Digital Materials

SCIENCES

Mechanics, Field Director – Gilles Pijaudier-Cabot

Numerical Methods in Mechanics, Subject Head – Sylvain Drapier

Digital Materials

Continuum Numerical Methods at the Mesoscopic Scale

Coordinated by

Marc Bernacki

Samuel Forest

iSTE

WILEY

First published 2024 in Great Britain and the United States by ISTE Ltd and John Wiley & Sons, Inc.

ISTE Ltd
27-37 St George's Road
London SW19 4EU
UK

www.iste.co.uk

John Wiley & Sons, Inc.
111 River Street
Hoboken, NJ 07030
USA

www.wiley.com

Library of Congress Control Number: 2024943341

British Library Cataloguing-in-Publication Data
A CIP record for this book is available from the British Library
ISBN 978-1-78945-197-9

ERC code:
PE1 Mathematics
PE1_13 Probability
PE1_14 Statistics
PE8 Products and Processes Engineering
PE8_3 Civil engineering, architecture, maritime/hydraulic engineering, geotechnics, waste treatment
PE8_7 Mechanical and manufacturing engineering (shaping, mounting, joining, separation)

Contents

Preface ix
Marc BERNACKI and Samuel FOREST

Chapter 1. Dislocation-based Mechanics: The Various Contributions of Dislocation Dynamics Simulations 1
Sylvain QUEYREAU

1.1. Introduction 1
1.2. Overview of discrete dislocation dynamics 2
1.2.1. Initial configurations and periodic boundary conditions 4
1.2.2. Mobility functions 5
1.2.3. Forces on dislocations 7
1.2.4. Topological changes 11
1.2.5. Boundary conditions 12
1.3. Mesoscale plasticity 15
1.3.1. Forest interactions 17
1.3.2. Statistical investigations of dislocation mechanisms 23
1.3.3. Lattice friction 29
1.3.4. Combination of multiple strengthening mechanisms 32
1.3.5. Toward polycrystalline plasticity 33
1.3.6. Cyclic deformations 40
1.4. Conclusion and future work 43
1.5. Acknowledgments 44
1.6. References 44

Chapter 2. Statistical Approach to the Representative Volume Element Size of Random Composites 55
Dominique JEULIN and Samuel FOREST

2.1. Introduction 55
2.2. Elements of numerical homogenization of heterogeneous media 56
2.2.1. Examples of physical properties of heterogeneous media 56
2.2.2. Change of scale in heterogeneous media 57
2.2.3. Principle of calculation of apparent properties 58
2.2.4. Homogenization of random media by numerical simulations . . . 63
2.3. Statistical definition of the RVE 64
2.3.1. Variance of estimation and integral range 65
2.3.2. Definition of the statistical RVE of apparent properties 67
2.3.3. Practical determination of the size of the RVE 69
2.3.4. Scaling of the bias, or systematic error 70
2.3.5. Determination of the size of the RVE in the nonlinear case 72
2.3.6. Remarks on the reduction of variance 73
2.4. Examples of application 73
2.4.1. Linear elastic properties and thermal conductivity 74
2.4.2. Elastic and viscoelastic properties of nanocomposites 89
2.4.3. Optical properties of nanocomposites 90
2.4.4. Fluid flow and permeability 94
2.4.5. Acoustic properties and wave propagation in fibrous media 95
2.4.6. Nonlinear properties 97
2.5. Conclusion 104
2.6. References 105

Chapter 3. Analytical Micromechanical Methods for Elasto-Viscoplastic Composites and Polycrystals 113
Stéphane BERBENNI and Samuel FOREST

3.1. Introduction 113
3.2. Translated field method 116
3.2.1. Field equations and integral equation 116
3.2.2. TF-based method: theoretical foundations 118
3.2.3. TF method for elasto-viscoplastic two-phase composite materials 119
3.2.4. TF method for elasto-viscoplastic polycrystals using the SC scheme 125
3.3. The β-model dynasty for multiphase elastoviscoplastic polycrystals . . 127
3.4. Applications of the analytical micromechanical methods 131
3.4.1. Applications of the TF method to composites and polycrystals . . 131
3.4.2. Applications of β-models to elastoviscoplastic polycrystals at finite deformations 141

3.5. Concluding remarks 147
3.6. References 148

Chapter 4. Vertex and Front-Tracking Methods for the Modeling of Microstructure Evolution at the Solid State 155
Marc BERNACKI

4.1. Introduction 155
4.2. Vertex frameworks 156
4.3. From Vertex to front-tracking then to enriched Vertex frameworks . . . 160
4.4. Enrichment of the Vertex approach 163
4.5. Other front-tracking frameworks for the modeling of microstructure evolution 169
4.6. Conclusion 173
4.7. References 173

Chapter 5. Phase Field: Theory, Numerical Implementation and Applications 177
Ingo STEINBACH and Oleg SHCHYGLO

5.1. Introduction 177
5.2. The soliton solution of a propagating wave front 179
5.3. Thermodynamically consistent derivation of the phase-field equations 183
5.4. The multi-phase fields approach 184
5.5. Multi-component alloy transformation 187
5.6. Multi-phase field with elasticity 191
5.7. Numerical treatment and applications 192
5.8. Examples 193
5.8.1. Martensitic transformation 193
5.8.2. Recrystallization and coarsening 194
5.8.3. Multicomponent solidification 198
5.9. References 199

Chapter 6. Level-Set Method for the Modeling of Microstructure Evolution 203
Marc BERNACKI

6.1. Introduction 203
6.2. Kinetics equations of microstructure evolution at the mesoscopic scale 205
6.3. Level-set function, description of polycrystal and meshing adaptation 208
6.4. Isotropic framework for LS modeling of ReX and GG 213
6.5. Anisotropy of GB properties and CDRX modeling 218

6.5.1. LS formulations in context of anisotropic GB properties 218
6.5.2. CDRX modeling 225
6.6. Static/evolutive SPPs 228
6.7. Modeling of diffusive solid-state phase transformation 235
6.8. Solute drag aspect 243
6.9. Conclusion 246
6.10. References 248

Chapter 7. Resolution Methods for Digital Materials – Recent Developments of Cellular Automaton Method 261
Lukasz MADEJ and Mateusz SITKO

7.1. Introduction 261
7.2. Cellular automaton and applications 262
7.3. CA front tracking model within the static recrystallization case study 268
7.4. Calculation times and perspectives 281
7.5. References 284

List of Authors 291

Index 293

Preface

Marc BERNACKI[1] and Samuel FOREST[2]
[1] CNRS UMR 7635, Centre for Material Forming (CEMEF), MINES Paris, PSL Research University, Sophia Antipolis, France
[2] CNRS UMR 7633, Centre des Matériaux, MINES Paris, PSL Research University, Evry, France

Digital materials have now become an integral part of design methods for industrial components and structures. Their mechanical properties can be predicted from a description of the microstructure and behavior laws of the constituents. This topic is now well established through the concept of integrated computational materials engineering (ICME).

This book covers a wide range of material properties from transport phenomena to the mechanics of materials and microstructure changes in physical metallurgy. It is not, however, an exhaustive review of numerical simulation methods for these phenomena, but rather a current look at the continuum simulation of elasticity, viscoplasticity, phase changes and grain boundary migration in metals, alloys and composites.

The elementary mechanisms of deformation and damage in materials involve complex atomic processes, which are now well explored by appropriate numerical simulations such as molecular dynamics. In contrast, this book shows how these mechanisms can be integrated into a continuum approach to the physics and mechanics of materials at the mesoscopic scale. It therefore focuses on the mechanics of continuous media and the continuum thermodynamics of irreversible processes. These models take into account the heterogeneities of materials, namely, their polycrystalline and/or composite nature, and constitute an intermediate step toward the establishment of effective laws for engineers in the calculation of structures and manufacturing processes.

The scientific topics selected in the book focus on the homogenization theory developed for periodic or random composite materials. The mean field approach is based on an idealization of the microstructure and provides estimates of the mean stresses and strains in the various phases present. These models can be fed by the full-field approach, where the representative elementary volume is the key concept. Plasticity is also given a significant place in this book, particularly at the grain scale, thanks to crystal plasticity in its discrete form, namely, the dynamics of dislocations, or its continuous form, which evaluates the quantities of slip and strain hardening associated with each slip system. Another strong point is the treatment of moving surfaces within the microstructure of materials, whether for phase changes, grain boundary migration and recrystallization. Statistical methods are used extensively to take account of the random nature of the distribution of constituents, dislocations and cracks, and also to exploit the large volumes of data produced by these simulations.

The book is structured into two parts. The first three chapters are devoted to mean-field and full-field approaches to establish the constitutive laws of elastoviscoplastic heterogeneous materials. The next four chapters focus on the numerical simulation of microstructure evolution, with particular attention paid to the modeling of the motion of interfaces (grain boundaries, moving interfaces for phase changes, etc.) and a whole range of numerical front-tracking or front-capturing methods that are now well established or in full development.

The first chapter is a summary presentation by Sylvain Queyreau of the methods and results of discrete dislocation dynamics (DDD) simulations. It begins with a description of the simulation technique, which is based on the elastic theory of dislocations in crystals, as well as on information relating to the core of dislocations and dislocation interaction rules derived from molecular dynamics analyses and microscopic observations. The method is capable of making predictions about the behavior of the single crystal, in particular the strain hardening laws. Recent developments address the response of polycrystals to complex loading, particularly cyclic loading conditions.

In the second chapter, Dominique Jeulin immediately places the question of the representative volume element (RVE) at the heart of the debate on the construction of the effective behavior of heterogeneous materials. He promotes statistical analysis of the morphology of phase distribution within the microstructure and the construction of random morphological models on which to base simulations of physical and mechanical properties. He proposes a precise statistical definition of the size of the RVE according to the precision required for the targeted property estimates. It emphasizes the biases associated with the choice of boundary conditions applied to the boundaries of the domain and stresses the importance of the concept of integral range for the definition of the RVE. The approach is highly effective for predicting linear material properties such as thermal and electrical conductivity, permeability and elasticity, but also acoustics and wave propagation. The statistical definition of the RVE, initially developed for linear properties, is extended here to the nonlinear case by using the space average of the local energy. Finally, the approach is heuristically extended to the elastoplastic behavior of heterogeneous materials, in particular metallic polycrystals and porous media.

The micromechanics of materials is addressed in Chapter 3 within the framework of the mean field approach. In a discipline whose history, outlined in the introduction, has been particularly rich over the last 60 years, the chapter focuses on the thorny issue of homogenization in the case of elastoviscoplastic behavior of the phases involved. The translated field method is highlighted and compared with other models available in the literature. It is first applied to the case of two-phase composites and then to metallic polycrystals. The second part of the chapter concerns the pragmatic approach developed by Cailletaud and Pilvin, which has proven its worth in the prediction of complex loading paths, particularly cyclic loading. It is extended here to the case of large deformations of metallic polycrystals in order to predict the evolution of the crystallographic texture and internal stresses for different deformation paths encountered particularly during metal forming.

The second part of the book, devoted to the simulation of evolving microstructures during thermomechanical treatments, begins with a presentation of the Vertex method in Chapter 4. The migration of grain boundaries and the growth of new grains are modeled by tracking grain boundaries, triple lines and vertices whose movements are dictated by surface energy, grain boundary curvature and, possibly, the energy stored in the grains in the form of dislocation densities assumed homogeneous per grain. Topological changes in the microstructure represent a major challenge in the tracking of these geometric objects within representative elementary volumes. Robust methods are now available and have proved their worth in simulating numerous metallurgical phenomena such as dynamic recrystallization or grain growth with second-phase particles and anisotropic properties of grain boundaries.

Predicting microstructural evolution during complex transformation processes, including mechanical, thermal and metallurgical aspects, is now possible thanks to the thermodynamics of continuous media and irreversible processes. This energy-based approach combines mechanical contributions (elasticity, viscoplasticity and strain hardening), chemical contributions (multi-constituent diffusion) and phase changes (oxidation, diffusive and displacive transformations). Electromagnetic coupling is also possible. The implementation of this holistic modeling is enhanced by the phase field method, which enables the kinetics and dynamics of interface motion to be calculated by formulating coupled free energy potentials and substituting diffuse interfaces for ideal surfaces. This is the subject of Chapter 5 by Ingo Steinbach and Oleg Schchyglo. The theoretical part of the chapter presents the multi-phase and multi-component formulation, which enables realistic alloy compositions to be addressed and the evolution of phase diagrams to be taken into account in detail. The phase field method excels in predicting changes in phase morphology, both coalescence and separation, but is often unsuitable for nucleation processes. Illustrations concern multi-component solidification, martensitic transformation and recrystallization phenomena.

The level-set method is described in Chapter 6 for applications to diffusive phase transformations (Oswald ripening and precipitate coalescence), grain growth and static or dynamic recrystallization in polycrystals. The driving pressures for interface migration are described in detail, as is the finite element numerical implementation. The crucial ingredient of the underlying model is the interface mobility law linking its velocity to the driving pressures of curvature and the stored energy jump in the case of recrystallization, or the Gibbs free energy jump in the case of diffusive phase changes. The technical subtleties of implementing these methods in finite element codes are explained, and the applications concern both 2D and 3D simulations, with original methods for meshing grain boundaries and interfaces. Once again, particular attention is paid to the treatment of multiple junctions. Large-scale computations of polycrystalline aggregates are used to study the influence of grain boundary types on their growth in relation to the anisotropy of surface energies.

The last chapter describes the latest advances in cellular automata methods applied to the computational mechanics of materials, and more specifically to recrystallization in metals and alloys during thermo-mechanical treatments. These are discrete approaches as opposed to the continuous approaches described in Chapters 4–6. Cellular automata, based on the consideration of a finite number of cells to which a certain number of variables are attached, are particularly suitable for introducing the stochastic aspects of the phenomena involved. They use the notion of cell neighborhood and transition rules to update the state of each cell. In particular, the method makes it possible to track the moving fronts within evolving microstructures during static recrystallization. It incorporates the influence of curvature, stored energy, thermal activation of the various phenomena and nucleation probabilities. Examples based on complex morphologies of polycrystalline microstructures from 3D electron backscattered diffraction (EBSD) analysis are presented.

The book is far from being an exhaustive snapshot of the numerical mechanics of materials in 2024, since only a very limited number of topics have been covered. The choice of topics is not only related to their importance in current research, but also to the tastes of the editors of this book, whom the reader should forgive. Other themes could have been explored, for example, a more in-depth study of the mechanics of composite materials in relation to their complex 3D microstructures, mechanical and electromagnetic couplings, the dynamics of microstructures (dispersion of elastic waves in metamaterials or high-speed deformation and damage), elastodynamics in polycrystalline materials, etc. Finally, the numerical mechanics of materials is now opening up to the methods of artificial intelligence, which is revolutionizing its algorithms. There is no doubt that this combination will produce astonishing successes, provided that the physics of the phenomena remains a requirement in the development of these methods.

We would like to thank Sylvain Drapier for his initiative in proposing this subject in the book series on computational mechanics, as well as all the authors of the chapters for their ambitious contributions and for their unfailing patience during the long preparation of the book. The quality of the result is all the more appreciated!

September 2024

1

Dislocation-based Mechanics: The Various Contributions of Dislocation Dynamics Simulations

Sylvain QUEYREAU
CNRS UPR 3407, Laboratoire des Sciences des Procédés et des Matériaux, Université Sorbonne Paris Nord, Villetaneuse, France

1.1. Introduction

Plastic deformation in crystalline materials is controlled by the motion and interactions of dislocations (Anderson et al. 2017). Discrete dislocation dynamics (DDD) simulations have now been available for about 30 years to investigate plastic flow at the mesoscale, which lies at a critical position between two very different scales. On a small scale, atomistic simulations are particularly well suited to investigate the dislocation core structure or dislocation interaction with the lattice that explains some elementary dislocation properties (such as mobility, cross-slip systems and reactions). On the other side of the spectrum, continuous approaches can simulate the complete and continuous mechanical behavior at the scale of the components, including microstructure heterogeneities and boundary conditions. The purpose of DDD is to simulate the evolution of large dislocation ensembles in reaction to an external load and to assess the collective behavior of dislocations, and

For a color version of all figures in this chapter, see www.iste.co.uk/bernacki/digital.zip.

Digital Materials,
coordinated by Marc BERNACKI and Samuel FOREST. © ISTE Ltd 2024.

thus act as a means of connecting the micro- and macroscales of plastic behavior, these scales would remain largely disconnected without DDD.

DDD accounts for both short- and long-ranged interactions between dislocations occurring within an evolving microstructure discretized in connected segments. The discretized microstructure is embedded in a usually homogeneous matrix. The numerical efficiency of DDD simulations resides in the fact that stresses are not evaluated in the entire domain but rather only on the degrees of freedom of the dislocation microstructure. The basic properties of dislocations are provided from dislocation theory and/or atomistic data. In the past decade, a number of remarkable technical advances have been achieved by the DDD community such as, for example, the development of non-singular dislocations theories (Cai et al. 2006; Po et al. 2014), a fast implementation of dislocation remote interactions in the frame of anisotropic elasticity which has been a long standing issue (Aubry and Arsenlis 2013; Aubry et al. 2014; Lazar et al. 2019), and tighter connections with atomistic simulations and robust hybrid methods to solve plasticity problems in finite domains including the case of the polycrystals (Weygand et al. 2002; Deng et al. 2008; Vattré et al. 2014; Jamond et al. 2016; Bertin and Capolungo 2018), to name a few. These technical advances lead to significant advances in the understanding of dislocation-based plasticity of metals.

This chapter aims to present a review of DDD technical developments and investigations from the past decade or so. We will mostly focus also on 3D approaches, as 2D simulations are somewhat associated with a different dislocation physics. In the context of this book, special attention will also be provided to the link toward larger scales. Readers interested in the rich and fascinating physics of dislocations should refer to Hull and Bacon (2001); Kubin (2013) and Anderson et al. (2017).

1.2. Overview of discrete dislocation dynamics

The dislocation dynamics technique has been designed and improved over the years to investigate the collective behavior and interactions of complex dislocation microstructures in response to an applied external loading. It therefore appears as a tool of choice to establish the link between microstructures and mechanical properties. The numerical competitiveness of DDD is twofold. First, since defect free regions of the materials play a small role in the plastic behavior of – elastically homogeneous – crystalline materials, modeling dislocations as an idealized network of connected lines contained in a continuous elastic medium has been recognized as an efficient strategy. Second, there are a number of analytical expressions and models that allow a rapid and accurate assessment of dislocation interactions[1]. The derivation of analytical or semi-analytical expressions for dislocation interactions is indeed an active field of research in the DDD community (see section 1.2.3). An exhaustive list of DDD codes is provided in Kubin (2013) (at the end of Chapter 4).

In brief, the DDD technique simulates the evolution of discretized dislocations in reaction to the mechanical stresses and interactions that set them in motion. A general mobility function M for a dislocations segment consists of the following form:

$$\mathbf{v} = M\left(\sigma(\mathbf{x}), \mathbf{b}, \xi, T, material\right)$$

The mobility depends on the type of the dislocation, its character and the temperature T but will above all be very material dependent even if there are only few different types of behavior (and corresponding functionals, see section 1.2.2). The motion of dislocations also depends upon the local stress field that is applied $\sigma(\mathbf{x})$ on them. While glide is often constrained in a finite set of predefined crystallographic planes, the number of degrees of freedom (DOF) remains virtually infinite. As the numerical cost of force evaluations scales with the number of DOFs, a compromise must be found between the precise description of the dislocation curvature and the number of discretized elements. This is why several discretization strategies have been proposed in the literature.

– In node-based strategies (Weygand et al. 2002; Arsenlis et al. 2007), the equations of motion are solved at the *nodes* as dislocations are described by straight *segments* connected by *nodes*. The forces experienced by dislocations are integrated along the segments and transformed in a nodal force by means of shape functions. The gliding mobility of the node is calculated as the average of the mobility of the two connected segments. A drag coefficient accounting for the node displacement along dislocation line – without physical meaning – must be defined and must preserve the flexibility of the dislocation.

– In *segment*-based codes (Devincre et al. 2011), equations of motion are solved at the center of straight segments typically. To further reduce the number of possible DOFs, segment orientations may also be limited in number and are often chosen as special physical orientations such as, pure screw, pure edge and junctions orientations. The nonlinear mobility functions can be easily accounted for but the management of topology is more complex than, for example, nodal approaches.

– In *parametric* approaches (Ghoniem et al. 2000; Po et al. 2014), dislocations are described by splines using a Galerkin approach. The use of non-straight segment captures well the flexibility of dislocations. The number of DOFs required to describe

1 As we will see further, analytical closed form expressions for energies and interactions of dislocations, in particular, exist for a handful of fairly simple configurations, and other solutions must be proposed for finite domain plasticity problems (see section 1.2.4).

a given dislocation curvature can potentially be smaller than other approaches. However, the development of analytical expressions for dislocation interactions is somewhat more cumbersome.

Furthermore, the so-called Burgers law: $\Sigma_i \mathbf{b}_i = \mathbf{0}$ states that the sum of Burgers vectors $\mathbf{b}_i$ of segments entering a node should balance the sum of Burgers vectors of outgoing segments. The microstructure should therefore consist of closed-loop geometries as dislocations ending freely in a crystal are unphysical and are associated with stress divergences[2]. Ultimately, the choice of discretization strategy is a matter of taste (Bulatov and Cai 2006) and should have a weak impact on the DDD results in comparison to critical choices regarding the physics included in the simulation (e.g. mobility functions).

1.2.1. *Initial configurations and periodic boundary conditions*

As metastable defects, dislocations are – almost – always present initially in crystalline materials obtained by conventional processes. Initial dislocation microstructures are therefore typically introduced at the beginning of DDD simulations. Since the so-called Frank network initially adopted by dislocations is poorly known, initial dislocation microstructures are arbitrary. The original solution based on Frank–Read sources – which are associated with unphysical pinning points and divergent stress fields – is gradually abandoned in favor of closed geometries such as planar or prismatic loops. Prismatic loops consist of four successive edge segments, two of which belong to a primary system and two of which belong to a cross-slip system. Interestingly, the length of the cross-slip segment has a huge impact on the initial strengthening, due to the very strong collinear annihilation interaction existing between primary and cross-slip systems. When starting from prismatic loops of various geometries (i.e. in terms of sizes and shapes), the mechanical response nevertheless converges after some deformation as the density on the cross-slip system will naturally be regulated by cross-slip systems events and collinear annihilations. Some other authors (Aubry et al. 2019) prefer to use relatively small loops that will operate akin to sources. Depending on the strain rate, a marked yield point may be observed on the deformation curve, and the microstructure will quickly be very different from the initial one.

In large-scale DDD simulations, dislocation density increases naturally, without the recourse to artificial sources or nucleation processes, but through the temporary or permanent pinning of dislocations by obstacles. In real systems, the nucleation of fresh dislocations can occur under special conditions from the surface or under

2 On a theoretical standpoint, the stress divergence of an open loop configuration can however be removed by using rational stress expressions as in Asano (1968) in the case of linear elasticity.

shock. Conventional DDD is not particularly suitable for the investigation of dislocation nucleation as it would require accounting for nonlinear elasticity and accurate description of the dislocation cores (Aubry et al. 2011a). However, practical solutions have been proposed to address these issues. Ultimately, confidence in the simulated dislocation microstructures can be found in the ability of DDD simulations to reproduce flow stresses, hardening and deformed dislocation patterns experimentally obtained on deformed single crystals (see section 1.3).

Since the mean free path of dislocations can be much larger than the simulation dimensions, extended gliding planes are enabled by the use of periodic boundary conditions (PBC). The total dislocation density tensor over the whole simulation domain must be null, and PBC usage enforces a balanced flux of dislocations across the boundaries of the domain. To further lengthen the extended gliding plane offered to dislocations, the simulation domain can be rotated with respect to the crystal lattice. For segment-based simulations, the size and orientation of the simulation basis must be carefully defined to respect the underlying simulation crystal lattice. Several approaches have been proposed to solve the associated Diophantine equations relating the simulation dimension and the projection of the boundary normals with respects to the lattice (Madec et al. 2004; Monnet 2006). It is interesting to note that similar – coupled – Diophantine equations in atomic simulations when two crystals are rotated relative to the simulation box to form grain boundaries have only recently been solved (Banadaki and Patala 2015).

1.2.2. *Mobility functions*

Mobility functions relate the effective shear stress felt by a given segment of dislocation to their gliding velocity (Kubin 2013). We will mention two prototypical mobility laws, one for the high temperature behavior associated with the athermal regime and one for the low temperature behavior or thermal regime of plastic flow. The effective stress τ_{eff} is the total shear stress contribution felt by the dislocation in the gliding direction. In the athermal regime and for conventional strain rates where relativistic effects are not felt, the velocity of dislocations v simply reads: $|\mathbf{v}| = |\mathbf{b}|\tau_{eff}/B(T)$ where $B(T)$ is a dragging coefficient describing the phonon scattering induced by dislocation motion. $B(T)$ generally increases with T. In materials, for example, bcc or hcp materials at low temperature and relatively low stress, dislocations exhibit a much more complicated mobility. Taking bcc Fe as an example, the thermal regime is known to be controlled by the slow motion of screw dislocations due to the large lattice friction. Screw dislocations move from one Peierls valley to the next (see Figure 1.1), separated by a distance h_0, by the so-called nucleation and propagation mechanism of a double kink along the dislocation line, if kinks are successful. The nucleation rate is generally the controlling mechanism. The resulting thermally activated behavior is commonly modeled as an Arrhenius equation at a mesoscale as:

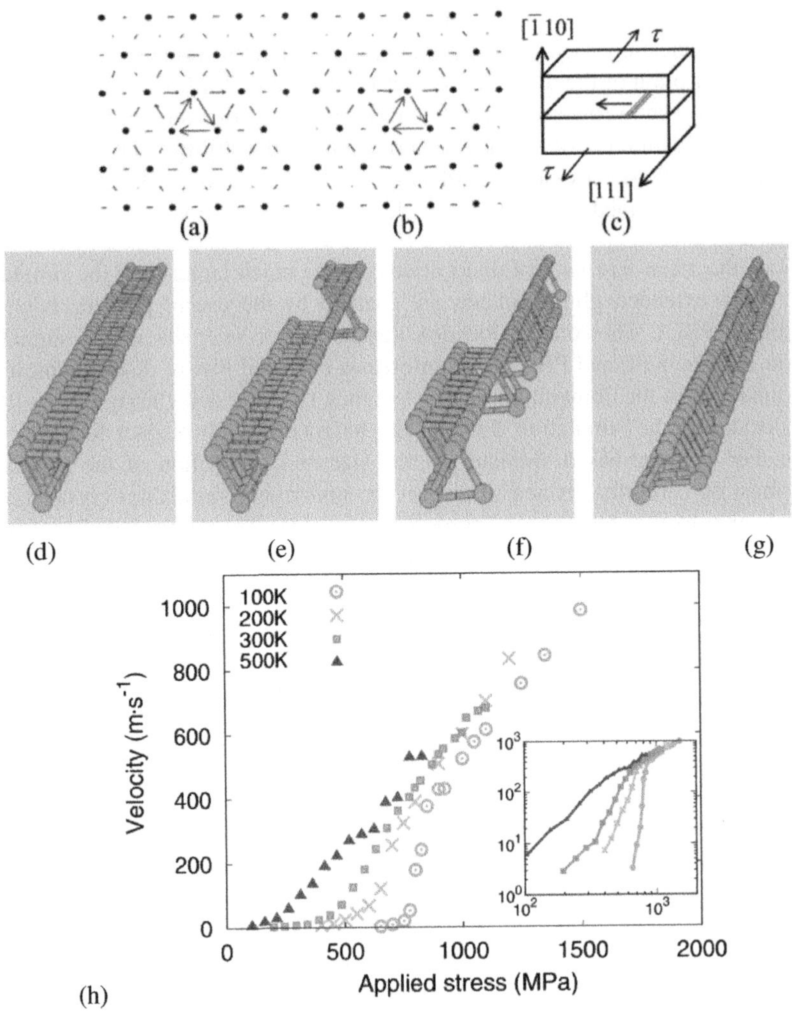

Figure 1.1. *(a and b) Compact core structure obtained by ab initio and by an EAM semi-empirical potential, for screw dislocations in pure bcc Fe. The displacement map shows that the core distortion is localized at the three closest <111> strings of atoms. (c) Typical geometry used to determine the mobility of dislocations with atomistics where a shear stress* τ *is applied on the free surfaces of a semi-infinite crystal. (d–g) Sequence of motion of a screw dislocation from one Peierls valley to the next through nucleation and propagation of double kinks (a–g from Monnet et al. (2004)). (h) Velocity of a screw dislocation obtained from molecular dynamics in bcc Fe. At low stress and low T, the velocity follows an Arrhenius type behavior consistent with equation [1.1]. At large applied stresses, the mechanical work is sufficient to overcome the activation barrier, and a linear regime is recovered, while differing from the linear drag regime observed for edges (from Gilbert et al. (2011))*

$$|\mathbf{v}| = h_0 \frac{v_D b}{w} \frac{L}{w} \exp\left(-\frac{\Delta H(\tau*)}{k_B T}\right) \quad [1.1]$$

where k_B is the Boltzmann constant and the attempt frequency of a kink-pair of width w is $v_D|\mathbf{b}|/w$ with v_D the Debye frequency. L/w, where L is the screw length, represents the ratio of competing nucleation sites. The activation enthalpy ΔH replaces here the Gibbs activation energy as entropy is commonly neglected. ΔH is typically constant with temperature $\Delta H = Ck_BT$. ΔH strongly decreases with the thermal stress $\tau*$, following the phenomenological expression (Kocks et al. 1975):

$$\Delta H = \Delta H_0 \left(1 - \left(\frac{\tau*}{\tau_P}\right)^p\right)^q \quad [1.2]$$

where τ_P is the Peierls stress and p, q are exponents. Atomistic simulations are particularly well suited to defined mobility law (Monnet et al. 2004; Gilbert et al. 2011). Note also that dislocations with other that screw character may be thermally activated (Monnet and Terentyev 2009; Queyreau et al. 2011).

It must be emphasized that the dependence upon the dislocation length L appearing in the exponential prefactor, which has been observed in situ using transmission electron microscopy (TEM) during deformation (Louchet et al. 1979; Caillard 2010, 2013), is crucial in capturing the proper low temperature behavior of bcc metals.

1.2.3. *Forces on dislocations*

The force evaluation on all segments of the dislocation microstructure is arguably the most demanding part of DDD simulations, despite the constant development of analytical and semi-analytical models. In this section, we will focus mostly on stress and force expressions that have been developed in the context of DDD simulations. Readers interested in a more complete list of analytical models for dislocations are referred to Anderson et al. (2017). Beyond the core region, dislocations generate only linear elastic distortions in the surrounding matrix. Therefore, the superposition principle applies. For the sake of simplicity, we start with the dislocation-dislocation interactions. When considering an infinite and homogeneous linear elastic medium, a first dislocation loop generates a stress field $\sigma^{\infty}(\mathbf{x})$ at a point $\mathbf{x}$ defined as a contour integral along the loop (Mura 1987):

$$\sigma_{ij}^{\infty}(\mathbf{x}) = C_{ijkl} \oint \epsilon_{lnh}\, C_{pqmn}\, G_{kp,q}(\mathbf{x} - \mathbf{x}')\, b'_m\, dx'_h \quad [1.3]$$

$$G_{ij}(\mathbf{x} - \mathbf{x}') = \frac{1}{8}\pi\mu \left[\delta_{ij}\partial_p\partial_p R - \frac{1}{2(1-\nu)}\partial_i\partial_j R\right], \quad [1.4]$$

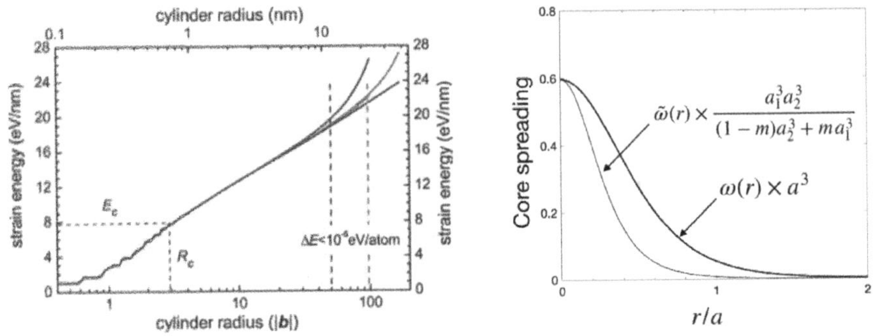

(a)

(b)

Figure 1.2. *(a) Strain energy integrated in a cylinder around an edge dislocation in bcc Fe as function of the cylinder radius from atomistics (Osetsky and Bacon 2003). The nonlinear domain corresponds to the core domain, and the linear behavior (in log scale) corresponds to the realm of linear elasticity. The red curve corresponds to the infinite medium. A natural core energy E_c can be defined for a given core radius R_c, but note that any couple of (E_c, R_c), could be used in a continuum framework as long as the rest of the curve is recovered. Not also that for conventional crystal sizes and dislocation spacings, core energy is perfectly negligible to elastic energy. (b) Comparison of the single and double convoluted core distribution of the non-singular theory by Cai et al. (2006) from Queyreau et al. (2019). a is the isotropic core spreading dimension, which should be close to few $|b|$*

where C_{ijkl} is the elastic stiffness matrix of the medium and ϵ_{lnh} is the permutation operator. $\mathbf{b}'$ is the Burgers vector of the loop and $\mathbf{x}'$ is the coordinate that runs along the dislocation. R is the metric: $||\mathbf{x} - \mathbf{x}'||$. $G_{kp}(\mathbf{x} - \mathbf{x}')$ are the fundamental Green's functions of elasticity, here given for isotropic elasticity. They correspond to the displacement component in the x_k direction at point $\mathbf{x}$ induced by a body force located at $\mathbf{x}'$ in the x_p direction (Mura 1987). If $\mathbf{x}$ now spans a second dislocation, this gives rise to the so-called Peach and Koehler force:

$$\mathbf{F}^{PK} = \oint N(\mathbf{x})(\sigma_{ij}^{\infty}(\mathbf{x}) \cdot \mathbf{b}) \times \xi \, dL, \qquad [1.5]$$

where $N(\mathbf{x})$ are potential shape functions to distribute the force on nodes connecting segments, and dL is a length along the second dislocation segment. $\mathbf{b}$ and ξ are the Burgers and line vectors of the second loop, respectively. Starting from the simplest case of isotropic elastic medium, closed form analytical expressions for the force integrals can found in Arsenlis et al. (2007) for straight segments and spline segments in Po et al. (2014).

The conventional theory of dislocations is associated with a well-known singularity of the stress field close to the dislocation core (see Figure 1.2). This is attributed to the sharp (Dirac) distribution of Burgers vectors implicitly assumed to calculate the elastic fields. While other regularization methods have been proposed to remove the singularity (Anderson et al. 2017; Brown 1964; Gavazza and Barnett 1976), the solution provided by Cai et al. (2006) and Arsenlis et al. (2007) leads to fully analytical and compact force expressions for the double integral above. The strategy proposed by these authors consists of replacing the sharp Burgers distribution by an isotropic smooth spreading function, as it is assumed that linear elasticity still applies at the dislocation core. The metric R is replaced by $R_a = R * w(\mathbf{x})$ its convolution with the core distribution $\mathbf{b}w(\mathbf{x})$. An appropriate choice of distribution leads to a very simple equation for $R_a = \sqrt{\mathbf{R}\cdot\mathbf{R}+a^2}$ and few additional higher order terms appearing in the stress expressions when compared to the singular theory. As the model was derived for dislocation–dislocation interaction, $w(\mathbf{x})$ is in fact a double convolution of the Burgers distribution, and only an approximate distribution is provided for the single convoluted distribution. This could potentially lead to smoother fields close to dislocation core and double convolution is, in principle, not required when calculating stress at a point in the elastic matrix. An approximated solution was proposed in Cai et al. (2006) as the weighted sum of two core sizes for the single convolution of the Burgers distribution. Difference in the elastic fields between single and double convolutions is large within the dislocation core, but limited outside (Queyreau et al. 2019). It must be noted that nodal force expressions identical to the ones derived from equation [1.5] can be obtained by a different route starting from the mechanical work of interaction existing between the two segments (Mura 1987):

$$
\begin{aligned}
E_{el} = & \frac{\mu}{16\pi}\oint\oint b_i b'_j \partial_k\partial_k R_a dx_i dx'_j - \frac{\mu}{8\pi}\oint\oint \epsilon_{ijq}\epsilon_{mnq} b_i b'_j \partial_k\partial_k R_a dx_m dx'_n \\
& + \frac{\mu}{8\pi(1-\nu)}\oint\oint \epsilon_{ikl}\epsilon_{jmn} b_k b'_m \partial_i\partial_j R_a dx_l dx'_n \qquad [1.6]
\end{aligned}
$$

A conservative force corresponding to mutual interactions can be then defined at dislocation node $\mathbf{q}$ as:

$$
\mathbf{F}_i = -\frac{\partial E_{el}}{\partial \mathbf{q}_i} \qquad [1.7]
$$

Another force contribution is the so-called self interaction of dislocation segments (Brown 1964; Gavazza and Barnett 1976). As for distinct dislocation interactions, the self force can be defined in a self-consistent fashion as the negative of gradient of self energy with respect to one of the segments node position. The self energy corresponds to the double line integral over the loop or equally the volume integral of the free energy and is therefore considered as a nonlocal contribution as well. In the theory put forward by Cai et al. (2006), when taking the gradient of the self energy, a part of the energy is coming from contribution within the spread radius a. This is a core contribution of the model, but it is not per se the core energy of a dislocation as there is no information from atomistics (see for example Clouet et al. (2011)) and linear elasticity is used within the core. Since line tension constitutes (de Wit and Koehler 1959) the leading term in the self force, line tension expressions or the heuristic-based approach from Brown (1964) are also commonly employed in DDD simulations. The case of anisotropic elasticity is typically more difficult to tackle as closed form analytical solutions for the stress field around dislocations do not exist. Anisotropic elasticity has thus been a long standing challenge for massive DDD simulations. Elastic anisotropy affects remote dislocation interactions and may lead to strong changes in equilibrium shape of dislocation loops (Bacon et al. 1979; Aubry et al. 2011b). The derivative of the required Green's functions can however be expressed as angular integrals around dislocation segments (Mura 1987). Aubry and Arsenlis (2013) approximated these integrals by means of spherical harmonics expansion, the order of which is truncated and adapted to the targeted precision. The truncation order must be increased as the anisotropy coefficient increases. Nevertheless, this semi-analytical approach is a significant improvement over double integration by brute force to obtain forces for interacting dislocations and has reduced the additional cost of complete anisotropy from more than two orders of magnitude to about an order more expensive than the classical isotropic solution (Aubry and Arsenlis 2013). The hybrid DDD-FFT solution (see section 1.2.5) is a promising method and is suggested as a strong competitor to fully anisotropic calculations, however the stress field is obtained on a volume grid and the same control of numerical errors is still to be seen. When not looking for the full anisotropic machinery, effective isotropic elastic constants and tabulated line tension can provide estimates of the behavior of curved dislocations in anisotropic medium (Bacon et al. 1979; Aubry et al. 2011b; Kubin 2013; Madec and Kubin 2017). Lazar et al. (2012, 2013); and Po et al. (2014) have formulated a non-singular theory for dislocations based on strain gradient elasticity of Helmholtz type as an alternative to Cai et al. (2006). A characteristic internal length l is introduced and associated with higher stress terms. The displacement and stress fields are non-singular thanks to the use of an isotropic spreading function $A(R)$ of the Burgers vector simply defined as the convolution of $A(R) = R * G$, where G is now the isotropic Green's function of the Helmholtz operator. It yields $A(R) = R + \frac{2l^2}{R}(1 - \exp^{-R/l})$. Expectedly, the internal length l was found to be slightly below $1|\mathbf{b}|$ from atomistics in W, which is close to the ideal isotropic material. In contrast to Cai et al. (2006), the Cauchy stress is defined from a single convolution with the core spreading function. The usage of strain gradient elasticity allows for capturing nonlinear elastic behavior at the dislocation core, the core distribution remains arbitrary. The improved description of the core elastic fields comes at the cost of higher order stress terms and elastic constants. The latter can however be identified through fully independent atomistic simulations. Recently, Lazar et al. (2019) extended their work to the cases of nonlocal anisotropic elasticity of generalized Helmholtz type. The theory is now associated with six internal lengths and the dislocation core distribution itself is now anisotropic. The line integrals along dislocations loops and angular integrals appearing in the Green tensor need however to be evaluated by numerical quadrature.

Naturally, other interactions between dislocations with other crystal defects can be included starting from equation [1.5] and a stress field expression or from equation [1.7] and a mechanical work of interaction, and the superposition principle, whether preference is given to stress or interaction energy. This is the case for the external applied stress, or the interactions with Eshelby inclusions that can be introduced to model second phase particles (Aubry et al. 2019) or punctual defects (Cai et al. 2014; Gu and El-Awady 2018).

1.2.4. *Topological changes*

During the simulation, the dislocation microstructure evolves strongly and a number of a topological changes have to be handled by the simulation.

– Dislocation lines may be rediscretized to adapt the length of segments to the local curvature. A different criterion has been proposed for this aim (Devincre et al. 2011; Arsenlis et al. 2007).

– Different parts of the dislocation microstructure may collide during the motion of moving segments. If the Burgers vectors are different, this may lead to the formation of a reaction that typically minimizes the elastic strain energy and the resulting dislocation may well be sessile in any of the initial slip planes. The formation and destruction of the reactions occur through zipping and unzipping by the lateral movement of the mobile segments. This plays an important role in the deformation of crystalline materials as reactions are almost always operative and they even constitute the most part of hardening in pure materials in the athermal regime.

– Finally, screw dislocations have the possibility to change of slip plane among the planes in zone that are available to them. In fcc crystals, *cross slip* (CS) occurs through the so-called Escaig–Friedel mechanism, according to which dissociated partial dislocations spread in a primary plane must recombine prior to any change of plane. This core transformation is necessarily thermally activated and with a power law dependence of the activation energy with respect to the stress and a length dependence of the prefactor. In DDD, the probability for a CS event to occur is handled by a kinetic Monte Carlo algorithm. Recent studies, at the atomic scale in particular, have shown that the CS mechanism is certainly more complicated that initially though (Kang et al. 2014; Hussein et al. 2015; Longsworth and Fivel 2021). In bcc iron, dislocation cores are compact (Frederiksen and Jacobsen 2003; Monnet et al. 2004) and multiple slip planes are possible, CS is commonly considered as very easy in agreement with the wavy slip lines observed at deformed sample surface. A basic and purely mechanical criterion was thus used in Queyreau (2008).

1.2.5. *Boundary conditions*

The numerical efficiency of DDD simulations lies in its efficient and accurate evaluation of dislocation interactions. This is permitted by the use of analytical expressions for the dislocation-dislocation interactions, for example, which have been formulated for a few simple configurations, typically the case of closed loops contained in an infinite and homogeneous domain. The use of PBC conditions mentioned earlier corresponds perfectly to this situation. However, this is not the case when considering finite domain geometries or internal interfaces or inhomogeneities. To remedy this, hybrid solution associating DDD with full-field finite element (FEM) or boundary element method (BEM) has been proposed to correct – or partially replace – the DDD fields (Giessen and Needleman 1995; Martinez and Ghoniem 2002; Zbib et al. 2001; O'Day and Curtin 2004; Weygand et al. 2002; Deng et al. 2008; Tang et al. 2006; Weinberger et al. 2009; Crone et al. 2014; Vattré et al. 2014; Jamond et al. 2016; Bertin and Capolungo 2018). The original solution was proposed by Giessen and Needleman (1995) and is based on superposition principle. We start with presenting this approach for didactic reasons and to highlight some associated challenges. The stress field solution is decomposed into a "classical" (∞) field associated with dislocations in an infinite domain and a corrective ($\hat{}$) field that enforces the desired boundary conditions at the internal or external boundaries of the simulated domain (see Figure 1.3).

$$\mathbf{u} = \mathbf{u}^{\infty} + \hat{\mathbf{u}} \qquad [1.8]$$

$$\sigma = \sigma^{\infty} + \hat{\sigma} \qquad [1.9]$$

The (∞) infinite domain fields do not respect the desired boundary conditions, the corrective field are obtained from a FE or BE simulations carried over a non-dislocated medium with the following boundary conditions:

$$\hat{\sigma} \cdot \mathbf{n} = \hat{\mathbf{T}} = \mathbf{T_0} - \mathbf{T}^{\infty} \text{ on } S_T \qquad [1.10]$$

$$\hat{\mathbf{u}} = \mathbf{u_0} - \mathbf{u}^{\infty} \text{ on } S_u \qquad [1.11]$$

$$[\mathbf{T}] = \mathbf{0}, \text{ on } S_I \qquad [1.12]$$

$$[\mathbf{u}] = \mathbf{0} \text{ on } S_I \qquad [1.13]$$

An alternative hybrid solution called discrete continuum model (DCM) was proposed in a series of papers (Vattré et al. 2014; Jamond et al. 2016). As for the "superposition principle-based" approaches, the coupling of the two simulation techniques is also rather strong, with the two codes handling the same domain and similar timesteps. DDD manages short range interactions between dislocations contained inside the same element, while FEM handles long range dislocation interactions among dislocation in different elements and boundary conditions. For this, the long-range contribution of dislocation interactions is thus performed using a

– plastic – eingenstrain formalism (Mura 1987). The latter is defined from the discrete plastic activity events by the gliding area swept by dislocation coming from DDD and a regularization procedure to match the continuous fields of FEM. The eigenstrain approximation thus replaces classical multipole expansions used in pure DDD. This method can also be seen as an FEM calculation where the constitutive equations of crystalline plasticity are replaced by a synchronous DDD calculation. Finally, the DCM hybrid approach can be shown to be formally equivalent to the superposition principle[3]. As for DDD, the choice of algorithm is thus a matter of taste. For periodic problems, FFT solver may be preferred to FEM and may speed up calculations (Bertin and Capolungo 2018).

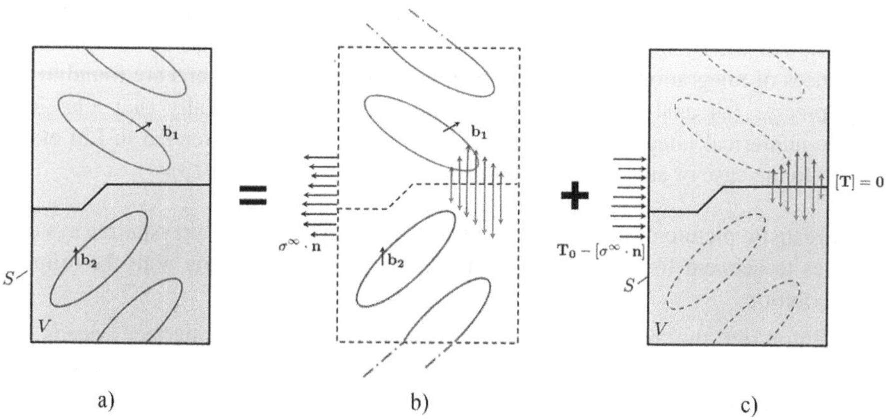

Figure 1.3. *Illustration of hybrid method solver from Queyreau et al. (2014). (a) Two slip families of Burgers vectors* $\mathbf{b}_1$ *and* $\mathbf{b}_2$ *are present in a bicrystal* V*, bounded by* S*. (b) The traction field due to the discrete dislocation ensemble in the infinite domain is evaluated on the domain internal and external boundaries. This calculation is performed on dislocations contained within the domain and on virtual segments used to close open dislocation loops. (c) Then, an FE or BE simulation of a non-dislocated elastic medium gives the corrective stress field that enforces the desired boundary conditions, that is, set traction values* $\mathbf{T}_0$ *at external surfaces* S_E *and no traction discontinuity at internal interfaces* S_I

The association of DDD with FEM methods, however, faces several challenges. In particular, the evaluation of the unbalanced tractions induced by dislocations at the boundaries of the domain (Weygand et al. 2002; Deng et al. 2008) is required prior to the FEM/BEM. The tractions $\sigma^{\infty} \cdot \mathbf{n}$ are expressed as the triple integration of the infinite domain stress field σ^{∞} along the dislocation line and over the surface element

3 Devincre (2019), personal communication.

of normal $\mathbf{n}$. Tractions are then projected and appropriately distributed as nodal forces $\mathbf{F}^{(n)}$ over the nodes of the surface element $S^{(E)}$ by means of shape functions N:

$$\mathbf{F}^{(n)} = \int_{S^{(E)}} [\sigma^{\infty}(\mathbf{x}) \cdot \mathbf{n}] \, N^{(n)}(\mathbf{x}) \, dS. \qquad [1.14]$$

The evaluation of unbalanced tractions induced by DDD stress fields is arguably the most demanding task when associating DDD to FE/BE as this operation must be carried out over all segments constituting the microstructure and on all surface elements delimiting the domain (Weygand et al. 2002; El-Awady et al. 2008). Queyreau et al. (2014, 2019) proposed an analytical solution in the case of the non-singular theory of Cai et al. (2006). These analytical solutions are obtained by sequences of integration by parts that exhibit recurrence relations and are found to be very precise (as analytical) and much more efficient numerically than classical double numerical integration strategies. This model was then extended in Liu et al. (2016) to the case of anisotropic elasticity (Aubry and Arsenlis 2013).

A realistic picture of the mechanical state of the dislocated polycrystalline system requires to account for several new contributions when comparing with the infinite body solution.

– Enforcing the proper boundary conditions at internal interfaces leads to significative modifications to the stress field solution of the infinite domain, leading to image and elastic incompatibility stresses felt by crystal dislocations. This particular boundary value problem in the presence of dislocation can only be assessed by hybrid DDD-FEM solver. These corrections are particularly important in the case of anisotropic elasticity, which explains why conventional DDD is still commonly employed when studying polycrystals in the simpler case of isotropic elastic materials.

– Second, plastic deformation is intrinsically anisotropic and different strains can result from different dislocation activity in adjoining crystals, which are rotated with respect to each other. As matter must remain compatible across the interface, this plastic incompatibility at the interface may be accommodated by elastic deformations in both grains (Gemperlová et al. 1989). Contrary to the elastic incompatibilities induced by different microstructures on both sides of the grain boundary (GB) (see above), plastic incompatibilities are uniform as the plastic deformation from dislocation motion concerns the entire surrounding grain, and they lead to long-range incompatibility stress fields (Gemperlová et al. 1989; Sutton and Balluffi 1995).

– Finally, a third ranged contribution may be found in the stress induced by the GB defect structure itself (Sutton and Balluffi 1995). From the seminal works of Read and Shockley (RS) and Frank (Sutton and Balluffi 1995), low angle grain boundaries are commonly described as arrays of dislocations. These arrays are particularly simple for symmetric pure tilt and twist grain boundaries. In the case of semi-coherent interfaces, misfit dislocations may accommodate the large lattice and elastic mismatch. As the spacing between GB dislocation decreases with misorientation angle according to Frank's equations, for HAGB, GB dislocations core overlap and cannot be individualized. This could also be the case for mixed tilt twist GB. Investigating the GB structure has motivated a wealth of atomistic study. It is however striking that the atomistic energy of so many GB can be fitted by a RS type equation in the 5D parametric space of GB (Bulatov et al. 2014; van Beers et al. 2015).

Simply put, the polycrystalline problem can be summed up for a grain G_1 as (Sutton and Balluffi 1995):

$$\sigma_{G1}^{tot} = \sigma^{dislo} + \sigma^{GB} + \sigma^{C} \quad [1.15]$$

$$\epsilon_{G1}^{tot} = \epsilon^{e} + \epsilon^{p} + \epsilon^{C} \quad [1.16]$$

where σ and ϵ are stresses and strains, respectively. σ^{dislo} is the stress field coming from the dislocation microstructure. σ^{GB} is the stress field induced by the GB presence and σ^{C} includes both elastic and plastic incompatibilities.

Knowing the stress field induced by a GB or an interface in general is uneasy. These planar defects can be modeled at the mesoscale as arrays of regularly spaced dislocations. Formally, it is possible to determine possible dislocation arrays from the transformation relating the two adjoining crystals. For planar interfaces, possible arrays of misfit dislocations are expected to comply with the so-called quantized Frank–Bilby equation: $\mathbf{B} = (\mathbf{F}_{\mathrm{M}}^{-1} - \mathbf{F}_{\mathrm{P}}^{-1}) \cdot \mathbf{k}$ that relates the displacement gradient $\mathbf{F}_{\mathrm{M}}$ and $\mathbf{F}_{\mathrm{P}}$ for the matrix M and particle P with respect to a reference interface state to the sum of interface Burgers vectors $\mathbf{B}$ crossed by an interface probing vector $\mathbf{k}$. In theory, the number of possible arrays of dislocations capable of accommodating is not unique and requires to find the reference state of the interface. Vattré and Demkowicz (2013, 2015) and Vattré et al. (2016) devised a strategy in a number of papers to define a unique solution for semi-coherent interfaces in a bicrystal by testing different reference states and using continuous mechanics and the Stroh anisotropic elasticity formulation (see Figure 1.4). The dislocation array solution is then identified as the one cancelling far field stress field induced by the interface and associated with the lowest energy. This strategy was successful in predicting the misfit dislocation structure obtained by MD.

1.3. Mesoscale plasticity

In this section, we review a number of recent contributions from the DDD community to the understanding of the plastic behavior of crystalline materials. The section focuses primarily on pure metallic systems of mostly cubic lattice structures. Other recent reviews can be found elsewhere regarding strengthening in alloys (Aubry et al. 2019), irradiated materials (Marian et al. 2018; Monnet 2018) or plasticity in small-scale systems (El-Awady 2015; Lavenstein and El-Awady 2019).

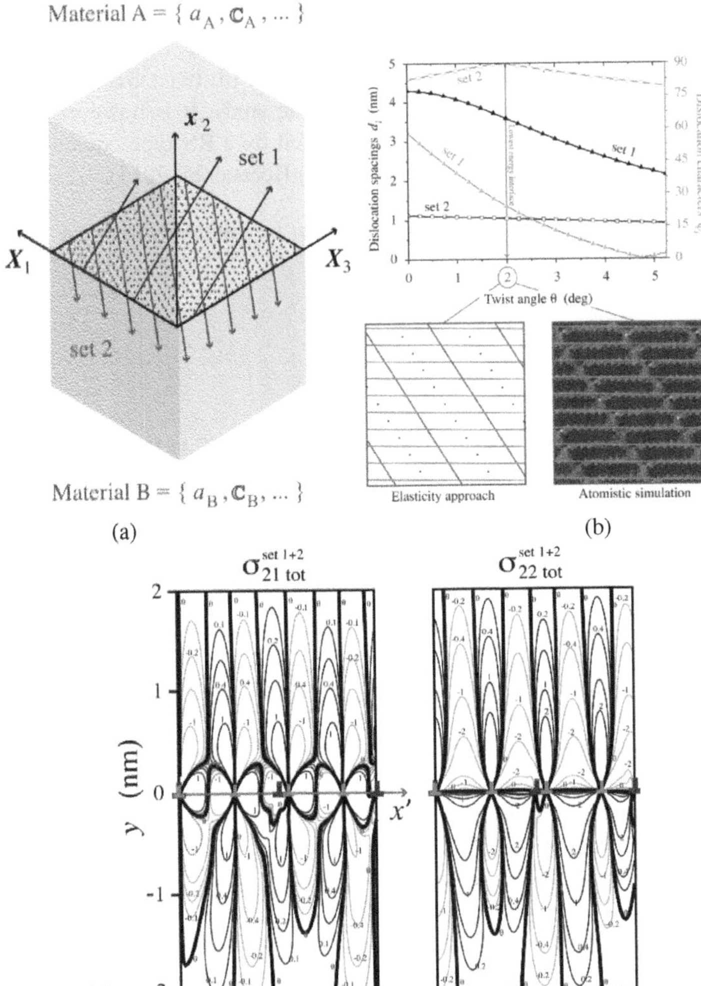

Figure 1.4. *(a) Schematic illustration of the modeling of an interface separating two materials by a semi-coherent interface containing two sets of misfit dislocations (Vattré and Demkowicz 2013). The dislocations separate regions of close coincidence between the two crystals, but the Burgers vector of the sets are undetermined in theory. (b) The resolution of the Frank–Bilby equations using anisotropic elasticity to remove this indeterminacy provides dislocation characters and spacings as function of the misorientation angle θ (Vattré and Demkowicz 2015), here illustrated for the Cu/Nb system. The dislocation arrays with the lowest energy, in red and for* $\theta = 2deg$*, matches the atomistic results. (c) Mappings of the stress fields induced by the corresponding two sets of misfit dislocations in the crystals*

1.3.1. *Forest interactions*

The so-called Forest interactions occurring between dislocations play a central role in the plastic flow of crystalline materials as dislocation-dislocation interactions are almost always operating and it helped shaping up strengthening models for other dislocation-obstacle interactions. Assuming that the Schmid law is respected, the critical stress τ_c required to activate a slip system 'i' is given as (Franciosi et al. 1980; Kocks and Mecking 2003):

$$\tau_c^i = \mu b \sqrt{\Sigma_j a_{ij} \rho_j} \qquad [1.17]$$

where ρ_j is the dislocation density attributed to system "j" and a_{ij} is the so-called interaction matrix that gathers the coefficients measuring the strength of the interaction between systems "i" and "j". a_{ij} are poorly known from experiments (Franciosi et al. 1980), and the determination of coefficients a_{ij} of the interaction matrix even from DDD is not an easy task a priori. Indeed, the length, density, character of the interacting dislocations and elastic properties can affect the reaction stability and these parameters should be averaged in a statistical manner. DDD simulations modeling the latent hardening test is now a common approach to achieve this, and consists of an active slip system that interacts under stress with one or several inactive slip systems leading to a single type of reactions. While having an inactive slip system may in principle lead to different dislocation patterns and therefore different sampling of dislocation characters and lengths, the coefficients determined this way are self-consistent with the DDD simulations where all reactions are included and robust enough for the prediction of the plastic anisotropy of single crystals (see below). This approach has been employed for fcc (Devincre et al. 2006; Madec and Kubin 2008, 2017), bcc (Madec et al. 2004; Madec and Kubin 2008; Queyreau and Devincre 2009; Madec and Kubin 2017) and hcp materials (Devincre 2013; Bertin et al. 2014) (see Figure 1.5). Note that the relative strength hierarchy among binary junctions in a fcc system was confirmed (except for the glissile junction that was found much stronger) in a very recent investigation (Akhondzadeh et al. 2020, 2021) through a different strategy where interaction coefficients are identified at once from a large number of different DDD simulations (see below). While no dislocation mechanism was thermally activated, a rate dependent term was added to the expression of τ_c^i above in order to improve the fit agreement with DDD data at large strain rates.

Since the largest part of the forest strengthening is related to the short range interactions, as opposed to remote elastic interactions, equation [1.9] finds its physical origin in the line tension to bow out the mobile segment pinned by a pair or reactions (Saada 1960). Therefore, the simple expression above neglects the logarithmic dependence of the line tension and a_{ij} are found to depend upon the dislocation density (Basinski and Basinski 1992; Devincre et al. 2006). Devincre et al. (2006) and Kubin et al. (2008) proposed to correct a_{ij} of this dependence as: $a_{ij}(\rho_j) = a^0_{ij}\left(\ln(1/b\sqrt{(a^0_{ij}\rho_j)})/\ln(1/b\sqrt{(a^0_{ij}\rho^0_j)})\right)^2$, where ρ^0_j is the reference density at which coefficient a^0_{ij} was determined, and $1/b\sqrt{(a^0_{ij}\rho_j)}$ is an approximation of the average dislocation spacing. This correction was later refined as the logarithmic correction should apply to the part of interaction strength related to contact reaction ($\approx$ 80%) and another contribution comes from long-range interactions ($\approx$ 20%) (Devincre et al. 2006; Kubin et al. 2008). Junctions – and their corresponding interaction coefficients – are typically weaker for bcc metals (Madec et al. 2004; Madec and Kubin 2008; Queyreau and Devincre 2009; Madec and Kubin 2017) than for fcc (Devincre et al. 2006; Madec and Kubin 2008, 2017). Among possible reactions, the collinear reaction that may occur between two distinct slip systems with same Burgers vector is of particular interest. First, while this reaction leads to the annihilation of the interacting segments, the effective strengthening obtained is by far the strongest among reactions as primary segments may move far from the collision point and many loops and debris are left behind. Contrary to other contact reactions, collinear annihilation does not lead to *direct* dislocation storage and can lead to unstable activity on active slip systems. This could explain the slip system selection observed for stable multislip conditions, where only noncollinear slip systems are activated. Second, collinear annihilation goes hand in hand with CS activity as screw dislocation have the possibility to deviate onto a plane of a collinear system. To the first order, reaction stability scales with the gain of elastic strain energy associated with the configurations where the reaction is formed when compared to the initially separated segments. Neglecting the character dependence and the logarithmic term, elastic strain energy of one segment roughly scales with $1/2\mu|\mathbf{b}|^2$. The cross-slip systems mechanism appears more ambivalent that the recovery mechanism it is commonly considered to be. At small to intermediate strains, CS along with collinear annihilations was instrumental in capturing the correct hardening of *stage I* in single crystals with an orientation such that a single slip system is activated (Devincre et al. 2007). The apparition of typical microstructures made up of entanglements is also intimately related to collinear annihilations. In bcc Fe, where CS can be considered to be very easy, the hardening effect of CS is even more apparent (Queyreau 2008) with the production of many debris and a large hardening rate in stage I. However, at larger stresses CS becomes an effective way to by-pass strong obstacles or stress concentrations, as soon as the external stress is sufficient to activate large segments in the cross-slip system. Only then is the recovery role of CS recovered (Kubin et al. 2009). CS activity will depend on applied stress, obstacle density and loading directions as it affects the Schmid factor on the cross-slip systems. Despite recent progresses in the description of this mechanism (Kang et al. 2014; Hussein et al. 2015; Longsworth and Fivel 2021), the full picture of CS impact on plastic deformation is yet to be provided.

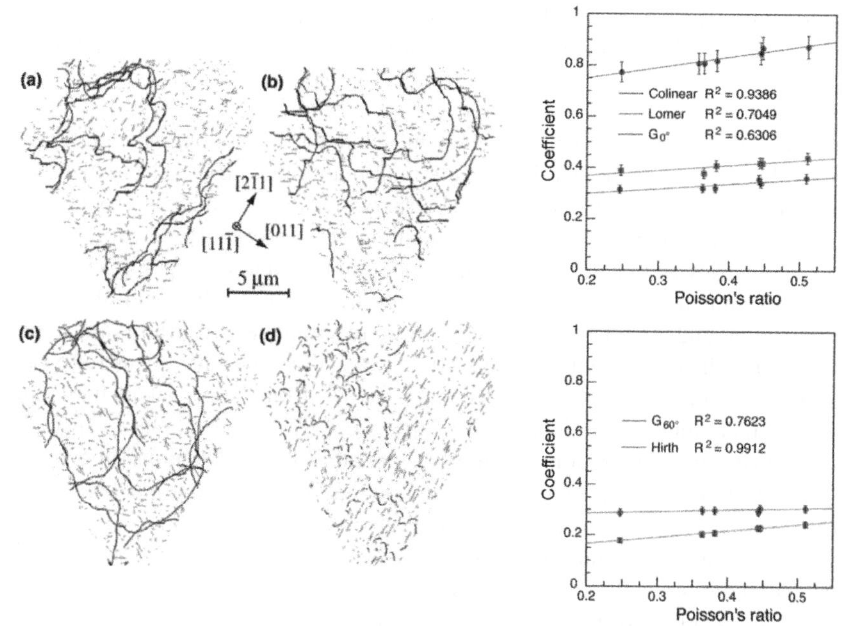

Figure 1.5. *(a–d) A thinfoil of $0.5\,\mu\mathrm{m}$ is extracted from sequences of latent hardening simulations in Cu (Devincre et al. 2006). Dislocations from the primary slip system (black) interact with mostly immobile forest dislocations (gray) leading to binary reactions (thick black). The observation plane corresponds to the primary slip plane. Note the various microstructures and junction length obtained for glissile, Lomer, Hirth and collinear reactions, in subfigures a, b, c and d, respectively. Linear scaling of the coefficients for binary reactions with elastic anisotropy of materials (Madec and Kubin 2017). The glissile reaction now exhibits two different configurations $G0$ and $G60$. The effective Poisson's ratio corresponds from left to right to Ge, Al, Ni, Cu, Ag and Au*

With an average length of mobile segment scaling as $k/\sqrt{a_{ij}\rho_j}$ ($\approx 75\%$ of the total length of dislocations, see below), collisions will occur mostly between pairs of mobile segments and binary junctions certainly represent the main portion of reactions occurring at least in the beginning of the deformation. More complex configurations exist and have drawn attention recently. For example multijunctions, may occur between a mobile segment and an already existing binary junction, as illustrated in Figure 1.6. As the reaction stability scales with the amount of elastic strain energy gained by the junction formation, these ternary objects may be very stable and strong pinning points. Bulatov and Cai (2006) demonstrated that multijunctions could be found in atomistic simulations and were confirmed by TEM observations. These ternary objects can also be found in spades in large-scale DDD simulations in bcc Mo, and they may act as strong anchoring points to the dislocation microstructures and operate as FR-like permanent sources leading to a noticeable extra hardening to binary junctions, at least for later deformation stages (Bulatov and Cai 2006).

Madec and Kubin (2008) investigated in detail the crystallography of ternary reactions. In the case of the bcc lattice, the multiplicity of $\{110\}$ and $\{112\}$ planes leads to an incommensurable number of ternary reactions, which is too large to tackle. Instead, the authors focused their attention on the set of 16 reactions – with various geometries and characters – that can be found when loading single crystals along high symmetry axes. Ten of these are *axial* reactions, in line with the parent binary junction, and six others are called *zigzag* reactions, which exhibit a different line orientation than the parent binary junction. Using the same equation as for binary junctions, strength of ternary reactions was assessed from model latent hardening simulations including an additional active slip system. Ternary reactions – especially delocalized zigzag configurations – can be stronger than binary junctions. In contrast, with a lower number of slip planes in fcc and putting aside coplanar reactions, three ternary zigzag reactions were considered and involve a glissile segment. As ternary reactions in fcc are very similar to the structure of binary reactions, they also slightly exhibit the same strength. Formally, we can imagine rewriting the critical stress (Franciosi and Zaoui 1982) as $\tau_c^i = \mu b\sqrt{\Sigma_j a_{ij}\rho_j + \Sigma_j\Sigma_l h_{ijl}\rho_j\rho_l}$ to capture higher order interactions. h_{ijl} are now ternary interaction coefficients between systems i, j and k.

When considering isotropic elasticity, interaction coefficients depend solely on the crystal structure and nature of the reaction formed. This is however not the case when considering anisotropic elasticity. Madec and Kubin (2017) recently probed this effect in the case of binary reactions in several fcc (Ge, Al, Ni, Cu, Ag, Au) and bcc (Mo, Nb, W, Ta, Fe) materials at the onset of the athermal regime (see Figure 1.5). Anisotropic elasticity was approximated through effective isotropic elastic constants and line tension expression from Bacon et al. (1979). For fcc materials, glissile junctions are now associated with a geometrically asymmetric configuration regarding the mobile segments orientation, which leads to an asymmetry in the junction mechanical stability. a_{ij} and a_{ji} now differ when active- and forest-slip systems are swapped. a_{ij} coefficients differ slightly from previous calculation with isotropic elasticity and they scale linearly with the Poisson's ratio. The picture is somewhat much more complicated for bcc materials with no such simple scaling of the coefficients. More reactions existing between $\{110\}$ and $\{112\}$ slip systems are now asymmetrical.

The glissile reaction has again recently attracted some attention. Stricker and Weygand (2015) recently assessed the impact of glissile junctions in the plastic behavior of micropilars. Similarly to CS segments or pairs of multijunctions, glissile reactions can also operate as FR-like sources if the local stress is sufficient.

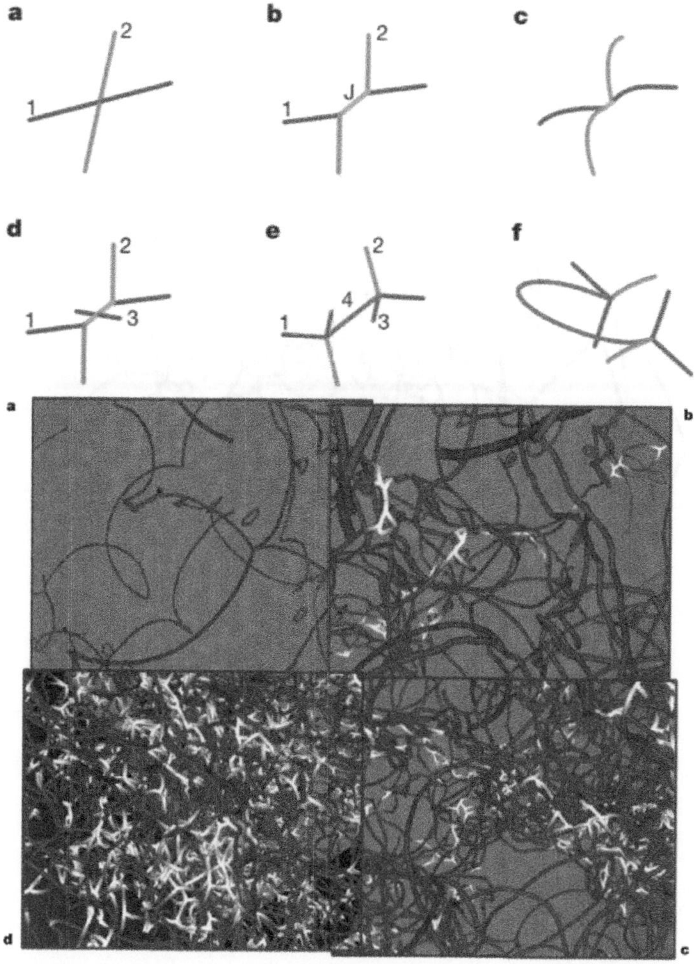

Figure 1.6. *Top panel: two dislocations from secant slip systems (a) collide and form a binary reaction (b) (Bulatov and Cai 2006). The destruction – unzipping – of the junction occurs though the lateral motion of parent segments. A third segment collides with the binary junction (d) leading to the formation of a ternary junction (e). The multijunction Burgers vector in bcc materials could be:* $1/2[\bar{1}11] + 1/2[1\bar{1}1] + 1/2[11\bar{1}] = 1/2[111]$. *If its length and local stress are sufficient, the multijunction can act as a dislocation source (f) around the quadruple nodes. Bottom panel: snapshot of the microstructure simulated during a [001] deformation in bcc Mo. Green segments are associated with binary collisions that are mostly observed initially (a) and at the onset of plastic flow (b). Then dislocations connected to four-nodes and mulltijunctions are seen in shades of white and their contribution increases with strain, up to* 2% *(d)*

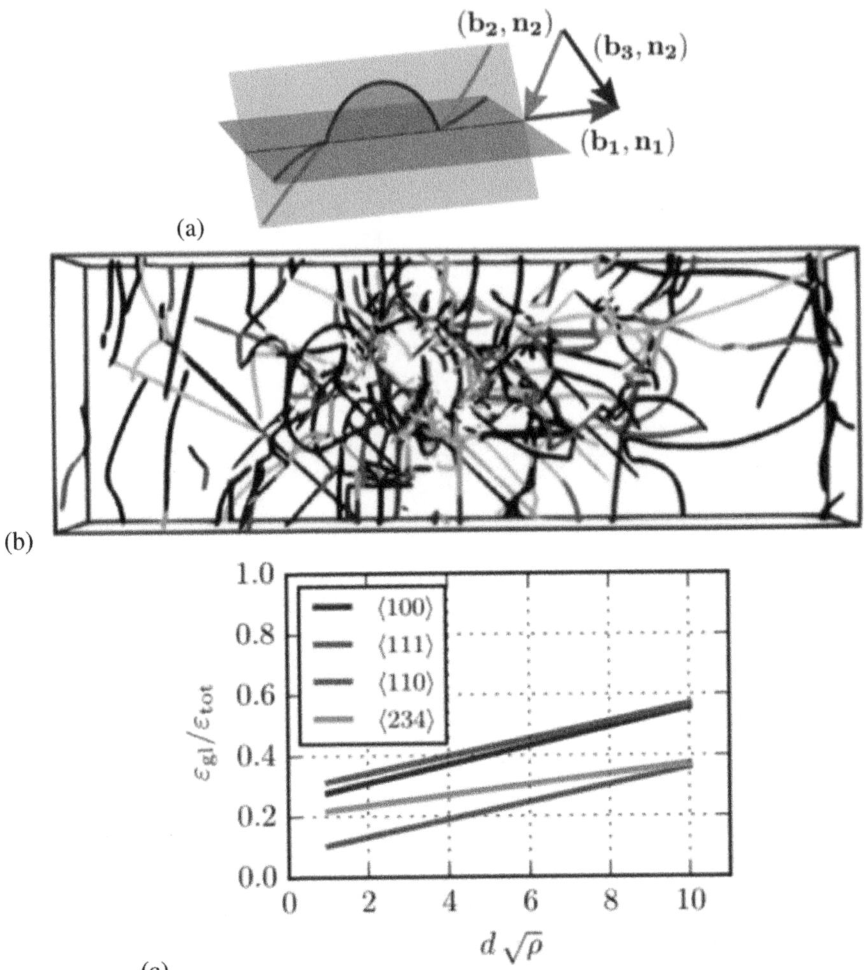

Figure 1.7. *(a) Schematic representation of a glissile junction formed in fcc, where the reaction leads to a Burgers vector b_2 that can glide along plane n_2 is the local stress is sufficient (Stricker and Weygand 2015). (b) Dislocation microstructure obtained in a single crystal deformed for 0.5% along <110> with a diameter of $1\,\mu\mathrm{m}$. Colored dislocations are coming from glissile reactions formed on different planes. Other dislocations are in black. Free surfaces are used. (c) Relative contribution of glissile activity ϵ_{gl} to the overall plastic activity ϵ_{tot} observed in average for various loading directions $< hkl >$ (Stricker and Weygand 2015)*

Since glissile junctions are virtually identical to mobile dislocations, the authors developed tools to track the plastic activity imputed to glissile junctions over a wide range of initial microstructure realizations and simulation sizes D up to 2 μm (see Figure 1.7). The contribution of glissile segments to the overall plastic deformation was found to be in the range of 10%–50% and roughly scaling with $D\sqrt{\rho}$. The glissile contribution is certainly much smaller in large crystals. Activation of a segment of length L akin to a FR source scales with $1/L$. Binary junctions, including glissile ones, typically exhibit an average length $1/\sqrt{a_{ij}\rho_j}$, which is much smaller than mobile *parent* segments (Devincre et al. 2006). Besides, the presence of free surface is strongly attractive to dislocations in the bulk leading possibly to a strong annihilation at the surfaces and to dislocation source starvation.

1.3.2. *Statistical investigations of dislocation mechanisms*

Most large-scale DDD simulations – including the references discussed in other sections of this chapter – aim to sample the wide variety of dislocation interactions and elementary mechanisms to provide an averaged and representative behavior of the simulated material. In this section, we highlight recent studies that aim to characterize and quantify the variety of dislocation configurations and corresponding distributions. A first example of this type of work was conducted by Devincre et al. (2006) who determined distributions of junctions lengths l_j and mobile segment lengths on the primary system l_p (see Figure 1.8). This was achieved on model simulations of latent hardening (see previous section) and [111] and [001] multislip simulations of fcc Cu with various forest dislocation densities ρ_f. In model simulations, junction length increases with the length of forest segments (Queyreau and Devincre 2009) until reaching a plateau of the order $< l_j >\approx 0.25\rho_f^{-1/2}$ (Devincre et al. 2006) that is independent of the nature of junctions. The average mobile segment length however was found to scale with the strength of the junction as $< l_p >\approx 1/\sqrt{a_f\rho_f}$ where a_f is the interaction coefficient. Recently, Sills et al. (2018) proposed a similar analysis for a [001] multislip simulation in Cu. These authors observed a distribution of primary segments that is mostly consistent with an exponential distribution of the form: $p(l_p) = (1/ < l_p >)\exp(-l_p/ < l_p >)$ (see also Figure 1.8). The averaged $< l_p >$ was also found to be proportional to $1/\sqrt{\rho}$. The authors explain the exponential distribution of l_p by the fact that the probability to obtain a collision scales with the given mobile segment length l. Neglecting the junction length, the two new mobile segments created around the junction splits the length into l_1 and $l - l_1$. The junction can then be seen as a 1D Poisson process. The authors define two dimensionless coefficients: $\phi = N^2/\rho$ with N the number of segments, and β the rate of stable junction formation, and recovered the classical rate equation for dislocation storage (Kocks and Mecking 2003) (see below). Another key ingredient entering in the modeling of materials hardening with deformation thus lies in the storage rate of dislocations $d\rho$ with deformation $d\gamma$. Following the seminal work of Kocks and Mecking (2003), storage rate on system "i" in pure fcc materials in the athermal regimes is:

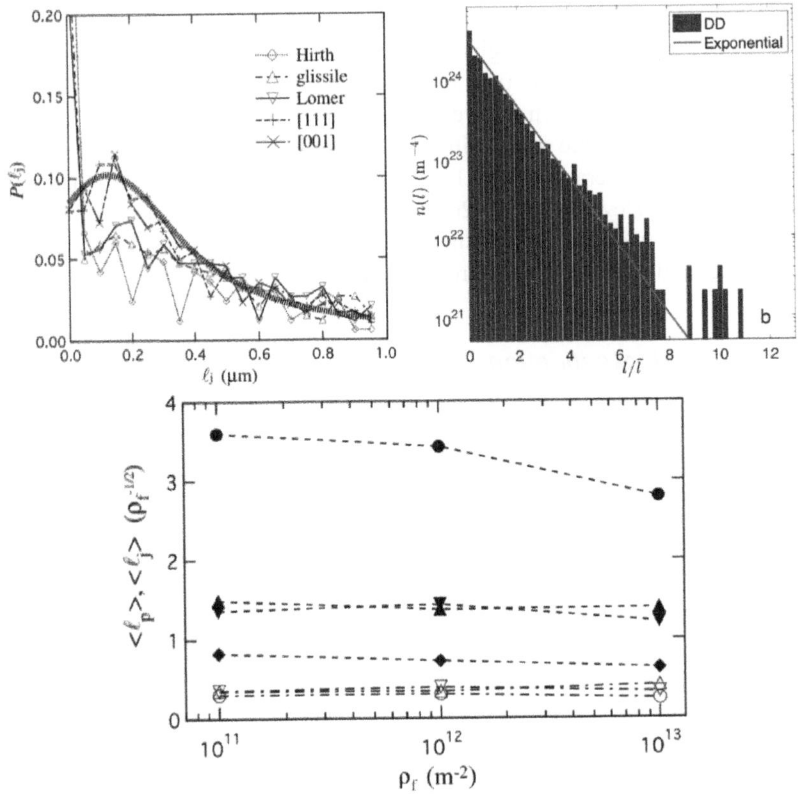

Figure 1.8. *Top left: distribution of junction lengths l_j under stress and at a density of 10^{12} m^{-2}. Empty symbols correspond to latent hardening simulation and crosses to simulations in multislip conditions (Devincre et al. 2006). Top right: distribution of primary segments – link between two reactions – determined during a [001] deformation simulation and about 1 shear strain (Sills et al. 2018). Bottom: from the distribution (a), scaling of the average primary $< l_p >$ $(= \bar{l})$ segments and junction lengths $< l_j >$ with forest dislocation density ρ_f (Devincre et al. 2006)*

$$\frac{d\rho_i}{d\gamma_i} = \frac{1}{b}\left(\frac{1}{L} - y_c\rho_i\right) \qquad [1.18]$$

where L is the mean free path of mobile dislocation before being temporary or permanently locked in the microstructure and y_c is the annihilation distance for screws controlling in part the dynamic recovery of $stage\,III$ (see Kocks and Mecking 2003 and Kubin et al. (2009) for more details). The MFP of dislocations is typically poorly known from experiments and is commonly expressed as $L = \sqrt{\Sigma a_{ij}\rho_j}/K_{hkl}$. This physical quantity depends on the number of active slip systems and thus the loading direction. In Devincre et al. (2008), the MFP is broken into three different physically based parameters that are defined from averages on large-scale simulations on few specific crystal orientations. These parameters correspond namely to (i) the rate of formation of stable junctions p_o close to 10%, and the two average lengths in the shape of (ii) free segments $< l >= k_0/\sqrt{\bar{a}\rho}$ and (iii) the junction segment $< l_j >= \kappa < l >$ that are stored in the process, with $k_0 = 1.08$ and $\kappa = 0.29$. Interestingly, Sills et al. arrived to the same storage functional as Kocks and Meckin from a different route. For this, the authors also define a rate of formation of stable junction and a pinning point spacing as $\lambda = 1/\sqrt{0.55\rho}$. The storage rate is then defined from exponential distribution of segment length described above. The complete CP model proposed by Devincre et al. (2008) was capable of reproducing most of the features of single-crystals deformed monotonously, while it was parameterized solely from DDD.

Recently, Akhondzadeh et al. (2020, 2021) performed a large-scale study of the deformation of Cu single crystals over about 120 different orientations within and at the border of the standard triangle in the absence of CS. Their simulations confirmed the importance of the glissile reaction (Stricker and Weygand 2015) to the strain hardening of fcc systems. Unexpectedly, significant dislocation multiplication was observed on slip systems with no applied stress or strain rate. This was also observed in recent ultra-scale MD simulations (Zepeda-Ruiz et al. 2021). This phenomenon referred to as *slip-free multiplication* was particularly linked to the glissile and coplanar interactions. Indeed, a binary glissile junction may act as a source of dislocation in the glissile slip plane (akin to Stricker and Weygand (2015)) by a direct glissile mechanism. These authors proposed the following modification of the storage rate (Akhondzadeh et al. 2020, 2021):

$$\dot{\rho}_i = \dot{\gamma}_i \left(\frac{c_1}{b} \sqrt{a_{ij}\rho_j} - c_2 \rho_i \right) + \frac{c_3}{b} \left(\dot{\gamma}_{i'} \sqrt{\rho_{i''}} + \dot{\gamma}_{i''} \sqrt{\rho_{i'}} \right) \quad [1.19]$$

where the first term in the rhs is similar to equation [1.18], and the second term accounts for the additional storage induced by slip systems *i'* and *i''* coplanar of system *i*. The constants c_1, c_2 and c_3 are dependent on the loading orientation and were fitted on the DDD data. It is interesting to note that the ratio of c_1 corresponding to [135], [112], [111] and [001] are identical to those in Devincre et al. (2008), while y_c is not defined from DDD. In Akhondzadeh et al. (2021), the authors proposed an analysis of the node types found in the set of $\approx$ 120 simulations. A large part of the nodes could not be categorized in terms of classical binary reactions, but correspond to higher order reactions. In particular, the authors identified two new mechanisms named *glissile + coplanar* and *assisted glissile mechanisms* that could contribute to the slip-free multiplication observed in the simulation (see Akhondzadeh et al. (2021) for more details).

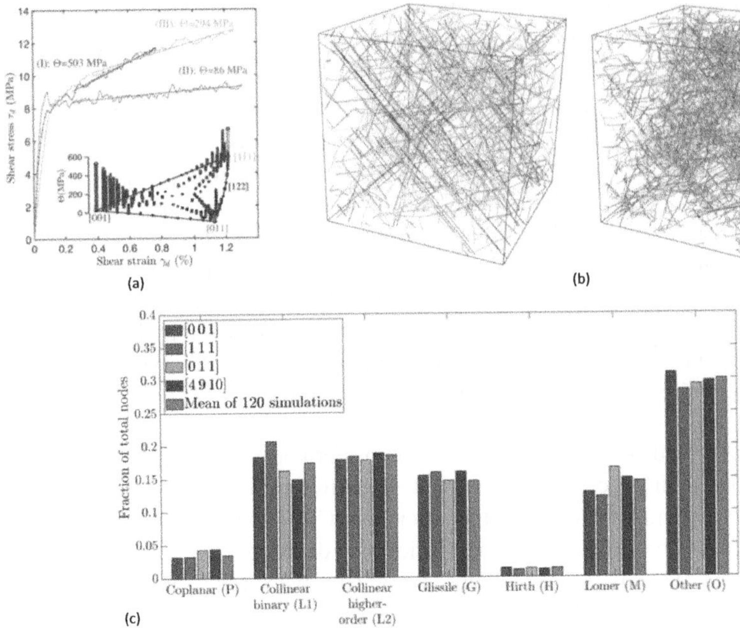

Figure 1.9. *(a) Simulated strain curves for Cu single crystals by DDD for three high symmetry loading directions (Akhondzadeh et al. 2021). The inset figure shows the smooth evolution of the simulated strain hardening rates for about 120 different loading directions spanning the standard triangle. (b) Illustration of the complexity that develops rapidly in the dislocation microstructure after* $\gamma = 0.5\%$ *for a* $[4\,9\,10]$ *loading direction. (c) Distribution of the type of dislocation nodes found in the simulated network for various loading directions at* $\gamma = 0.5\%$ *(Akhondzadeh et al. 2021). Nodes are categorized as the five* $\mathit{classical}$ *binary junctions, higher order interactions from a gliding reaction (L2), and all other configurations are grouped into the other type*

In a series of papers, El-Azab and coworkers performed an extensive statistical analysis of time and spatial correlations existing in dislocation microstructures (Deng and El-Azab 2007, 2010; Xia et al. 2016; Anderson and El-Azab 2021) with the ambition in mind to provide a continuum dislocation dynamics (CDD) descriptions based on DDD results. Similarly to crystal plasticity, CDD aims at providing a continuous description of a dislocation system and its evolution by means of dislocation densities, but contains additional physical ingredients, for example, lines orientation or line curvature to capture more naturally the increase in line length and possibly reproduce dislocation patterns. To this aim, time and spatial coarse grainings must be carefully performed. Interested readers are referred to other chapters on continuous dislocation fields theories in the present book. Deng and El-Azab (2007) determined from DDD the pair correlations that exist during the deformation of Mo single-crystals along [001]. The radial correlation existing between pairs of segments of the same slip systems is peaked at $\mathbf{r_{ij}}|- > 0$ due to the line connectivity and displays oscillation indicating preferred spacing distances (see Figure 1.10). The repartition in these preferred spacings decreases with the separation distance and these amplitude oscillations in the correlation functions decrease with strain. The pair correlation is also strongly anisotropic as it depends upon the two angles necessary to orient the dislocation line. Dislocations are correlated at short distance, again due to connectivity and can be correlated or anti-correlated at medium distance, and uncorrelated or anti-correlated at long range. The authors suggest that the behavior at an intermediate range can be due to the distance interactions between dislocations that may be attractive or repulsive. Spatial correlations among pairs of segments belonging to different slip systems exhibit similar features, with a peak at close range and oscillations at preferred distances that decrease in intensity with strain. The correlation patterns are found to be very anisotropic. Rationalizing pair correlations remain however a difficult task due to the large number of slip systems and geometric and physical parameters affecting correlations (Deng and El-Azab 2010).

Within the framework of CDD, density evolution is controlled by the rate of reactions R and cross-slip systems events, which are described as stochastic processes. Defining a relevant mesoscopic time scale at which these processes can be seen as continuous becomes key. Deng and El-Azab (2010) investigated the temporal statistics and the rates of reactions (junctions and annihilations) and cross-slip systems events. For this, the authors used $marked\,point\,process$ and $time\,series$ to model the statistics observed in the DDD data. The correlation times for junctions and annihilation are found to be of the same magnitude and much smaller that correlation time for CS events. The rate of junction events was found to exhibit strong fluctuations that are superposed to a low background activity slowly increasing with strain. As two successive events within a correlation time are seen as correlated, a shorter correlation time means that events are independent. Interestingly, the probability distributions $p(R)$ of rate parameter also follows a power law scaling as $\approx R^{-k}$ with $k \approx 2 \pm 0.5$. The analysis of CS events was then refined in Xia et al. (2016) where CS rates are not considered as stationery but rather increase with strain. As mentioned earlier, the local stress fluctuations building up with accumulated strain will eventually favor CS events (Queyreau 2008; Xia et al. 2016). The coarse grained rate for CS events now includes an additional stochastic component to reproduce fluctuations, and is accounted for in CDD by a Monte Carlo scheme. This solution ultimately made possible to reproduce the $\mu b/\tau_c$ scaling of the deformation cells (see Figure 1.10), which is a long-standing challenge for macroscopic approaches (while failing in other quantitative areas) (Xia et al. 2016).

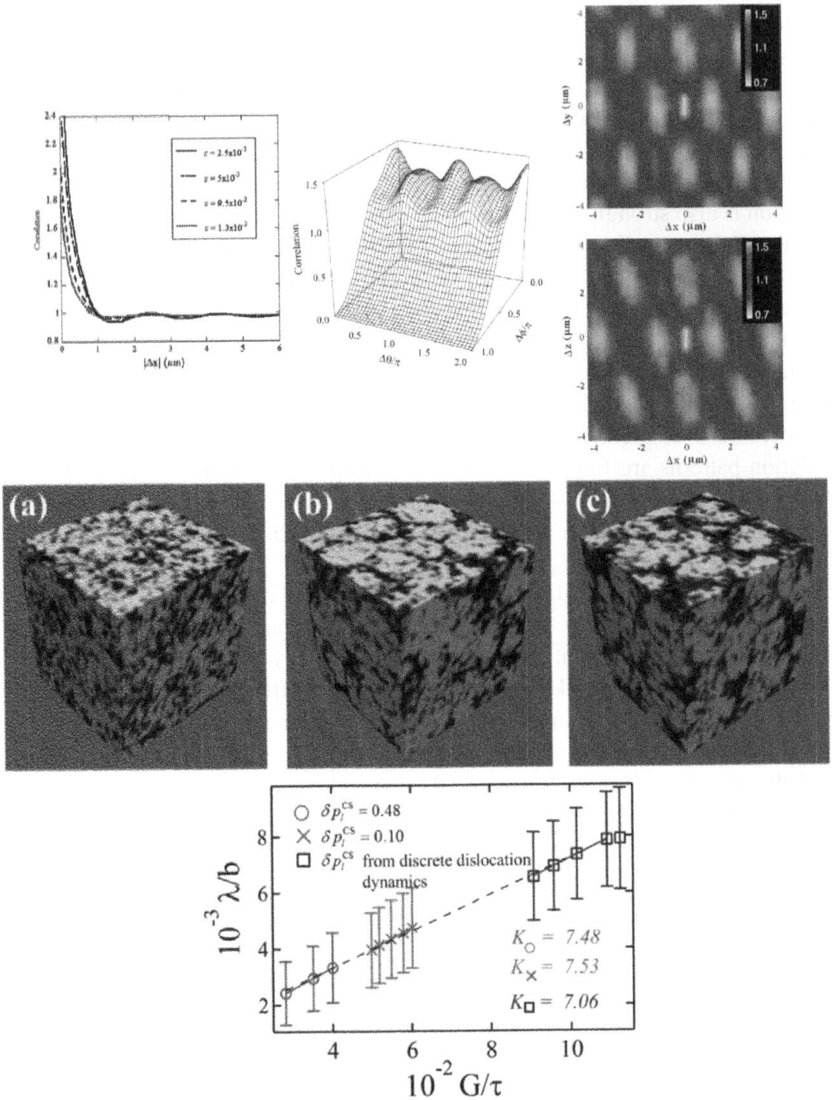

Figure 1.10. *Top panel: (a–d) pair correlations* $\zeta^{(\alpha_1\alpha_1)}$ *observed during a DDD simulation of Mo deformed along [001] among same type of dislocations* α_1 *(*$\mathbf{b} = a_0/2[111]$*) (Deng and El-Azab 2007). Pair correlations can be described in a 5D parametric space with* $\mathbf{\Delta x} = \mathbf{x_2} - \mathbf{x_1}$ *the position vector and* $\mathbf{\Delta}\theta = \theta_2 - \theta_1$ *and* $\mathbf{\Delta}\phi = \phi_2 - \phi_1$ *two spherical angles. See text for more details. Middle panel: CDD simulation of the patterning of the dislocation density during a [001] deformation at 0.5, 1 and 1.28% of strain (Xia et al. 2016). Scaling of the dislocation cell size* λ *as function of shear modulus* G *and shear stress* τ*. CDD results are compared with DDD results from Xia and El-Azab (2015)*

1.3.3. *Lattice friction*

A number of DDD investigations of the thermal regime of materials with large lattice friction exist in the literature. While many of them include a thermally activated character of screw mobility corresponding to DK nucleation and propagation, only a few of them account for the dislocation length dependence that is instrumental in capturing the thermal regime of plastic flow (Tang et al. 1998; Monnet et al. 2004; Naamane et al. 2010; Monnet et al. 2011). In Naamane et al. (2010), such a mobility law was parameterized from experiments conducted on single crystals of pure α-Fe or α-Fe with small amount of impurities and focusing on $\{110\}$ systems. Naamane et al. paid great care in the definition of the onset of plastic flow, which can be very different from the engineering estimate owing to the presence of an extended $stage\,0$. During this initial stage, edge dislocations that are very mobile produce most of the plastic deformation leading to an imperceptible hardening until they annihilate, become blocked or exit the sample. At the end of this stage, the microstructure is almost essentially composed of screw dislocations with a slow and thermally activated mobility (see Figure 1.11), fitting the classical picture of low temperature plasticity (Louchet et al. 1979; Caillard 2013; Kubin 2013). At the timescale of DDD simulations, the backward jumps of DK cannot be neglected and the exponential form of equation [1.1] may transform into hyperbolic sine function. For reasons of numerical efficiency, mobility functions of non-screw dislocations in Naamane et al. (2010) are set to be proportional to the mobility of screw dislocation (and is independent of non-screw length). The ratio between non-screw and screw characters is set such as $k >> 1$, with k increasing as the temperature decreases. These ingredients allowed the authors to reproduce the onset of plastic flow of Fe single crystals and the evolution of dislocation microstructures with decreasing temperatures (Naamane et al. 2010) (see Figure 1.11).

Monnet et al. (2011) investigated the Orowan strengthening at low temperature using the mobility laws described above, first in the case of periodic arrays of impenetrable precipitates. This step is necessary to rationalize the results observed in more complex configurations. In the case of edges, the authors found that the additional stress to the lattice friction remain controlled by line tension in agreement with the so-called Bacon Kocks Scattergood (BKS) model (Bacon et al. 1973), as the mobility of edge dislocations is independent of length. This quasistatic behavior contrasts with the case of screw dislocations that exhibit a strain rate dependence. The Orowan strengthening still depends upon spherical size D and spacing L of the precipitates but now the extra stress is significantly decreased with temperature when compared to the BKS model. Due to the length dependence in their mobility, screw dislocations remain mostly straight when forced against precipitates and non-screw arc is formed closed to precipitates. The authors identified two different configurations depending on the relative ratio of the border-to-border spacing $L - D$ and twice the local curvature adopted 2κ. For large precipitate spacings such that $L - D > 2\kappa$, the Orowan stress corresponds to the change of stress induced by the change of screw length. This first result is consistent to earlier model proposed in Louchet et al. (1979); Kubin (2013) from in situ TEM observations. The arc formed by the non-screw segments is not controlled by line tension (Kubin 2013) but by the production of DK. Monnet et al. (2011) provided an analytical model assuming that screw and non-screw velocities are identical. For small particle spacing $L - D < 2\kappa$, the initially screw dislocation bows out so much that the screw length is considered to be close to the critical DK length, and line tension effects need to be included.

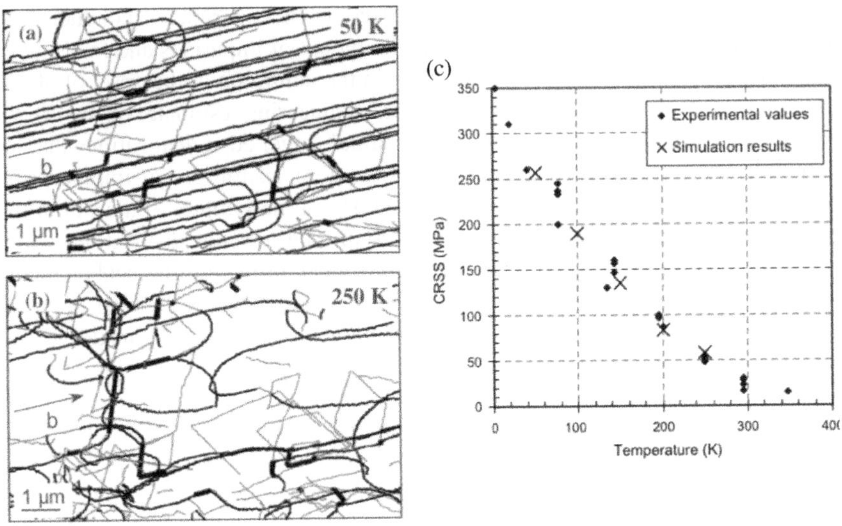

Figure 1.11. *Evolution of the simulated dislocation microstructures with temperature in bcc Fe from Naamane et al. (2010). (a) and (b) thin foils of 1 μm are extracted from the simulation lying in the plane of the primary slip system. Primary dislocations (in black) form junctions (thick black) with forest dislocation (gray). At 50K (a), the dislocation microstructure is formed of long screw dislocations controlling the plastic flow, fewer and shorter junctions are observed. (c) The critical resolved shear stress (CRSS) obtained by DDD is in good agreement with experiments (after correction of the stage 0 existing at low T) (Naamane et al. 2010)*

In the same study, the authors investigated the low-temperature strengthening induced by now random distributions of precipitates on initially infinite screws. In steady-state conditions, the area swept by dislocation equals $A = \lambda^2 = \bar{L}(\kappa + D/2)$ where $\bar{L}$ is average particle spacing in the length direction. In these massive calculations, the precipitate microstructure was such that $\bar{L} - D > 2\bar{\kappa}$. The additional stress induced by Orowan particle was therefore derived from the periodic raw model as:

$$\Delta\tau_{Orowan} = \tau_P \frac{2k_B T}{\Delta G_0} \ln\left(\frac{l_0}{\bar{l}}\right)\left(2 + \frac{k_B T}{\Delta G_0} \ln\left(\frac{v^2}{H^2 l_0 \bar{l}}\right)\right) \qquad [1.20]$$

with τ_0 and ΔG_0 the parameters first appearing in the screw mobility in equation [1.1] and $\bar{l} = 1/\left(D\rho\left(\frac{\mu b}{\tau_{app}} + \frac{D}{2}\right)\right)$.

Naamane (2008) investigated the evolution of forest strengthening at low temperature making use of model latent hardening simulations. To mimic low-temperature microstructures, long mobile screw dislocations interact with forest dislocations made of screws and may form symmetrical mixed junction. When a junction is formed, one of the ternary node is zipping the reaction in the direction of screw motion while the other is zipping in the opposite direction. This may help the motion of one of the mobile arm of the junction, and ultimately leading to an asymmetry in the unzipping process depending on which of the two arm is the most mobile (longest). Interestingly, the average junction length was found to decrease with temperature (Naamane et al. 2010). The additional strengthening due to forest interactions $\Delta\tau_F$ to the lattice friction was found to decrease as the temperature decreases. Finally, the additional $\Delta\tau_F$ at low temperature was found in qualitative agreement with the model proposed by Louchet et al. (1979). Recalling that the screw velocity is function of the applied stress and screw length, its expression can be used to determine the effective stress to set in motion screws of a given length.

Monnet et al. (2013) proposed a bit later a unified crystal plasticity formalism based on these DDD results that encompasses both the low and high temperature behaviors of bcc metals (Monnet et al. 2013). For this, the shear rate equation associated with screw gliding on slip system "i" is obtained from the so-called Orowan equations as:

$$\dot{\gamma}^i_{friction} = 2\rho^i_{screw} b^i v^i_{screw}, \qquad [1.21]$$

where ρ^i_{screw} and v^i_{screw} are the density and average density for screw dislocations on "i", respectively. v^i_{screw} can be defined from the screw mobility equation and the average obstacle spacing. In the athermal regime, the shear strain rate equation exhibit the so-called power expression:

$$\dot{\gamma}^i_{drag} = \dot{\gamma}_0 \left(\frac{|\tau^i_{app}|}{\tau^i_c}\right) \mathrm{sign}(\tau^i), \qquad [1.22]$$

with $\dot{\gamma}_0$ a constant and m the strain rate sensitivity coefficient. The strain rate on system "i" is thus the harmonic mean of these two strain rate:

$$\dot{\gamma}^i = \left((\dot{\gamma}^i_{friction})^{-1} + (\dot{\gamma}^i_{drag})^{-1}\right)^{-1} \qquad [1.23]$$

Note that interaction coefficients and recovery parameters appearing in the equation for the dislocation density evolution must be altered due to the presence of the lattice friction (Monnet et al. 2013). This model captures most features of the plastic deformation of bcc single crystals. More details on the peculiar dislocation physics observed at low temperature can be found in Kubin (2013); Monnet et al. (2013).

1.3.4. *Combination of multiple strengthening mechanisms*

In real materials, several strengthening mechanisms are often operating simultaneously, such as several reactions and obstacles pinning dislocations (e.g. elements in solid solution or second-phase particles). The question of how to combine the various strengthening associated with define the overall flow stress τ_{tot} from the separated mechanism strengths τ_c^m becomes important and exhibits generally the following form: $(\tau_{tot})^k = \Sigma_m(\tau_c^m)^k$ (Brown and Ham 1972). The relative strengths and densities of the dislocation obstacles may dictate the choice of exponent k. The quadratic law of mixture where $k = 2$ is only exact when the mechanisms have identical strengths but densities are different, while $k = 3/2$ is exact when the two have identical densities but strengths are different. Interestingly, these two extreme cases are rarely very far from each other for a wide range of relative densities and strengths. The laws of mixture have been identified from DDD (Monnet 2006; Queyreau et al. 2010). Monnet (2006) has determined the law of mixture associated with the strengthening associated with two different populations of incoherent precipitates in Zr Alloy system. Only a mixture law close to the one proposed by Brown was capable of predicting the set of DDD results: $\tau_{total} = \left(\frac{\rho_1}{\rho_1+\rho_2}\right)\tau_1 + \left(\frac{\rho_2}{\rho_1+\rho_2}\right)\tau_2$, where ρ_i are area densities of the two families of obstacles. In Queyreau et al. (2010), different laws of mixture have been compared and a quadratic law provided the best agreement with various strengthenings ranging from forest interaction to spherical impenetrable precipitates. Linear superposition is often considered to be justified for the combination of relatively large or strong obstacles among a population of weak obstacles, say as in the case of solute atoms with forest dislocations or precipitates.

Finally, when considering most of possible strengthening mechanisms (Monnet 2018), a generic law of mixture can be found in the following form:

$$\tau_{total} = \tau_{Sol.Sol.} + \tau_{HP} + \sqrt{\tau_{forest}^2 + \tau_{precipitates}^2} \qquad [1.24]$$

where $\tau_{Sol.Sol.}$ is the strengthening induced by solid solution when present, $\tau_{precipitates}$ is the strengthening associated with strong second-phase particles (i.e. Orowan by-passing) and τ_{HP} is the Hall–Petch contribution (see next section).

1.3.5. *Toward polycrystalline plasticity*

Accounting for the formation and presence of grain boundaries (or interfaces) is certainly the new frontier for DDD simulations. This is associated with a number of challenges and/or open physical issues. (i) The presence of internal interfaces makes the resolution of the stress and displacement fields much more difficult in comparison to the case of the single crystals, as already discussed in section 1.2.4. (ii) In addition to the remote interaction existing between the stress fields of crystals dislocations and interfaces, a set of contact interactions are possible. These range from impingement, partial absorption or recombination with the interface defects to direct or indirect reemission of dislocations on the other side of the interface. (iii) The plastic behavior may then differ from the one observed in single crystal with additional dislocation accumulation at grain boundaries, different strain and stress localizations, modified hardening and size effects. However, since the overall behavior of (iii) depends on the stress state in (i) and elementary mechanisms that are operating in (ii), the full picture of the polycrystalline plasticity is certainly still out of reach for many years to come. At small (atomic) scale, the possible interactions between a single dislocation and an interface are rather difficult to rationalize and thus to model at the mesoscale. Remote interactions between a GB and dislocations can lead to the repulsion or attraction of the dislocations. At close range, attracted dislocations may be blocked close to the GB, but they may also be partially absorbed by GB and/or recombined with GB dislocations. These mechanisms involve the core structure of the dislocation and of the interface. However, reproducing and characterizing the atomic structure of GB is in its own right a complex issue, except for some simple cases (i.e. symmetrical pure tilt coincidence-site lattice (CSL) GB of low misorientation angle and/or Sigma value). Burgers vectors of GB dislocations may belong to DSC lattice (Sutton and Balluffi 1995), which is defined from the dichromatic structure of the GB. After absorption, a new dislocation can eventually be emitted in the adjoining crystal, leaving behind a residual dislocation. The mechanical picture seems a priori difficult to assess once more, but in de Koning et al. (2003), a qualitative line tension model was proposed to evaluate the chance of a given slip system to be transmitted across several simple CSL GBs. If direct transmission is associated with complex dislocation rearrangements and large stresses to operate, indirect transmission could be a more effective mechanism to plastic transmission through activation of another slip system in the adjoining grain by local stress concentrations. Finally, for polycrystals with micrometer grain sizes, additional mechanisms such as GB source nucleation and GB sliding are suggested.

In the absence of generic rules to describe short-range interactions between crystal dislocations and interfaces, different modeling strategies have been proposed. Interfaces have been modeled in DDD as impenetrable boundaries (de Sansal et al. 2009; Aifantis et al. 2009; Zhang et al. 2014; Stricker et al. 2016), partially impenetrable boundaries but with direct transmission of dislocation permitted according to a simplified criterion (Fan et al. 2015), explicit GB structure as simple dislocation arrays in a continuous matrix (Liu et al. 2011, 2012; Burbery et al. 2017) or gamma/gamma prime coherent interfaces using DCM modeling in Ni-based super alloys (Vattré et al. 2014; Jamond et al. 2016). As was the case for single-crystal simulations, isotropic elastic framework is mostly employed here. The choice of modeling thus directly impacts the elementary mechanisms of interaction allowed with the GB, but this choice will also impact the overall plastic response of the simulated polycrystal. The simulated plastic behavior will also be impacted by other parameters such as the geometry of grains, relative grain orientation with respect to neighboring grains and grains dimensions. Simulation configurations thus cover a wide range of geometries from bi- or tricrystals (Aifantis et al. 2009; Zhang et al. 2014; Devincre and Gatti 2015), perfectly periodic cubic grains (Fan et al. 2015; Jiang et al. 2019), aggregates made up of four cubic or rectangular grains (Jiang et al. 2019) to 16 dodecahedral grain aggregates (de Sansal et al. 2009). Due to the usage of PBC and limitations in computing resources, these configurations are very regular and idealized when compared to polycrystalline microstructures existing in real materials. One of the first impacts of the presence of GB, is the reduction of dislocation mean free path. A number of studies have investigated the accumulation of dislocations at GB and the resulting backstresses. In Daveau (2012) and Devincre and Gatti (2015), a few sets of large and semi-infinite bicrystals have been investigated with DDD, where one or the two grains are plastically deformed in single glide. The selected grain orientations matched the actual orientation found in a real tricrystal of Ni (see Figure 1.12). The accumulation of geometrically necessary dislocations (GND) at the GB was found to differ from the textbook pile-ups obtained from FR sources as the initial microstructures are made of prismatic loops and cross-slip systems are operating. The authors proposed a modified pile-up expression reproducing DDD results including the GND density decrease with normal distance and slip-plane orientation. The total GND density is then included in modified CP model following (Kocks and Mecking 2003): (i) this density is removed from the Taylor expression; (ii) an additional storage is expressed as $d\rho^i_{GND}/d\gamma^i = 0.5/br$, with r the distance from the GB and; (iii) the GND presence is also at the origin of a backstress that is included in the yield criterion. In Aifantis et al. (2009) and Zhang et al. (2014), DDD simulations are employed for the deformation of a tricrystal in Al. Two impenetrable, parallel grain boundaries of pure twist around [001] separate the three crystals arranged in series (see Figure 1.12). Deformation is imposed along [001] that corresponds to a multiple slip condition. The initial microstructure is made of FR sources and different lengths were tested. The distribution of GND at GB is analyzed in terms of dislocation density tensor calculated over slices of the simulation box. Alternatively to a CP model, DDD results are used in this study to implement a gradient plasticity (GP) model (see corresponding chapter in this book). Few parameters are fitted from the confrontation of DDD simulations with the 1D resolution of the GP. The backstress induced by GND dislocations is a natural outcome of the model. Interestingly, the internal length associated with the higher order term is not constant but scales with the controlling length scale, for example, the FR source length or the forest spacing. Besides, while

GB is impenetrable in the DDD simulations, a better agreement is found when the GB is modeled as deformable interface in the GP with an efficient GB strength defined from dislocation activity within the grains (see also Stricker et al. 2016). The same DDD simulation has been also compared to a different GP model in Bayerschen et al. (2015). Both Crystal Plasticity (CP) (Daveau 2012; Devincre and Gatti 2015) and GP (Zhang et al. 2014; Bayerschen et al. 2015) models are self-consistent in the sense that they match rather well the DDD simulations. It would be interesting to test their predictability to other cases.

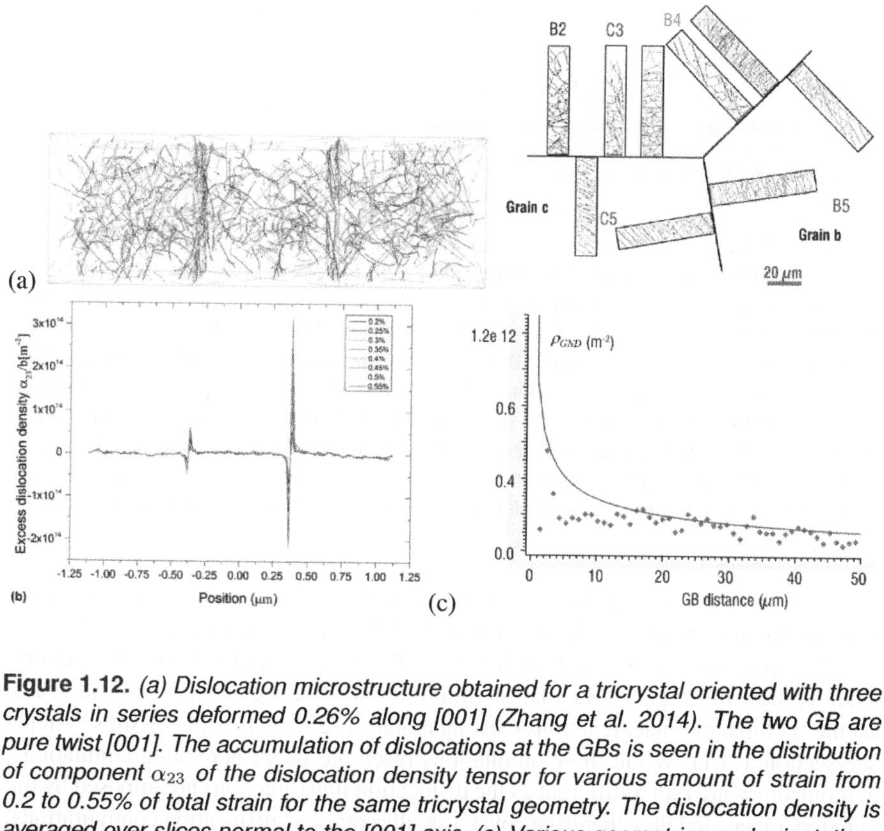

Figure 1.12. *(a) Dislocation microstructure obtained for a tricrystal oriented with three crystals in series deformed 0.26% along [001] (Zhang et al. 2014). The two GB are pure twist [001]. The accumulation of dislocations at the GBs is seen in the distribution of component α_{23} of the dislocation density tensor for various amount of strain from 0.2 to 0.55% of total strain for the same tricrystal geometry. The dislocation density is averaged over slices normal to the [001] axis. (c) Various geometries and orientations of DDD simulations that probe the configurations existing in a real tricrystal of Cu (Devincre and Gatti 2015). The colors and numbers correspond to the active slip system following the Schmid and Boas notation. Corresponding average profile of the distribution of dislocation density with respect to GB distance is in red (Devincre and Gatti 2015)*

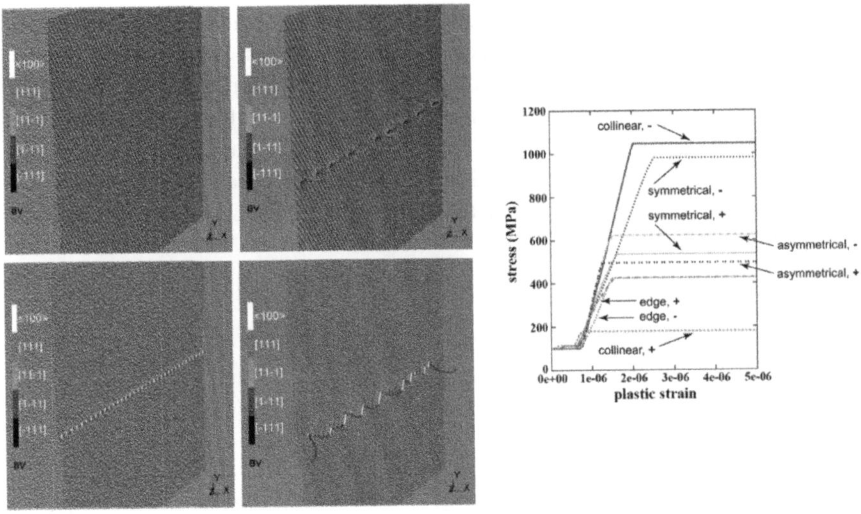

Figure 1.13. *Negative dislocation interacting with colinear GB edge dislocations (all in red) leading to a first partial annihilation of the incoming dislocation (a), then transmission through bowing out and recombination (b). Crystal and GB dislocations are color coded according to their Burgers vectors. The incoming dislocation of a different slip system now (in blue) reacts to form arrays mixed-symmetrical junctions (in white) with GB dislocations. When the applied stress is sufficient, transmission occurs through junction unzipping, and dislocation bowing out on the other side of the GB. (e) Stresses required for the direct transmission of an incoming dislocation for the various type of reactions that can occur with the GB dislocations. An asymmetrical behavior is observed depending on the sign ("+" or "-") of the incoming dislocations*

The direct transmission of crystal dislocations through a low-angle grain-boundary (LAGB) was investigated by Liu et al. (2011, 2012) in bcc Fe. In Liu et al. (2011), a $< 111 >$ pure tilt symmetric LAGB is modeled by edge dislocations separated by $100b$ (as shown in Figure 1.13). Different infinite mobile segments are considered to test the six binary contact reactions existing among $1/2 < 111 > \{110\}$ slips systems (see section 1.3.1). While these simulations resemble latent hardening simulations, their outcomes are rather different as the dislocation line directions are imposed by the boundary plane, and they therefore cover a smaller range of dislocation configurations. As a result, no weak reaction (i.e. edge and asymmetrical mixed junctions) are formed leading to an easy transmission of the mobile segment, and the strongest interaction is found, as expected, with the negative collinear mobile segment and a symmetrical edge junction formation. For the latter case, a recombination may be observed as a hexagonal dislocation network consistent with a mixed GB, depending on the applied stress. Following results from atomistics (Terentyev et al. 2010), mobility of binary $< 100 >$ junctions was enabled and this could potentially impact the stability of ternary junctions. A follow-up study, Liu et al. (2012) investigated the interaction between mobile segments with mixed tilt-twist LAGB still in bcc Fe. The LAGB is modeled as two sets of $1/2 < 111 > \{110\}$ dislocation relaxing into a hexagonal array including a $[010](101)$ binary reaction. The interaction with an infinite segment now leads to two possible binary junctions and a ternary reaction, which can be axial or delocalized zig-zags (see section 1.2.1). Different loading conditions of the GB were considered. Various levels of transmission stress were observed as a function of the nature of the interaction, and the destruction of the GB was sometimes observed before the penetration of the dislocation. It is interesting to note the already large stress level ($\approx$ 0.2–1 GPa) associated with the penetration of these LAGB of and 0.3° of misorientation (see Figure 1.13). It is expected that GB dislocation spacing decreases with an increase in the misorientation angle, leading to possible higher stress level for dislocation transmission.

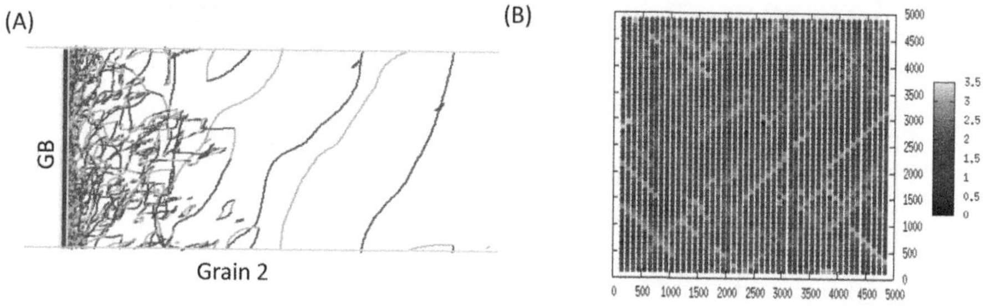

Figure 1.14. *Cross view of dislocation accumulation at impenetrable GB in a fcc Al bicrystal after 0.2 plastic strain. The GB is a $\Sigma 5$ pure twist. Corresponding mapping of the stresses σzz induced by dislocations at the GB with $\mathbf{z}$ the normal to the GB plane. Stresses are in GPa and distance units in $|\mathbf{b}|$*

In the context of crack initiation at GB, Shi and Queyreau (2023) investigated the stress induced by crystal dislocations at impenetrable GB in fcc Al. Initial microstructures are made up of prismatic loops. The stresses at the GB are evaluated using the non-singular dislocation theory (Cai et al. 2006) on various microstructures and few GB planes. The stress is not per say the Cauchy stress due to the double convolution used in Cai et al. (2006), and stress distributions constitute an upper bound as elastic relaxations are neglected in pure DDD (as opposed to the elastic relaxation allowed by DDD-FEM methods). As expected stress concentrations are rapidly going up with the accumulation of dislocations at the boundaries (see Figure 1.14). From a fine mapping of the stresses at the GB, the distribution of stress values seem to converge toward a log-normal distribution. After a few percent of deformation, some locations at the GB reach a maximum stress of few GPa demonstrating that dislocations are the necessary link between atomistically informed debonding formalism (i.e. the so-called traction separation law) (Van der

Ven 2004; Tahir et al. 2013; Ehlers et al. 2016, 2017) and macroscopic description of fracture (i.e. cohesive zone models (Rice and Wang 1989)).

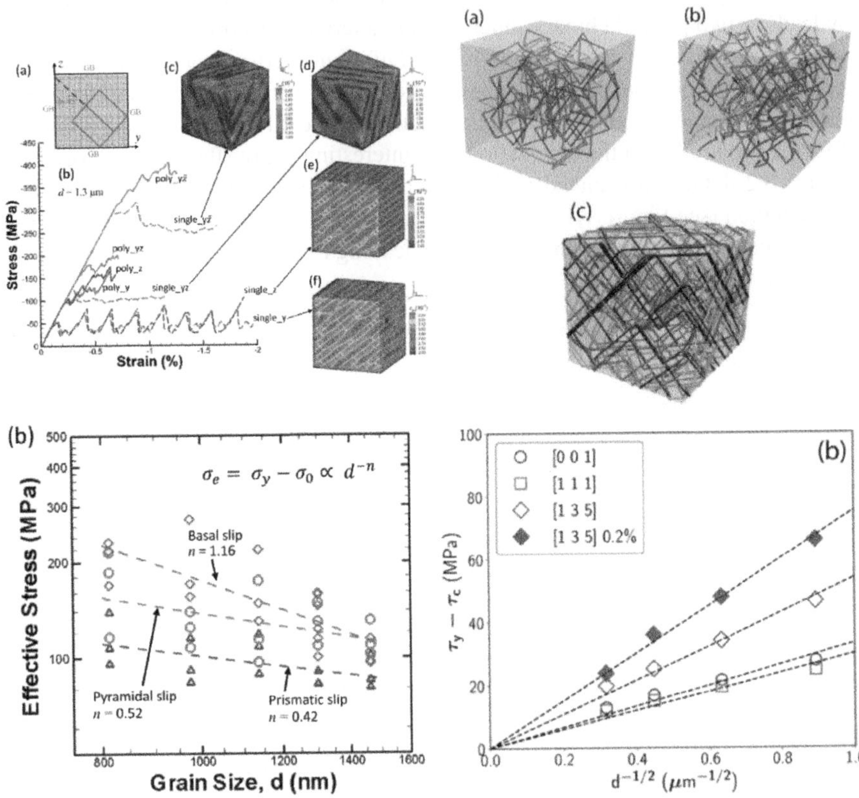

Figure 1.15. *Top left: stress strain curves and traces of slip activity corresponding to various orientations of a single grain in hcp Mg (Fan et al. 2015). GB is partially impenetrable to dislocations. See text for more details. Bottom left: scaling of the flow stress removed from the Periels contribution with the size of the grain d and the nature of active slip. Top right: evolution of the dislocation microstructure from the initial one, made up of prismatic loops (a) to the deformed microstructures obtained for transparent GB (b) or impenetrable GB (c) (Jiang et al. 2019). Bottom right: scaling of the Hall–Petch term as function of the crystal orientation as obtained after a plastic strain of 0.1%*

Finally, the presence of GB induces a size effect on the polycrystalline mechanical response. The so-called Hall–Petch has been the subject of several investigations. It states that the flow stress of a polycrystals σ scales as the grain size D as:

$$\sigma = \sigma_0 + \frac{k_{HP}}{D^n} \qquad [1.25]$$

where σ_0 is a reference stress, k_{HP} is a poorly known constant and exponent n is close to 0.5. The challenge is twofold here, as the precise physical origin of the HP is still uncertain while widely observed in experiments, besides k_{HP} and n are loosely defined and values can vary widely, that is, with the number of activated systems, grains geometry or the polycrystalline texture. Fan et al. (2015) investigated size effects in periodic grains in hcp Mg. A single cubic grain was included in the simulations with GB on all PBC faces to simulate a strong basal texture component (see Figure 1.15). The c-axis was rotated with respect to the z-axis of the simulation. GB are partially impenetrable with a simplified transmission criterion. The initial microstructure is made up of FR sources, the different slip systems existing for hcp structure and Mg: basal, prismatic, pyramidal I with <a> Burgers vector or pyramidal II <c+a> Burgers vector. Pyramidal and Prismatic slip systems experience a Peierls stress. Different loadings were considered to favor the various slip family. Surprisingly, almost no hardening was observed in absence of GB. A size effect was obtained when GBs are accounted, which is strongly impacted by the FR sources sizes and the type of activated systems. The effective flow stress $\sigma_y - \sigma_0$ scales as D^n, where σ_0 accounts for Peierls stress. n is found in the range $[-0.4, -1.16]$ depending on the nature of the activated slip.

de Sansal et al. (2009) investigated the plastic flow onset of 16 grain aggregates of 14 faces polyhedra with DDD (see Figure 1.15). Grains have random orientations with a Taylor factor of 3.05, and are separated by impenetrable grain boundaries. Consistent with early plastic deformation stages, grains are mostly deformed in single slip. The initial microstructure made of sources, leads to pile-ups at the GB but CS of screws rapidly redistributes dislocation activity within grains. A Voce law is used to detect the onset of plastic flow and a grains size dependence with an exponent $n \approx$ [-0.5, -1] is found for grain sizes in the range 1–2.5 μm. The authors defined a k_{HP} from DD simulation from an argument based on dislocation storage at GB and that was scaling with total deformation. Recently, Jiang et al. (2019) investigated the impact of the grain shape and grain environment on the HP effect in fcc Cu. Initial microstructures are made of prismatic loops in impenetrable grains. Simulations encompass varied configurations, with periodic rectangular grains with varied orientations including a single glide [135], and two multislip [001] and [110], four grain aggregates with two random orientations and a [001], and [110] orientations, and different grain shapes with different aspect ratios mimicking cube, lath and needle geometries. The reference stress σ_0 is defined as the yield stress on an ideal Taylor polycrystal. At plastic onset, the grain size dependence is found with an exponent of -0.5 with grain sizes D from 1.25 to 10 μm. For equiaxed periodic grains, k_{HP} is mostly independent of the number of slip systems with $\approx 4/bD$. k_{HP} depends strongly on the grains shape, and a shape factor is introduced by the authors to relate to the equiaxed grains. Finally, the behavior of a four grain aggregate is consistent with the average behavior of the constituents.

1.3.6. *Cyclic deformations*

Previous sections were concerned with monotonous loading, however plastic deformation is known to be strongly dependent upon deformation history. One way to assess this aspect of plastic deformation is through cyclic deformations and the corresponding hysteresis curves, starting from the simplest Bauschinger experiments. Initially, isotropic materials exhibit a rounded onset of plastic flow over a transient in compression and sometimes a permanent recovery, when a first loading in tension was applied. This hysteresis behavior can be found in a multitude of crystalline materials ranging from single crystals to polycrystals and in pure or alloyed materials (Mughrabi 1978, 2010; Li et al. 2011). While the ubiquity of this hysteresis behavior is clear, its origins are much less evident. Single crystals are an elementary part of polycrystals, the cyclic deformation in these systems is even more confusing. Many of the current explanations reside in the easy dissolution of existing dislocation microstructures formed during the initial loading, or in the existence of polarized long-range internal stresses. However, there was no clear explanation of the former in terms of elementary mechanisms until recently. The latter explanation is typically associated with the so-called composite model (Mughrabi 1988), according to which the microstructure is made of *hard* dislocation walls, mostly composed of edge dislocations, and *soft* interior cells where most of the plastic flow occurs. Such a polarized microstructure leads to long range internal stresses (LRIS), which shall oppose or help the forward or backward motion of dislocations. However, X-ray microdiffraction analyses question the very existence of polarized LRIS in single crystals (Kassner et al. 2009, 2013).

Experimental investigations have revealed the complex evolution and features of the materials response to cyclic deformations. When now focusing onto cyclic deformation of single crystals (see review in Li et al. (2011)), hysteresis curves evolve with the number of cycles as function of loading direction, the material and the amount of applied plastic strain per cycle. The maximum stress increases each cycle until it reaches a saturation value $|\tau_{sat}|$. In single glide condition, there is a range of intermediate strain per cycle γ_p^{cy}, for which the saturation stress is almost constant (or slightly increasing). For larger strain per cycle just below 1%, double slip eventually activated and saturation stresses increased with the γ_p^{cy}, while for multislip conditions, the saturation stress increases continuously over the range of typical γ_p^{cy}. In this context, tridimensional DDD simulations are certainly very attractive as they can provide elementary mechanisms and average the collective behavior of dislocations during cyclic deformation. However, despite the constant progress in computing performance and numerical algorithms, only a few dozens of cycles are typically accessible to DDD. Therefore, a number of strategies have been proposed to connect to latter stages of deformation by either extrapolating the dislocation patterns or through multiscale approaches.

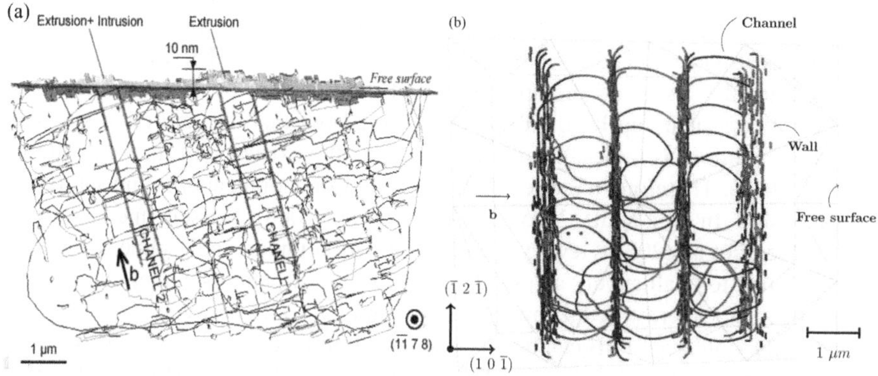

Figure 1.16. *(a) Simulated persistent slip bands simulated in Déprés et al. (2006) obtained after 23 cycles at a plastic increment of $\Delta\epsilon_p = 10^{-3}$ and a grain diameter of 5 μm. Observation view corresponds to the primary plane normal. (b) Initial PSB structure built artificially as a ladder structure (Erel et al. 2017) with walls made up of mostly long edge dislocation and cells swept mostly by screw dislocations*

Déprés et al. (2008) considered the cyclic response of a single dodecahedric grain in single slip conditions in AISI316L stainless steel to 20 cycles at 1.5×10^{-3} strain per cycle (see Figure 1.16). The initial microstructure is made of sources, with cross-slip systems enabled and the grain boundaries are impenetrable to dislocations, and will leave the place to the presence of many dipolar configurations after the 20 cycles. A distribution of dipole height is then obtained from the simulations and is assumed to represent the effective opposing or promoting effect to subsequent dislocations. The authors then modified a crystal plasticity model where the dipolar distribution appears in a backstress term. After fitting of a couple of additional parameters, the CP calculations qualitatively captures the evolution of hysteresis curves with cumulated strain for one strain increment in the case of single glide conditions. At the heart of BE and cyclic deformation, lies the question of how much of plastic deformation is reversible by an opposite deformation. This was discussed by Queyreau and Devincre (2009) in the case of very model simulations where an infinite dislocation interact with random particle of various strength, where the mean free path of dislocation is directly related to the number of precipitates being swept by dislocations.

More than in other applications, reproduction of the peculiar patterns observed during deformation have been the motivation to a number of DDD investigations, in an effort to directly relate microstructures to observed deformation behavior, effort that remains mostly unsuccessful due to the complexity of the problem at hand and the requirement of considering large deformation. In single glide conditions, the saturation stress in the plateau is rather close to the onset of dynamic recovery τ_{III} in monotonic conditions (Mughrabi 1979), where the motion and interactions among screw dislocations confined in cells seem evident key aspects. The existence of the plateau coincides also with the apparition and development of ladder structures is also called persistent slip band. However, it turned out that the saturation stress in multislip conditions is rather far from τ_{III}, despite the existence of marked patterns (Li et al. 2011). Therefore, the direct link between dislocation patterns and saturation remain unsure. In Shin et al. (2005); El-Awady et al. (2008); Hussein and El-Awady (2016); Meng et al. (2021), few cycles of deformation under single-glide conditions are performed to qualitatively assess the development of patterns. Other simulations have model the microstructure by long edge dipolar walls between which screws navigates and interacts. PSB can be made of long infinite and rigid edge dislocations (El-Awady et al. 2008) to more realistic and flexible dipolar loops (Erel et al. 2017) (see Figure 1.16). Another motivation to simulate the building up of dislocation microstructures is to investigate the specific surface roughening associated with cyclic experiments. Dislocation activity localization associated with PSD formation is correlated with dislocation exiting though out crystal surface in the form of band extrusions, or accumulating a grain boundaries, possibly leading to crack initiation. Slip band on surface has been investigated in Déprés et al. (2006); Hussein and El-Awady (2016); Meng et al. (2021).

Queyreau and Devincre (2020, 2021) revisited recently the Bauschinger effect (BE) and cyclic deformation of fcc single crystals (see Figure 1.17). It turned out that in DDD simulations a relatively large BE could be obtained but internal stresses are shown to be absent. Two original mechanisms are identified from the simulations: (i) the stability of junctions formed during the first loading is actually asymmetric, and easily destroyed in the beginning of compression deformation; (ii) the avalanches leading to the storage of dislocation segments now occur in region that have been swept during tension, leading to an unwinding of stored segments and loop collapses as in particle strengthened materials (Queyreau and Devincre 2009). The interaction matrix and MFP of dislocations are weakened by means of two sigmoidal functions r_a and r_{MFP} that state the competition between promoting effects (scaling with $\Delta\rho_p$ stored during forward motion) and finite amount of reversible strain (scaling with γ in backward motion). These functions are obtained from a comprehensive set of DDD simulations. These results are extended to a larger scale through a modified crystal plasticity model that is solely based on the DDD parameters. Other required parameters are provided by former DDD simulations or known experimental results. This continuous model was compared to most available data existing on fcc single crystals and was successful in predicting from the DDD results most of the features of the cyclic deformations. Saturation of the flow stress with the increase in cycle number occurs naturally. The saturation stress is known to depend upon the amount of plastic strain per cycle and the loading direction and materials. It is interesting to point out here that there is no requirement to back-stresses or need for the existence of a pattern. The controlling effects are the statistical behavior of short-range reactions, and to the first order, patterning does not seem to play major role in the behavior, as this is the case for monotonic deformations (Devincre et al. 2008). Ultimately, a comparison with data obtained on polycrystals with micrometer grains clearly shows that these effects are controlling the cyclic behavior in polycrystals as well.

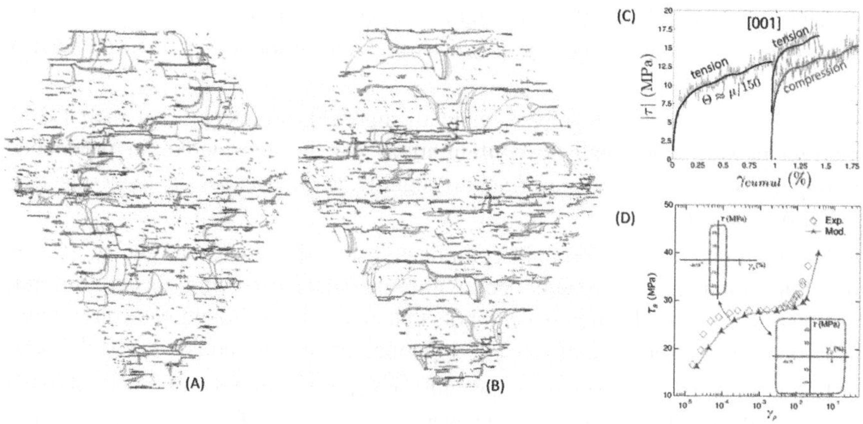

Figure 1.17. *Strong tension-compression asymmetry in the plastic deformation of pure fcc metals. (A) and (B) are sequences of dislocation motion during a Bauschinger experiment using DDD. A Ni single crystal is deformed along [112] direction. Primary dislocations in light blue form junctions in red with the second active slip system. Thinfoils that are 0.2 µm thick are extracted from the same region, slightly before and after strain reversal. Note the changes in the paths taken by dislocations and the less continuous propagation of dislocation during compressive deformation. (C) Typical Bauschinger curves obtained from DDD in the case of [001] deformation. (D) Comparison of the predictions obtained from DDD and the modified CP model with the saturation stress observed experimentally in Cu single crystals (Mughrabi 1978) (see text for more details)*

1.4. Conclusion and future work

At the end of the present review, a number of remarks become obvious. First, the growing number of applications clearly demonstrates that DDD has become a very versatile tool. Second, DDD can provide both qualitative and quantitative results, but quantitative data are unsurprisingly more difficult to obtain and often require other inputs or coarse graining at other physical scales, whether simulations or experiments. A recent example in Zepeda-Ruiz et al. (2017); Bertin et al. (2024) demonstrates that a one-to-one agreement between large-scale MD and DDD is possible. Third, an important part of the papers presented here are actually not DDD investigations, and they were required to justify DDD inputs or outcomes. As materials science is a mature field, DDD simulations should find a natural place within the field, between tradition and innovation.

Next, we highlight a number of objectives that we intend to follow:

– Simulations should continue to benefit from the constant progresses in terms of analytical or semi-analytical models (e.g. force evaluations) or more efficient numerical algorithms. The efficiency gain associated with these progresses should then be objectively shown and compared to previous solutions at fixed numerical errors.

– Thanks to the constant improvement of algorithm efficiency and increase in computing power, more realistic configurations can be considered, in particular, including elastic anisotropy may lead to significant differences in the collective behavior of dislocations.

– DDD simulations are certainly complicated scientific codes and the subject of constant developments. A growing need within the DDD community is certainly the definition of a set of well-posed benchmarks that could shared among the community. Indicators of the convergence of force of dynamics evaluations could also be defined.

– Large-scale DDD simulations provide a natural sampling of the collective behavior of dislocations. There is however still a lot of effort to provide on the statistical analysis of dislocation microstructures, dislocation kinetics and the link between the two, if any. This is particularly important when coarse graining DDD data to larger scales.

– With a number of success regarding the understanding of single crystal behavior, the new frontier to DDD simulations is certainly the polycrystal system. Due to the complexity of the GB configuration and structures, a close link with atomistic simulations will certainly be more necessary than ever.

1.5. Acknowledgments

The author would like to thank S. Aubry, B. Devincre and S. Forest for their careful reading and insightful comments on this chapter. Special thanks are also dedicated to T. Arsenlis, V. Bulatov, B. Devincre and G. Monnet for having taught the author in his younger years the tricks of the trade of DDD simulations.

1.6. References

Aifantis, K.E., Senger, J., Weygand, D., Zaiser, M. (2009). Discrete dislocation dynamics simulation and continuum modeling of plastic boundary layers in tricrystal micropillars. *IOP Conference Series: Materials Science and Engineering*, 3, 012025.

Akhondzadeh, S., Sills, R.B., Bertin, N., Cai, W. (2020). Dislocation density-based plasticity model from massive discrete dislocation dynamics database. *Journal of the Mechanics and Physics of Solids*, 145, 104152.

Akhondzadeh, S., Bertin, N., Sills, R.B., Cai, W. (2021). Slip-free multiplication and complexity of dislocation networks in FCC metals. *Materials Theory*, 5(1), 2.

Anderson, J.P. and El-Azab, A. (2021). On the three-dimensional spatial correlations of curved dislocation systems. *Materials Theory*, 5(1), 1.

Anderson, P.M., Hirth, J.P., Lothe, J. (2017). *Theory of Dislocations*. Cambridge University Press, Cambridge.

Arsenlis, A., Cai, W., Tang, M., Rhee, M., Oppelstrup, T., Hommes, G., Pierce, T.G., Bulatov, V.V. (2007). Enabling strain hardening simulations with dislocation dynamics. *Modelling and Simulation in Materials Science and Engineering*, 15(6), 553–595.

Asano, S. (1968). Stress tensor of a finite dislocation. *Journal of the Physical Society of Japan*, 25(1), 220–226.

Aubry, S. and Arsenlis, A. (2013). Use of spherical harmonics for dislocation dynamics in anisotropic elastic media. *Modelling and Simulation in Materials Science and Engineering*, 21(6), 065013.

Aubry, S., Fitzgerald, S.P., Dudarev, S.L., Cai, W. (2011a). Equilibrium shape of dislocation shear loops in anisotropic α-Fe. *Modelling and Simulation in Materials Science and Engineering*, 19(6), 065006.

Aubry, S., Kang, K., Ryu, S., Cai, W. (2011b). Energy barrier for homogeneous dislocation nucleation: Comparing atomistic and continuum models. *Scripta Materialia*, 64(11), 1043–1046.

Aubry, S., Fitzgerald, S.P., Arsenlis, A. (2014). Methods to compute dislocation line tension energy and force in anisotropic elasticity. *Modelling and Simulation in Materials Science and Engineering*, 22(1), 015001.

Aubry, S., Queyreau, S., Arsenlis, A. (2019). Dislocation dynamics modeling of the interaction of dislocations with Eshelby inclusions. In *Mechanics and Physics of Solids at Micro- and Nano-Scales*, Ionescu, I., Queyreau, S., Picu, C., Umut, S. (eds). ISTE Ltd, London, and John Wiley & Sons, New York.

Bacon, D.J., Kocks, U.F., Scattergood, R.O. (1973). The effect of dislocation self-interaction on the Orowan stress. *Philosophical Magazine*, 28(6), 1241–1263.

Bacon, D.J., Barnett, D.M., Scattergood, R.O. (1979). Anisotropic continuum theory of lattice defects. *Progress in Materials Science*, 23, 212.

Banadaki, A.D. and Patala, S. (2015). An efficient algorithm for computing the primitive bases of a general lattice plane. *Journal of Applied Crystallography*, 48(2), 585–588.

Basinski, Z. and Basinski, S. (1992). Fundamental aspects of low amplitude cyclic deformation in face-centred cubic crystals. *Progress in Materials Science*, 36, 89–148.

Bayerschen, E., Stricker, M., Wulfinghoff, S., Weygand, D., Böhlke, T. (2015). Equivalent plastic strain gradient plasticity with grain boundary hardening and comparison to discrete dislocation dynamics. *Proceedings of the Royal Society A: Mathematical, Physical and Engineering Sciences*, 471(2184), 20150388.

van Beers, P.R.M., Kouznetsova, V.G., Geers, M.G.D., Tschopp, M.A., McDowell, D.L. (2015). A multiscale model of grain boundary structure and energy: From atomistics to a continuum description. *Acta Materialia*, 82, 513–529.

Bertin, N. and Capolungo, L. (2018). A FFT-based formulation for discrete dislocation dynamics in heterogeneous media. *Journal of Computational Physics*, 355, 366–384.

Bertin, N., Tomé, C., Beyerlein, I., Barnett, M., Capolungo, L. (2014). On the strength of dislocation interactions and their effect on latent hardening in pure Magnesium. *International Journal of Plasticity*, 62, 72–92.

Bertin, N., Cai, W., Aubry, S., Arsenlis, A., Bulatov, V.V. (2024). Enhanced mobility of dislocation network nodes and its eject on dislocation multiplication and strain hardening. *Acta Materialia*, 271(1), 19884.

Brown, L.M. (1964). The self-stress of dislocations and the shape of extended nodes. *Philosophical Magazine*, 10(105), 441–466.

Brown, L.M. and Ham, R.K. (1972). Dislocation-particle interactions. In *Strengthening Methods in Crystals*, Kelly, A. and Nicholson, R. (eds). Elsevier, Amsterdam.

Bulatov, V.V. and Cai, W. (2006). *Computer Simulations of Dislocations*. Oxford University Press, Oxford.

Bulatov, V.V., Reed, B.W., Kumar, M. (2014). Grain boundary energy function for fcc metals. *Acta Materialia*, 65, 161–175.

Burbery, N.B., Po, G., Das, R., Ghoniem, N., Ferguson, W.G. (2017). Dislocation dynamics in polycrystals with atomistic-informed mechanisms of dislocation–grain boundary interactions. *Journal of Micromechanics and Molecular Physics*, 2(1), 1750003.

Cai, W., Arsenlis, A., Weinberger, C., Bulatov, V. (2006). A non-singular continuum theory of dislocations. *Journal of the Mechanics and Physics of Solids*, 54(3), 561–587.

Cai, W., Sills, R., Barnett, D., Nix, W. (2014). Modeling a distribution of point defects as misfitting inclusions in stressed solids. *Journal of the Mechanics and Physics of Solids*, 66, 154–171.

Caillard, D. (2010). Kinetics of dislocations in pure Fe. Part I. In situ straining experiments at room temperature. *Acta Materialia*, 58(9), 3493–3503.

Caillard, D. (2013). A TEM in situ study of alloying effects in iron. II – Solid solution hardening caused by high concentrations of Si and Cr. *Acta Materialia*, 61(8), 2808–2827.

Clouet, E., Ventelon, L., Willaime, F. (2011). Dislocation core field. II. Screw dislocation in iron. *Physical Review B*, 84(22), 224107.

Crone, J.C., Chung, P.W., Leiter, K.W., Knap, J., Aubry, S., Hommes, G., Arsenlis, A. (2014). A multiply parallel implementation of finite element-based discrete dislocation dynamics for arbitrary geometries. *Modelling and Simulation in Materials Science and Engineering*, 22(3), 035014.

Daveau, G. (2012). Interaction dislocations – joints de grains en déformation plastique monotone : étude expérimentale et modélisations numériques. PhD Thesis, École Centrale Paris, Gif-sur-Yvette.

Deng, J. and El-Azab, A. (2007). Dislocation pair correlations from dislocation dynamics simulations. *Journal of Computer-Aided Materials Design*, 14(S1), 295–307.

Deng, J. and El-Azab, A. (2010). Temporal statistics and coarse graining of dislocation ensembles. *Philosophical Magazine*, 90(27–28), 3651–3678.

Deng, J., El-Azab, A., Larson, B.C. (2008). On the elastic boundary value problem of dislocations in bounded crystals. *Philosophical Magazine*, 88(30–32), 3527–3548.

Déprés, C., Robertson, C.F., Fivel, M.C. (2006). Low-strain fatigue in 316L steel surface grains: A three dimension discrete dislocation dynamics modelling of the early cycles. Part 2: Persistent slip markings and micro-crack nucleation. *Philosophical Magazine*, 86(1), 79–97.

Déprés, C., Fivel, M., Tabourot, L. (2008). A dislocation-based model for low-amplitude fatigue behaviour of face-centred cubic single crystals. *Scripta Materialia*, 58(12), 1086–1089.

Devincre, B. (2013). Dislocation dynamics simulations of slip systems interactions and forest strengthening in ice single crystal. *Philosophical Magazine*, 93(1–3), 235–246.

Devincre, B. and Gatti, R. (2015). Physically justified models for crystal plasticity developed with dislocation dynamics simulations. *AerospaceLab Journal*, 9, 1–7.

Devincre, B., Kubin, L., Hoc, T. (2006). Physical analyses of crystal plasticity by DD simulations. *Scripta Materialia*, 54(5), 741–746.

Devincre, B., Kubin, L., Hoc, T. (2007). Collinear superjogs and the low-stress response of fcc crystals. *Scripta Materialia*, 57(10), 905–908.

Devincre, B., Hoc, T., Kubin, L. (2008). Dislocation mean free paths and strain hardening of crystals. *Science*, 320(5884), 1745–1748.

Devincre, B., Madec, R., Monnet, G., Queyreau, S., Gatti, R., Kubin, L. (2011). Modeling crystal plasticity with dislocation dynamics simulations: The ‘microMegas’ simulation code. In *Mechanics of Nano-Objects*, Forest, S., Ponchet, A., Thomas, A. (eds). Presses des Mines, Paris.

Ehlers, F.J.H., Seydou, M., Tingaud, D., Maurel, F., Charles, Y., Queyreau, S. (2016). Ab initio determination of the traction – Separation curve for a metal grain boundary: A critical assessment of strategies. *Modelling and Simulation in Materials Science and Engineering*, 24(8), 085014.

Ehlers, F.J.H., Seydou, M., Tingaud, D., Maurel, F., Queyreau, S., Charles, Y. (2017). Ab initio studies of two Al grain boundaries subjected to mixed tension/shear mode loading: How shear may promote breakage. *Modelling and Simulation in Materials Science and Engineering*, 25(6), 064001.

El-Awady, J.A. (2015). Unravelling the physics of size-dependent dislocation-mediated plasticity. *Nature Communications*, 6(1), 5926.

El-Awady, J.A., Bulent Biner, S., Ghoniem, N.M. (2008). A self-consistent boundary element, parametric dislocation dynamics formulation of plastic flow in finite volumes. *Journal of the Mechanics and Physics of Solids*, 56(5), 2019–2035.

Erel, C., Po, G., Crosby, T., Ghoniem, N. (2017). Generation and interaction mechanisms of prismatic dislocation loops in FCC metals. *Computational Materials Science*, 140, 32–46.

Fan, H., Aubry, S., Arsenlis, A., El-Awady, J.A. (2015). Orientation influence on grain size effects in ultrafine-grained magnesium. *Scripta Materialia*, 97, 25–28.

Franciosi, P. and Zaoui, A. (1982). Multislip in fcc crystals a theoretical approach compared with experimental data. *Acta Metallurgica*, 30(8), 1627–1637.

Franciosi, P., Berveiller, M., Zaoui, A. (1980). Latent hardening in copper and aluminium single crystals. *Acta Metallurgica*, 28(3), 273–283.

Frederiksen, S.L. and Jacobsen, K.W. (2003). Density functional theory studies of screw dislocation core structures in bcc metals. *Philosophical Magazine*, 83(3), 365–375.

Gavazza, S. and Barnett, D. (1976). The self-force on a planar dislocation loop in an anisotropic linear-elastic medium. *Journal of the Mechanics and Physics of Solids*, 24(4), 171–185.

Gemperlová, J., Paidar, V., Kroupa, F. (1989). Compatibility stresses in deformed bicrystals. *Czechoslovak Journal of Physics*, 39(4), 427–446.

Ghoniem, N.M., Tong, S.-H., Sun, L.Z. (2000). Parametric dislocation dynamics: A thermodynamics-based approach to investigations of mesoscopic plastic deformation. *Physical Review B*, 61(2), 913–927.

Giessen, E.V.D. and Needleman, A. (1995). Discrete dislocation plasticity: A simple planar model. *Modelling and Simulation in Materials Science and Engineering*, 3(5), 689–735.

Gilbert, M.R., Queyreau, S., Marian, J. (2011). Stress and temperature dependence of screw dislocation mobility in α-Fe by molecular dynamics. *Physical Review B*, 84(17), 174103.

Gu, Y. and El-Awady, J.A. (2018). Quantifying the effect of hydrogen on dislocation dynamics: A three-dimensional discrete dislocation dynamics framework. *Journal of the Mechanics and Physics of Solids*, 112, 491–507.

Hull, D. and Bacon, D.J. (2001). *Introduction to Dislocations*, 4th edition. Butterworth-Heinemann, Oxford.

Hussein, A.M. and El-Awady, J.A. (2016). Quantifying dislocation microstructure evolution and cyclic hardening in fatigued face-centered cubic single crystals. *Journal of the Mechanics and Physics of Solids*, 91, 126–144.

Hussein, A.M., Rao, S.I., Uchic, M.D., Dimiduk, D.M., El-Awady, J.A. (2015). Microstructurally based cross-slip mechanisms and their effects on dislocation microstructure evolution in fcc crystals. *Acta Materialia*, 85, 180–190.

Jamond, O., Gatti, R., Roos, A., Devincre, B. (2016). Consistent formulation for the discrete-continuous model: Improving complex dislocation dynamics simulations. *International Journal of Plasticity*, 80, 19–37.

Jiang, M., Devincre, B., Monnet, G. (2019). Effects of the grain size and shape on the flow stress: A dislocation dynamics study. *International Journal of Plasticity*, 113, 111–124.

Kang, K., Yin, J., Cai, W. (2014). Stress dependence of cross slip energy barrier for face-centered cubic nickel. *Journal of the Mechanics and Physics of Solids*, 62, 181–193.

Kassner, M., Geantil, P., Levine, L., Larson, B. (2009). Backstress, the Bauschinger effect and cyclic deformation. *Materials Science Forum*, 604–605, 39–51.

Kassner, M., Geantil, P., Levine, L. (2013). Long range internal stresses in single-phase crystalline materials. *International Journal of Plasticity*, 45, 44–60.

Kocks, U.F. and Mecking, H. (2003). Physics and phenomenology of strain hardening: The FCC case. *Progress in Materials Science*, 48(3), 171–273.

Kocks, U.F., Argon, A.S., Ashby, M.F. (1975). *Thermodynamics and Kinetics of Slip*. Pergamon Press, Oxford.

de Koning, M., Kurtz, R., Bulatov, V., Henager, C., Hoagland, R., Cai, W., Nomura, M. (2003). Modeling of dislocation – Grain boundary interactions in FCC metals. *Journal of Nuclear Materials*, 323(2–3), 281–289.

Kubin, L. (2013). *Dislocations, Mesoscale Simulations and Plastic Flow*. Oxford University Press, Oxford.

Kubin, L., Devincre, B., Hoc, T. (2008). Modeling dislocation storage rates and mean free paths in face-centered cubic crystals. *Acta Materialia*, 56(20), 6040–6049.

Kubin, L., Hoc, T., Devincre, B. (2009). Dynamic recovery and its orientation dependence in face-centered cubic crystals. *Acta Materialia*, 57(8), 2567–2575.

Lavenstein, S. and El-Awady, J.A. (2019). Micro-scale fatigue mechanisms in metals: Insights gained from small-scale experiments and discrete dislocation dynamics simulations. *Current Opinion in Solid State and Materials Science*, 23(5), 100765.

Lazar, M. (2012). Non-singular dislocation loops in gradient elasticity. *Physics Letters A*, 376(21), 1757–1758.

Lazar, M. (2013). The fundamentals of non-singular dislocations in the theory of gradient elasticity: Dislocation loops and straight dislocations. *International Journal of Solids and Structures*, 50(2), 352–362.

Lazar, M., Agiasofitou, E., Po, G. (2019). Three-dimensional nonlocal anisotropic elasticity: A generalized continuum theory of Ångström-mechanics. *Acta Mechanica*, 231, 743–781.

Li, P., Li, S., Wang, Z., Zhang, Z. (2011). Fundamental factors on formation mechanism of dislocation arrangements in cyclically deformed fcc single crystals. *Progress in Materials Science*, 56(3), 328–377.

Liu, B., Raabe, D., Eisenlohr, P., Roters, F., Arsenlis, A., Hommes, G. (2011). Dislocation interactions and low-angle grain boundary strengthening. *Acta Materialia*, 59(19), 7125–7134.

Liu, B., Eisenlohr, P., Roters, F., Raabe, D. (2012). Simulation of dislocation penetration through a general low-angle grain boundary. *Acta Materialia*, 60(13), 5380–5390.

Liu, B., Arsenlis, A., Aubry, S. (2016). Computing forces on interface elements exerted by dislocations in an elastically anisotropic crystalline material. *Modelling and Simulation in Materials Science and Engineering*, 24(5), 055013.

Longsworth, M. and Fivel, M. (2021). Investigating the cross-slip rate in face-centered cubic metals using an atomistic-based cross-slip model in dislocation dynamics simulations. *Journal of the Mechanics and Physics of Solids*, 153, 104449.

Louchet, F., Kubin, L.P., Vesely, D. (1979). In situ deformation of b.c.c. crystals at low temperatures in a high-voltage electron microscope. Dislocation mechanisms and strain-rate equation. *Philosophical Magazine A*, 39(4), 433–454.

Madec, R. and Kubin, L.P. (2008). Second-order junctions and strain hardening in bcc and fcc crystals. *Scripta Materialia*, 58(9), 767–770.

Madec, R. and Kubin, L.P. (2017). Dislocation strengthening in FCC metals and in BCC metals at high temperatures. *Acta Materialia*, 126, 166–173.

Madec, R., Devincre, B., Kubin, L. (2004). On the use of periodic boundary conditions in dislocation dynamics simulations. In *IUTAM Symposium on Mesoscopic Dynamics of Fracture Process and Materials Strength*, volume 115, Gladwell, G.M.L., Kitagawa, H., Shibutani, Y. (eds). Springer, Dordrecht.

Marian, J., Fitzgerald, S., Po, G. (2018). Discrete dislocation dynamics simulations of irradiation hardening in nuclear materials. In *Handbook of Materials Modeling*, Andreoni, W. and Yip, S. (eds). Springer International Publishing, Cham.

Martinez, R. and Ghoniem, N.M. (2002). The influence of crystal surfaces on dislocation interactions in mesoscopic plasticity: A combined dislocation dynamics-finite element approach. *Comput. Modeling Eng. Sci.*, 3, 229–243.

Meng, F., Ferrié, E., Déprés, C., Fivel, M. (2021). 3D discrete dislocation dynamic investigations of persistent slip band formation in FCC metals under cyclical deformation. *International Journal of Fatigue*, 149, 106234.

Monnet, G. (2006). Investigation of precipitation hardening by dislocation dynamics simulations. *Philosophical Magazine*, 86(36), 5927–5941.

Monnet, G. (2018). Multiscale modeling of irradiation hardening: Application to important nuclear materials. *Journal of Nuclear Materials*, 508, 609–627.

Monnet, G. and Terentyev, D. (2009). Structure and mobility of the edge dislocation in bcc iron studied by molecular dynamics. *Acta Materialia*, 57(5), 1416–1426.

Monnet, G., Devincre, B., Kubin, L. (2004). Dislocation study of prismatic slip systems and their interactions in hexagonal close packed metals: Application to zirconium. *Acta Materialia*, 52(14), 4317–4328.

Monnet, G., Naamane, S., Devincre, B. (2011). Orowan strengthening at low temperatures in bcc materials studied by dislocation dynamics simulations. *Acta Materialia*, 59(2), 451–461.

Monnet, G., Vincent, L., Devincre, B. (2013). Dislocation-dynamics based crystal plasticity law for the low- and high-temperature deformation regimes of bcc crystal. *Acta Materialia*, 61(16), 6178–6190.

Mughrabi, H. (1978). The cyclic hardening and saturation behaviour of copper single crystals. *Materials Science and Engineering*, 33(2), 207–223.

Mughrabi, H. (1979). Plateaus in the cyclic stress-strain curves of single- and polycrystalline metals. *Scripta Metallurgica*, 13(6), 479–484.

Mughrabi, H. (1988). Dislocation clustering and long-range internal stresses in monotonically and cyclically deformed metal crystals. *Revue de physique appliquée*, 23(4), 367–379.

Mughrabi, H. (2010). Fatigue, an everlasting materials problem – Still en vogue. *Procedia Engineering*, 2(1), 3–26.

Mura, T. (1987). *Micromechanics of Defects in Solids*. Kluwer, Boston/The Hague.

Naamane, S. (2008). Étude de la déformation plastique de la ferrite à basse température: simulations de dynamique des dislocations. PhD Thesis, Pierre and Marie Curie University, Paris.

Naamane, S., Monnet, G., Devincre, B. (2010). Low temperature deformation in iron studied with dislocation dynamics simulations. *International Journal of Plasticity*, 26(1), 84–92.

O'Day, M.P. and Curtin, W.A. (2004). A superposition framework for discrete dislocation plasticity. *Journal of Applied Mechanics*, 71(6), 805.

Osetsky, Y.N. and Bacon, D.J. (2003). An atomic-level model for studying the dynamics of edge dislocations in metals, 11, 427–446.

Po, G., Lazar, M., Seif, D., Ghoniem, N. (2014). Singularity-free dislocation dynamics with strain gradient elasticity. *Journal of the Mechanics and Physics of Solids*, 68, 161–178.

Queyreau, S. (2008). Étude des mécanismes d'écrouissage sous irradiation de la ferrite par dynamique des dislocations. PhD Thesis, Pierre and Marie Curie University, Paris.

Queyreau, S. and Devincre, B. (2009). Bauschinger effect in precipitation-strengthened materials: A dislocation dynamics investigation. *Philosophical Magazine Letters*, 89(7), 419–430.

Queyreau, S. and Devincre, B. (2020). On the origins of tension – Compression asymmetry in crystals and implications for cyclic behavior. *arXiv:2012.15540*.

Queyreau, S. and Devincre, B. (2021). A multiscale investigation of the physical origins of tension–compression asymmetry in crystals and their implications for cyclic behavior. *arXiv:2112.15481*.

Queyreau, S., Monnet, G., Devincre, B. (2010). Orowan strengthening and forest hardening superposition examined by dislocation dynamics simulations. *Acta Materialia*, 58(17), 5586–5595.

Queyreau, S., Marian, J., Gilbert, M.R., Wirth, B.D. (2011). Edge dislocation mobilities in bcc Fe obtained by molecular dynamics. *Physical Review B*, 84(6), 064106.

Queyreau, S., Marian, J., Wirth, B.D., Arsenlis, A. (2014). Analytical integration of the forces induced by dislocations on a surface element. *Modelling and Simulation in Materials Science and Engineering*, 22(3), 035004.

Queyreau, S., Hoang, K., Shi, X., Aubry, S., Arsenlis, A. (2019). Analytical integration of the tractions induced by non-singular dislocations on an arbitrary shaped triangular quadratic element. *Modelling Simul. Mater. Sci. Eng.*, 28, 075001.

Rice, J.R. and Wang, J.-S. (1989). Embrittlement of interfaces by solute segregation. *Materials Science and Engineering: A*, 107, 23–40.

Saada, G. (1960). Sur le durcissement dû à la recombinaison des dislocations. *Acta Metallurgica*, 8(12), 841–847.

de Sansal, C., Devincre, B., Kubin, L.P. (2009). Grain size strengthening in microcrystalline copper: A three-dimensional dislocation dynamics simulation. *Key Engineering Materials*, 423, 25–32.

Shi, X. and Queyreau, S. (2023). Quantification of the stresses induced by dislocation microstructures on grain boundaries from dislocation dynamics simulations.

Shin, C., Fivel, M., Verdier, M., Robertson, C. (2005). Dislocation dynamics simulations of fatigue of precipitation-hardened materials. *Materials Science and Engineering: A*, 400–401, 166–169.

Sills, R.B., Bertin, N., Aghaei, A., Cai, W. (2018). Dislocation networks and the microstructural origin of strain hardening. *Physical Review Letters*, 121(8), 085501.

Stricker, M. and Weygand, D. (2015). Dislocation multiplication mechanisms – Glissile junctions and their role on the plastic deformation at the microscale. *Acta Materialia*, 99, 130–139.

Stricker, M., Gagel, J., Schmitt, S., Schulz, K., Weygand, D., Gumbsch, P. (2016). On slip transmission and grain boundary yielding. *Meccanica*, 51(2), 271–278.

Sutton, A.P. and Balluffi, R.W. (1995). *Interfaces in Crystalline Materials*. Clarendon Press, Oxford.

Tahir, A.M., Janisch, R., Hartmaier, A. (2013). *Ab initio* calculation of traction separation laws for a grain boundary in molybdenum with segregated C impurites. *Modelling and Simulation in Materials Science and Engineering*, 21(7), 075005.

Tang, M., Kubin, L.P., Canova, G.R. (1998). Dislocation mobility and the mechanical response of b.c.c. single crystals: A mesoscopic approach. *Acta Materialia*, 46(9), 3221–3235.

Tang, M., Cai, W., Xu, G., Bulatov, V.V. (2006). A hybrid method for computing forces on curved dislocations intersecting free surfaces in three-dimensional dislocation dynamics. *Modelling and Simulation in Materials Science and Engineering*, 14(7), 1139–1151.

Terentyev, D.A., Osetsky, Y.N., Bacon, D.J. (2010). Effects of temperature on structure and mobility of the <100> edge dislocation in body-centred cubic iron. *Acta Materialia*, 58(7), 2477–2482.

Van der Ven, A. (2004). The thermodynamics of decohesion. *Acta Materialia*, 52(5), 1223–1235.

Vattré, A. and Demkowicz, M. (2013). Determining the Burgers vectors and elastic strain energies of interface dislocation arrays using anisotropic elasticity theory. *Acta Materialia*, 61(14), 5172–5187.

Vattré, A. and Demkowicz, M. (2015). Partitioning of elastic distortions at a semicoherent heterophase interface between anisotropic crystals. *Acta Materialia*, 82, 234–243.

Vattré, A., Devincre, B., Feyel, F., Gatti, R., Groh, S., Jamond, O., Roos, A. (2014). Modelling crystal plasticity by 3D dislocation dynamics and the finite element method: The discrete-continuous model revisited. *Journal of the Mechanics and Physics of Solids*, 63, 491–505.

Vattré, A., Jourdan, T., Ding, H., Marinica, M.-C., Demkowicz, M.J. (2016). Non-random walk diffusion enhances the sink strength of semicoherent interfaces. *Nature Communications*, 7, 10424.

Weinberger, C.R., Aubry, S., Lee, S.-W., Nix, W.D., Cai, W. (2009). Modelling dislocations in a free-standing thin film. *Modelling and Simulation in Materials Science and Engineering*, 17(7), 075007.

Weygand, D., Friedman, L.H., Giessen, E.V.D., Needleman, A. (2002). Aspects of boundary-value problem solutions with three-dimensional dislocation dynamics. *Modelling and Simulation in Materials Science and Engineering*, 10(4), 437–468.

de Wit, G. and Koehler, J.S. (1959). Interaction of dislocations with an applied stress in anisotropic crystals. *Physical Review*, 116(5), 1113–1120.

Xia, S. and El-Azab, A. (2015). Computational modelling of mesoscale dislocation patterning and plastic deformation of single crystals. *Modelling and Simulation in Materials Science and Engineering*, 23(5), 055009.

Xia, S., Belak, J., El-Azab, A. (2016). The discrete-continuum connection in dislocation dynamics: I. Time coarse graining of cross slip. *Modelling and Simulation in Materials Science and Engineering*, 24(7), 075007.

Zbib, H.M., Diaz de la Rubia, T., Bulatov, V. (2001). A multiscale model of plasticity based on discrete dislocation dynamics. *Journal of Engineering Materials and Technology*, 124(1), 78–87.

Zepeda-Ruiz, L.A., Stukowski, A., Oppelstrup, T., Bulatov, V.V. (2017). Probing the limits of metal plasticity with molecular dynamics simulations. *Nature*, 550(7677), 492–495.

Zepeda-Ruiz, L.A., Stukowski, A., Oppelstrup, T., Bertin, N., Barton, N.R., Freitas, R., Bulatov, V.V. (2021). Atomistic insights into metal hardening. *Nature Materials*, 20(3), 315–320.

Zhang, X., Aifantis, K.E., Senger, J., Weygand, D., Zaiser, M. (2014). Internal length scale and grain boundary yield strength in gradient models of polycrystal plasticity: How do they relate to the dislocation microstructure? *Journal of Materials Research*, 29(18), 2116–2128.

2

Statistical Approach to the Representative Volume Element Size of Random Composites

Dominique JEULIN[1] **and Samuel FOREST**[2]
[1] *Centre de Morphologie Mathématique, MINES Paris, PSL Research University, France*
[2] *CNRS UMR 7633, Centre des Matériaux, MINES Paris, PSL Research University, Evry, France*

2.1. Introduction

The usual theoretical prediction of effective properties of random materials concerns media with an infinite extension, modeled by ergodic random structures. On the other hand, more and more 3D digital images of real microstructures have become available from efficient observations techniques, such as X-ray microtomography. Images are obtained on large-scale instruments like synchrotrons, and even on the laboratory scale, since commercial microtomographs are becoming as common as scanning electron microscopy (SEM). Electron microtomography in the transmission electron microscope (TEM) provides images with a much higher magnification, while optical confocal microscopy can be used for this purpose on media transparent to light, as, among others, biological tissues.

In parallel with these experimental breakthroughs, more and more powerful computational facilities have become available, even on desktop computers, making

Digital Materials,
coordinated by Marc BERNACKI and Samuel FOREST. © ISTE Ltd 2024.

it easier to handle very large data sets like 3D images and their processing to improve their quality or to extract objects that they contain.

A faithful representation of microstructures by flexible 3D random models of microstructures, relying on identification of models and their parameters from partial data or from prior assumptions, can often be obtained with a low computational cost.

Starting from images, efficient numerical solvers and parallel computation techniques allow us to get fields as accurate numerical solutions of physical problems driven by PDE, from which apparent physical properties can be estimated by homogenization.

Combining these novelties together, numerical approaches for microstructural physics and mechanics are rapidly expanding in the area of digital materials. However, these approaches are irrelevant and questionable if the statistical representativity of the results cannot be mastered. The purpose of this text is to remind through physical examples the principle of numerical computation of apparent properties. This numerical approach involves a statistical definition of a representative volume element (RVE) to provide intervals of confidence of the estimated apparent properties in the cases of linear and nonlinear constitutive equations. It is illustrated by examples of application to linear and nonlinear micromechanics, optics, fluid flows and wave propagation for various types of microstructures of industrial materials, as well as of models of random media.

2.2. Elements of numerical homogenization of heterogeneous media

2.2.1. *Examples of physical properties of heterogeneous media*

Consider a continuum embedded inside a domain (usually bounded) B within the Euclidean space $\mathbb{R}^n$. In each point x of B, define a set of physical properties $P(x)$ (for instance, a stress or strain tensor, a velocity, a temperature, etc.). $P(x)$ builds a field defined on the domain B, solution of a problem defined by the following conditions:

– application of conservation principles (for instance, momentum and energy) from a local balance involving systems of partial differential equations (PDEs). Equilibrium condition imposes that for a given flux $F(x)$:

$$\operatorname{div} F = 0 \quad [2.1]$$

– appropriate choice of boundary conditions (BCs) given on the boundaries ∂B of the domain B;

– use of constitutive equations. For instance, the following linear laws are commonly used:

- Hooke's law between the stress tensor σ and the strain tensor e in linear elasticity through the fourth rank tensor of elasticity $\mathbf{C}$:

$$\sigma = \mathbf{C}e \tag{2.2}$$

- Navier linearized law linking for a fluid the pressure gradient (with components $\partial_i p$) to the Laplacian of the velocity u through the kinematical viscosity μ:

$$\partial_i p = \mu \triangle u_i \tag{2.3}$$

- Darcy's law between the macroscopic flow rate (with components Q^i) and the macroscopic pressure gradient (with components $\partial_i P$) for a flow in a porous medium, K being the second rank tensor of permeability:

$$Q^i = -\frac{1}{\mu} K^{ij} \partial_j P \tag{2.4}$$

- for electrostatics of a dielectric medium, proportionality between the electric displacement D and the electric field E, ϵ being the second rank dielectric permittivity tensor, and ϕ the potential:

$$D = \epsilon E = -\epsilon \text{ grad}\, \phi \tag{2.5}$$

Laws [2.2–2.5] are defined for each point x, where only local information is involved in the relations between the variables. For a homogeneous medium, the variables L, μ, K, ϵ (which are positive definite tensors) in equations [2.2]–[2.5] remain uniform in space. The solution to the problem is a field $P(x)$ in B, compatible with the BCs. For the various instances given above, equation [2.1] is formulated as:

$$\begin{aligned} \operatorname{div} \sigma &= 0 \text{ for continuum mechanics} \\ \operatorname{div}(\operatorname{grad} p) &= 0 \text{ for fluid flows} \\ \operatorname{div} Q &= 0 \text{ for macroscopic flows} \\ \operatorname{div} D &= 0 \text{ for electrostatics} \end{aligned} \tag{2.6}$$

in the static case and in the absence of source terms.

2.2.2. *Change of scale in heterogeneous media*

In the case of linear constitutive equations, existence and uniqueness of the solution for a given geometry are proved for appropriate BCs. In addition, on a macroscopic scale, the same linear constitutive laws apply, with so-called effective properties. More generally, nonlinear constitutive equations can be used.

In a heterogeneous medium, the constitutive equations can change in space from:

– different types of constitutive equations (for instance, linear for some places and nonlinear for other places within the domain B);

– variations in space of the coefficients of the equations. They can be modeled either by random functions (Matheron 1967; Sanchez-Palencia and Zaoui 1987) or by periodic fields (Sanchez-Palencia and Zaoui 1987).

In the case of heterogeneous media, the change in scale problem can be formulated as follows:

– Is it possible to replace a heterogeneous medium (with support B) with an equivalent homogeneous medium from a macroscopic point of view?

– If yes, what is the macroscopic constitutive equation and what are the values of its macroscopic coefficients (or effective properties)?

The probabilistic study of the fluctuations of apparent properties estimated in realizations of a random medium in a bounded domain B with possible scale effects is far less advanced. To correctly handle scale depending on fluctuations of apparent properties, the notion of statistical RVE has to be introduced (section 2.3). Presently, this problem can be only studied by numerical methods of simulations on realizations of random media (like, among others, finite elements (FEs) (Cailletaud et al. 1994; Kanit et al. 2003, 2006) or fast Fourier transform (FFT) (Willot and Jeulin 2009)).

The interest of change of scale methods is straightforward for applications in material sciences:

– They help to account for the microstructure in the description of the overall behavior of heterogeneous media. As a result, they can contribute to the optimization of materials. Another important aspect concerns the implementation of constitutive equations on a mesoscale containing microscopic information for introduction into computation codes used for the design of parts, or of structures.

– They connect the microstructure to physical properties of materials useful for their applications. The latter can combine several properties, possibly studied by the same theoretical approach from the single micro-geometrical information.

2.2.3. *Principle of calculation of apparent properties*

Definition of apparent properties from space averages

First consider the case of a vector field, namely, the electric field E in a dielectric medium with a real permittivity. The average of any field $P(x)$ in B with volume V is defined as:

$$\langle P \rangle = \frac{1}{V} \int_B P(x) dx \tag{2.7}$$

For instance, the average of the electric field E and the displacement field D is:

$$\langle E\rangle = \frac{1}{V}\int_B E(x)dx$$
$$\langle D\rangle = \frac{1}{V}\int_B D(x)dx = \frac{1}{V}\int_B \epsilon(x)E(x)dx \qquad [2.8]$$

The apparent dielectric tensor ϵ^* of the equivalent homogeneous medium contained in B is defined in such a way that equation [2.5] is satisfied on a macroscopic scale:

$$\langle D\rangle = \epsilon^* \langle E\rangle \qquad [2.9]$$

Boundary conditions

In general, ϵ^* depends on the applied boundary conditions on ∂B, local permittivity of the components and the geometry of the medium. The most current applied BCs are as follows:

– Dirichlet BC: a uniform electric field E_0 is applied on ∂B, or equivalently, introducing the potential $\phi(x)$ so that $E_0 = -\operatorname{grad}\phi$, $\phi(x) = -E_0.x$, $\forall x \in \partial B$.

– Neumann BC: $D.n = D_0$ for $x \in \partial B$ with normal n.

– Periodic BC: $\phi(x) = -E_0.x + \phi_P(x)$, where $\phi_P(x)$ is a periodic fluctuation, with $\phi_P(x^+) = \phi_P(x^-)$, x^+ and x^- being two homologous points on ∂B. D must be antiperiodic: on two homologous boundaries, $D^+.n^+ + D^-.n^- = 0$.

Furthermore, conditions must be satisfied on the interface between a pair of components. For perfect interfaces, the potential is continuous on the interface, but its normal derivative makes a jump, so that the normal component of E makes a jump on the interface, but its tangential component is continuous. In the remaining parts of this text, the interfaces are assumed to be perfect.

The apparent dielectric tensor can equivalently be defined from the energy $U(x) = \frac{1}{2}E(x)\epsilon(x)E(x) = \frac{1}{2}E(x)D(x)$ (the energy is a quadratic form in the case of a linear constitutive equation, as for linear electrostatics):

$$\langle U\rangle = \frac{1}{2}\langle E\rangle \epsilon^* \langle E\rangle \qquad [2.10]$$

The relations given by equations [2.8]–[2.10] can be applied to any heterogeneous medium in a bounded region B, provided the field $E(x)$ is expressed from the applied BCs and from the PDE expressing the Gauss' law:

$$\operatorname{div} D = \sum_i \frac{\partial}{\partial x_i}(D_i(x)) = 0 \qquad [2.11]$$

This procedure may be used for a periodic medium from the knowledge of the cell building the period (Jikov et al. 1994; Sanchez-Palencia and Zaoui 1987). When applying periodic BC on the period or homogeneous BC on a finite domain B, the averages of the fields are given by (Jikov et al. 1994; Sanchez-Palencia and Zaoui 1987):

$$\langle E\rangle = E_0 \qquad [2.12]$$

or:

$$\langle D\rangle = D_0 \qquad [2.13]$$

The Hill–Mandel lemma

Starting from the local energy $U(x) = \frac{1}{2}E(x)\epsilon(x)E(x) = \frac{1}{2}E(x)D(x)$ and averaging over the domain B, the following relation is satisfied for the homogeneous or periodic applied BCs to the bounded domain B, from which the apparent dielectric tensor can equivalently be defined (in linear elasticity of heterogeneous media, this result is called the Hill–Mandel lemma for heterogeneous media (Hill 1963; Kröner 1972; Sanchez-Palencia and Zaoui 1987)):

$$\langle U\rangle = \frac{1}{2}\langle ED\rangle = \frac{1}{2}\langle E\rangle\langle D\rangle = \frac{1}{2}\langle E\rangle\epsilon^*\langle E\rangle \qquad [2.14]$$

Equation [2.14] shows that the random fields E and D are uncorrelated for the homogeneous or periodic applied BCs to the bounded domain B. It holds for any pair (E, D) where $E = -\operatorname{grad}\phi$ and $\operatorname{div}(D) = 0$ satisfying these BCs. This relation is independent on the microstructure contained in a domain B, and does not rely on the linear (or nonlinear) constitutive equation [2.5] connecting D and E. The Hill–Mandel lemma is the key point for the estimation of the apparent (or effective in an infinite medium) properties and their fluctuations with the scale of observation, as detailed in section 2.3.

When a constant electric field E_0 is applied on ∂B, from equation [2.14] is deduced:

$$\langle D\rangle = \epsilon^* E_0 \qquad [2.15]$$

When a constant displacement field D_0 is applied on ∂B, equation [2.14] yields:

$$\langle E\rangle = \left(\epsilon^{-1}\right)^* D_0 \qquad [2.16]$$

Therefore, in the case of periodic or of homogeneous BC applied on ∂B, the apparent dielectric permittivity ϵ^* (or the apparent inverse $\left(\epsilon^{-1}\right)^*$) is obtained from the average $\langle D\rangle$ (or $\langle E\rangle$). In practice, it can be implemented from a numerical solution of the electrostatic problem, knowing the geometry of the medium and the BCs, as will be seen in section 2.3. Notice that for a bounded domain B in general $(\epsilon^*)^{-1} \neq \left(\epsilon^{-1}\right)^*$, the equality being obtained asymptotically when $B \to \mathbb{R}^n$.

If the field $P(x)$ is modeled by an ergodic random multivariate function, for domains B converging to $\mathbb{R}^n$ the spatial averages converge toward mathematical expectations with the following meaning for the variance D^2:

$$D^2\left[\langle P\rangle\right] \to 0 \text{ when } B \to \mathbb{R}^n$$

For an ergodic random field of permittivity $\epsilon(x)$, the random field $E(x)$ (and then $D(x) = \epsilon(x)E(x)$) is stationary and ergodic. It can be shown that in these conditions, the averages $\langle.\rangle$ in equations [2.9], [2.14], [2.15] and [2.16] become mathematical expectations $E\{.\}$, and the Hill–Mandel lemma can be applied for the infinite random medium (Matheron 1967, 1968; Jeulin 2021) (p. 631):

$$\begin{aligned} &E\{D\} = \epsilon^* E\{E\} \\ &E\{U\} = E\left\{\tfrac{1}{2}E(x)D(x)\right\} = \tfrac{1}{2}E\{E\}E\{D\} = \tfrac{1}{2}E\{E\}\epsilon^* E\{E\} \end{aligned} \qquad [2.17]$$

The two definitions of ϵ^* given in equation [2.17] are then equivalent.

Apparent elastic properties

For the apparent elastic properties of a random medium, a similar approach is made, where the potential is now replaced by the displacement field $u(x)$, the electric field by the deformation field $e(x)$ and D by the stress field $\sigma(x)$. Equation [2.14] corresponding to the Hill–Mandel lemma becomes:

$$\langle \sigma e\rangle = \langle \sigma\rangle\langle e\rangle \qquad [2.18]$$

Equation [2.18] holds for any pair (e, σ) where **e** derives from a displacement field $u(x)$:

$$e_{ij}(x) = \frac{1}{2}\left(\frac{\partial u_i}{\partial x_j} + \frac{\partial u_j}{\partial x_i}\right) \qquad [2.19]$$

and $\text{div}(\sigma) = 0$ satisfying BCs given below. It does not rely on the linear constitutive relation [2.2] connecting σ and e. For micromechanics in a domain B, the common applied BCs for which the Hill–Mandel lemma is satisfied are as follows (Kanit et al. 2003):

– Dirichlet BC: A uniform deformation e is applied on ∂B, or equivalently $u(x) = e.x, \forall x \in \partial B$. It is also called kinematic uniform BC (KUBC).

– Neumann BC: For the normal stress components on the boundary ∂B with normal n, $\sigma.n = t$: a uniform traction t is applied on ∂B. This corresponds to static uniform BC (SUBC).

– Periodic BC: $u(x) = \overline{e}.x + u_P(x)$, where $u_P(x)$ is a periodic fluctuation vector, with $u_P(x^+) = u_P(x^-)$, x^+ and x^- being to homologous points on ∂B. The traction t must be anti-periodic: on two homologous boundaries, $\sigma^+.n^+ + \sigma^-.n^- = 0$.

As for the electrostatics case, homogenization tools presented in this text are valid for perfect interfaces between components, insuring the continuity of the displacement and of the traction vector components.

Effective and apparent properties in the nonlinear case

The procedure based on the average energy can be extended to nonlinear constitutive equations, for which the function U is not any more quadratic. This happens for elastoplastic porous media or elastoviscoplastic polycrystals in micromechanics. This is studied by means of variational techniques to work out bounds in Talbot and Willis (1985); Ponte Castaneda and Willis (1988); Willis (1991, 2001), among others. The definition of apparent properties becomes more difficult in that case, since the average $\langle U \rangle$ or its mathematical expectation $E\{U\}$ can have various expressions as a function of physical properties of the various components A_i of the random medium and their constitutive equations. In contrast to the linear case, the composition of nonlinear properties by spatial averaging can be more complex. However, in the nonlinear case, the expressions [2.14], [2.17], [2.18] resulting from the Hill–Mandel lemma still apply and can be used to compute the macroscopic constitutive equations followed by the random medium. For instance, in the case of nonlinear mechanical properties, imposing an average deformation $E\{e\}$, use can be made of the local potential energy $W(x, e)$ and its space average $\langle W(x, e) \rangle$. For a domain B with infinite extension, the spatial averages of variables are replaced by their mathematical expectations in the ergodic case, so that an effective potential energy W^* is obtained:

$$E\{W(x, e)\} = W^*(E\{e\})$$

from which the macroscopic constitutive equation relating $E\{\sigma\}$ to $E\{e\}$ is obtained:

$$E\{\sigma\} = \frac{\partial}{\partial E\{e\}} W^*(E\{e\}) \qquad [2.20]$$

Alternatively, when imposing an average stress $E\{\sigma\}$, use can be made of the local complementary energy $U(x, \sigma)$ and of its space average $\langle U(x, \sigma) \rangle$. When $B \to \mathbb{R}^n$, the spatial averages of variables are replaced by their mathematical expectations, so that an effective energy U^* is obtained:

$$E\{U(x,\sigma)\} = U^*(E\{\sigma\})$$

from which the macroscopic constitutive equation relating $E\{e\}$ to $E\{\sigma\}$ is obtained:

$$E\{e\} = \frac{\partial}{\partial E\{\sigma\}} U^*(E\{\sigma\}) \quad [2.21]$$

When using parametric expressions relating $E\{e\}$ and $E\{\sigma\}$, effective material parameters can be estimated. Operating on a bounded domain B, apparent material parameters can be identified from the averages $\langle W(x,e)\rangle$ or $\langle U(x,\sigma)\rangle$, or directly from the curves relating $\langle\sigma\rangle$ and $\langle e\rangle$ obtained by spatial averages of the fields $\sigma(x)$ and $e(x)$ numerically computed from a range of values of applied fields $\langle\sigma\rangle$ or $\langle e\rangle$. Again, these average values (or apparent energies) are subject to statistical fluctuations, which have to be considered in order to provide intervals of confidence of the results of estimation, and to properly define a RVE.

2.2.4. *Homogenization of random media by numerical simulations*

An efficient way to solve the problem of homogenization makes use of digital materials and generates numerical solutions of the corresponding PDEs, namely, the required full fields. In a second step, the effective properties are estimated by spatial averaging of the solution using the Hill–Mandel condition (equations [2.14] and [2.18]). The input image can be a 3D image of the studied medium obtained by various techniques, for instance, microtomography. When such a 3D reconstruction is not available, use can be made of simulations of realizations of an appropriate model of random structure, after identification from 2D images obtained on sections. One of the techniques used to solve the homogenization problem is based on the FE method, as illustrated in Barbe et al. (2001a). A convenient way to numerically solve the homogenization problem is to use periodic BCs applied on a periodic cell. For this purpose, simulations of periodic random media were developed, like the Voronoi 3D polycrystal (Decker and Jeulin 2000). Usually, they start from a periodization of the Poisson point process. Beside the FE method, efficient iterative techniques operating by iterations of FFT on periodic media are available for micromechanics problems (Moulinec and Suquet 1994) and electrostatics (Eyre and Milton 1999). This numerical approach is based on the convolution form of the Lippmann–Schwinger equation (Kröner 1972), which is easily handled in the Fourier space after Fourier transformation of the involved fields. The main advantages of FFT computations is to avoid a laborious meshing procedure for complex microstructures, as required by the FE method, and to operate with periodic BCs, which are shown to provide a faster convergence of apparent properties with respect to the volume of numerical simulations (Kanit et al. 2003), as illustrated in section 2.4.1. This is directly applied to real or simulated 3D images. Up to $1,600^3$ images were used to estimate the apparent elastic moduli of granular materials in Escoda et al. (2016). Moreover, in the case of a two-component random medium, it is possible to handle separately the geometry and the physical properties of the components, so that a single calculation of fields has to be performed (see Chapter 19 in Jeulin 2021, pp. 725–726). As a result, a considerable reduction of the cost of computations can be expected when the impact of the change of contrast of properties, or when we look for the effect of the frequency ω through $\epsilon(x, \omega)$ in optics or through viscoelastic moduli in viscoelastic materials.

By implementation of such numerical techniques on realizations of models of random media, it is easy to provide parametric studies of the impact on the physical behavior of realistic microstructures. This provides guidelines toward the design of optimal microstructures with regard to their properties, as was done in the case of ice cream materials (Kanit et al. 2011).

2.3. Statistical definition of the RVE

When using numerical simulations as defined in section 2.2.4, a question immediately arises: how far are the effective properties estimated on a bounded domain B representative of the full microstructure? In other words, what is the size of a so-called "representative volume element" (RVE) (Kanit et al. 2003)? Given an estimation of apparent properties obtained from volume averages of computed fields (electric field in electrostatics, stress of strain field in elasticity), what is the statistical interval of confidence of this estimation? A similar question appears for engineering purposes, when working on parts with dimension of the same order as the microstructure; this occurs for some devices like microbeams in micro-electronics, where a few grains only are seen across the specimen. In that case, the effective properties of different parts (like the elastic moduli of the beam) fluctuate, so that a specific procedure may be required for the quality control in production, with a more or less important amount of rejected parts.

To answer to these questions, higher order statistical information than average values is required, but unfortunately, this is presently not available from a theoretical calculation without resorting to numerical simulations.

The second-order moment of the field in a domain with an infinite extension can be worked out when the effective property is known (Kreher and Pompe 1985; Bobeth and Diener 1986; Kreher 1990; Suquet and Ponte Castaneda 1993; Buryachenko and Kreher 1995; Buryachenko 2007), as detailed in Chapter 18 of Jeulin (2021), but it does not provide any information on the fluctuations of the average made over a finite domain B. The second moment of the field is obtained by partial derivation of the effective permittivity ϵ^* with respect to the permittivity of phases, starting from equation [2.17]. A first-order formal estimation of the variance of the average flow rate in porous media over a finite domain is given by Matheron (1967). It is derived from the first-order term of the perturbation expansion of the flow rate. Bounds of the variance are worked out in the same reference for radial flows and specific covariances of the permeability random function.

Developments for the statistical definition of the RVE in numerical homogenization of random media make use of a geostatistical approach based on the experimental determination of the integral range (Matheron 1971, 1989) from numerical simulations (Cailletaud et al. 1994; Kanit et al. 2003). This is a direct transposition of the approach followed in image analysis of microstructures to give a confidence interval of volume fractions, applied to steels and to iron ore sinters in Hersant and Jeulin (1976), and detailed in Chapter 3 of Jeulin (2021).

2.3.1. *Variance of estimation and integral range*

When estimating the average of an ergodic stationary random function $Z(x)$ over a domain B with volume V (briefly written $\bar{Z}(V) = \frac{1}{V}\int_V Z(x)dx$) from an average over a domain B' with volume V' ($\bar{Z}(V')$), the obtained result is known with a random error of estimation $\bar{Z}(V) - \bar{Z}(V')$. The variance of this error $\sigma_E^2(V, V')$ is given as a function of the central covariance function $\overline{C}(h)$ of $Z(x)$ by (Matheron 1965, 1971):

$$\begin{aligned} \sigma_E^2(V,V') &= E\left\{\left(\bar{Z}(V) - \bar{Z}(V')\right)^2\right\} \\ = \frac{1}{V^2}\int_B\int_B \overline{C}(x-y)\,dxdy &+ \frac{1}{V'^2}\int_{B'}\int_{B'} \overline{C}(x-y)\,dxdy \\ &-\frac{2}{VV'}\int_B\int_{B'} \overline{C}(x-y)\,dxdy \end{aligned} \quad [2.22]$$

where

$$\overline{C}(h) = E\{(Z(x) - E(Z))\,(Z(x+h) - E(Z))\}$$

In equation [2.22], a structural property (through the central covariance) and the geometry of implantation of the fields of measurements (through the domains of integration B and B') are present. In practical applications, simplified expressions can be used (for large domains B and B', they contain the integral ranges A_n in $\mathbb{R}^n$), as illustrated for solving sampling problems in image analysis applied to steels and to iron ore sinters in Hersant and Jeulin (1976) and below for homogenization. This enables us to define the size of a statistically representative volume element (RVE).

Consider now fluctuations of average values over different realizations of a random medium inside the domain B with the volume V. For an ergodic stationary random function $Z(x)$ with mathematical expectation $E(Z)$, the variance $D^2_Z(V)$ of its average value $\bar{Z}(V)$ in the domain B of volume V is derived from equation [2.22] by letting $B' \to \mathbb{R}^d$ and is given as a function of the central covariance function $\overline{C}(h)$ of $Z(x)$ (Matheron 1971) by:

$$D^2_Z(B) = \frac{1}{V^2} \int_B \int_B \overline{C}(x-y)\, dxdy \qquad [2.23]$$

For a large specimen (with $V \gg A_3$), equation [2.23] can be asymptotically expressed to the first order in $1/V$ as a function of the integral range in $\mathbb{R}^3$, A_3, by:

$$D^2_Z(B) = D^2_Z(V) = D^2_Z \frac{A_3}{V} \qquad [2.24]$$

with:

$$A_3 = \frac{1}{D^2_Z} \int_{R^3} \overline{C}(h)\, dh \qquad [2.25]$$

where D^2_Z is the point variance of $Z(x)$ and A_3 is the integral range of the random function $Z(x)$, defined when the integrals in equations [2.23] and [2.25] are finite. When $Z(x)$ is the indicator function of the random set A, equation [2.24] provides the variance of the local volume fraction (in 3D) as a function of the point variance $D^2_Z = p(1-p)$, p being the probability for a point x to belong to the random set A. When working in 2D, as was done to solve sampling problems in image analysis (Hersant and Jeulin 1976), the volume V is replaced by the surface area, and the integral range becomes A_2 after integrating the covariance in $\mathbb{R}^2$ in equation [2.25]. The asymptotic scaling law [2.24] is valid for an additive variable Z over the region of interest B and gives a decrease in the variance $D^2_Z(B)$ inversely proportional to its volume V. From equation [2.24], the volume V is equivalent to $k = \dfrac{V}{A_3}$ subvolumes in which the average values of the RF $Z(x)$ are uncorrelated random variables. This is why the integral range can be interpreted as a volumetric scale of the microstructure.

Considering now for B a subdomain of B' with volume $V < V'$. For large specimens with $V' > V \gg A_3$, the asymptotic expression of equation [2.22] becomes:

$$\sigma^2_E(V, V') \simeq D^2_Z A_3 \left(\frac{1}{V} - \frac{1}{V'} \right) \qquad [2.26]$$

When $V' \gg V$, the term $1/V'$ can be neglected in equation [2.26], so that equation [2.24] is recovered. This is usually the case in practice of numerical homogenization.

For infinite integral range, the variance $D_Z^2(V)$ decreases more slowly than V^{-1}. From empirical observations, a decrease in $V^{-\gamma}$ with $\gamma < 1$ was proposed (Lantuejoul 1991). In Jeulin (2016), the coefficient γ is theoretically computed in the case of Boolean random varieties (see Chapter 6 in Jeulin 2021). When $A_3 = 0$, the decrease in $D_Z^2(V)$ is much faster than V^{-1}, which induces a fast convergence of $Z(V)$ toward its expected value.

Equations [2.24] and [2.26] give access to an experimental estimation of the integral range by plotting the curve $D_Z^2(V)$ obtained for different sizes of samples, without the knowledge of the theoretical or the experimental covariance $\overline{C}(h)$ (Matheron 1989; Lantuejoul 1991). This multiscale method of estimation of the integral range is much more robust than using a numerical integration of the experimental covariance, often subject to uncontrolled spurious oscillations around its sill. Its equivalent form when using a $V^{-\gamma}$ scaling gives access to γ and to a test of the validity of the scaling law (Kanit et al. 2003).

2.3.2. *Definition of the statistical RVE of apparent properties*

To estimate the effective dielectric permittivity ϵ^* by homogenization from equations [2.15] and [2.16], it is required to compute the averages $\langle D\rangle$ or $\langle E\rangle$ in domain B:

$$\langle D\rangle = \frac{1}{V}\int_B D(x)dx = \frac{1}{V}\int_B \epsilon(x)E(x)dx = \langle \epsilon E\rangle$$

For an appropriate choice of the constant electric field E_0 applied on ∂B (namely of the constant displacement field D_0) and in the isotropic case, ϵ^* is obtained from the estimations of a single component of $\langle D\rangle$ or $\langle E\rangle$. Therefore, the variance of the apparent property ϵ^* (it would be the effective property for $B \to \mathbb{R}^n$) follows equations [2.24] or [2.26] when the integral range A_3 of the relevant field is known. This requires the stationarity of the fields D and E, obtained for appropriate BCs (homogeneous or periodic, as seen in section 2.2.3). A similar approach can be used when the local (or even the global) permittivity is described by a tensor, but numerical simulations must be done with various vectors D_0 or E_0 applied on ∂B. This approach is exactly the same as in the case of elastic properties of a random medium and was introduced from numerical simulations in 2D in Cailletaud et al. (1994). To be valid, the approach giving the variance of estimation of $\bar{Z}(V)$ based on equations [2.24] or [2.26] requires the stationarity and the ergodicity of the fields (E, D, or e, σ). This is satisfied for any infinite random medium with stationary and ergodic moduli (see Jeulin 2001, 2021, pp. 629–630). Large efforts have been recently made to prove the $1/V$ scaling of the variance of estimation corresponding to equation [2.24] by involving some central limit theorem proved only for some specific random fields of coefficients like the dielectric permittivity $\epsilon(x)$ (see Egloffe et al. 2015; Gloria et al. 2015; Armstrong et al. 2019; Duerinckx et al. 2020, and the references quoted therein, among others). In fact, the general geostatistical approach does not need such restrictions.

Since the theoretical or the experimental covariance of the fields (E, D, or e, σ) is usually not available, the integral range can be estimated in the case of homogenization according to the procedure proposed by G. Matheron for any RF (Matheron 1989): working with multiple realizations on domain a B with an increasing volume V, the parameter A_3 is estimated by fitting the obtained empirical variance according to the expression [2.24]. Alternatively, considering now average values in subdomains B_i with a wide range of sizes, or simulations in a large domain B', the parameter A_3 is estimated by fitting the obtained variance according to the expression [2.24] or [2.26] for volume $V \gg A_3$ where this expression is satisfied. This parameter does not depend on V, provided it is large enough to ensure the stationarity of the field by minimizing the effect of BCs. A first application of this methodology to elastic fields was developed in Cailletaud et al. (1994).

The point variance of the concerned field is directly estimated from its calculated map. In early studies on the statistical RVE, and even still in some recent papers, the point variance was estimated from the corresponding point variance of a given property. For instance, in the case of a two phase composite with a scalar property Z_1 for phase 1 (with the volume fraction p), and Z_2 for phase 2, the point variance D_Z^2 of the random variable Z is given by:

$$D_Z^2 = p(1 - p)(Z_1 - Z_2)^2 \quad [2.27]$$

However, the value D_Z^2 given in equation [2.27] is a rough estimate of the required point variance correctly deduced from the corresponding field, as illustrated in section 2.4.3. Furthermore, for an infinite contrast obtained when $Z_1 \rightarrow \infty$, D_Z^2 given by equation [2.27] goes to infinity and the empirical estimation of the integral range would give a zero value, which is non-physical.

Some typical microstructures with long range correlations, like dilated Poisson hyperplanes or dilated Poisson lines in 3D studied in Chapter 6 of Jeulin (2021), have an infinite integral range (Jeulin 1991a, 1991b), so that the computation of the variance $D_Z^2(V)$ of equation [2.24] cannot be used any more. In this situation, a scaling power law with an exponent $\gamma < 1$ was suggested (Lantuejoul 1991), and used in various applications, where a coefficient close to 1 was empirically estimated (Cailletaud et al. 1994; Kanit et al. 2003). With this scaling law, the variance becomes:

$$D_Z^2(V) = D_Z^2 \left(\frac{A_3}{V}\right)^\gamma, \quad [2.28]$$

where the volume A_3 is no more the integral of the central covariance function $\overline{C}(h)$, but is still homogeneous to a microstructural volume. In practice, the effect of the volume of samples given by equation [2.24] is correct for a restricted range of volumes (the largest ones), as a result of its asymptotic character. This has to be checked on data. In addition, the perturbation of boundaries when imposing homogeneous BCs introduces a lack of stationarity, which is more sensitive at small scale, and may induce some spurious power law of the variance (with $\gamma < 1$) on a small scale, according to equation [2.24]. Again, this has to be carefully checked in every situation, as it is done in the references.

2.3.3. *Practical determination of the size of the RVE*

The size of a RVE can be defined for a physical property Z, a contrast, and a given precision in the estimation of the effective properties, depending on the number n of realizations that are available. By means of a standard statistical approach, the absolute error ϵ_{abs} and the relative error ϵ_{rela} on the mean value $E\{Z\}$ obtained with n independent realizations of volume V are deduced from the 95% interval of confidence by:

$$\epsilon_{abs} = \frac{2D_Z(V)}{\sqrt{n}}; \epsilon_{rela} = \frac{\epsilon_{abs}}{E\{Z\}} = \frac{2D_Z(V)}{E\{Z\}\sqrt{n}}. \quad [2.29]$$

The size of the RVE can now conventionally be defined as the volume for which $n = 1$ realization (as a result of an ergodicity assumption on the microstructure) is sufficient to estimate the mean property $E\{Z\}$ with a relative error (for instance, $\epsilon_{rela} = 1\%$), provided the variance $D^2_Z(V)$ is known from the asymptotic scaling law [2.24], [2.26] or [2.28]. Alternatively, it can be decided to operate on smaller volumes (provided no bias is introduced by the edge effect resulting from the BCs), and to consider n realizations to obtain the same relative error. It is noticed from extensive results of numerical simulations in Kanit et al. (2003) that the mean value given by the periodic BCs varies slightly as a function of the size of the domain, as compared to the other BCs (SUBC or KUBC). This more or less significant bias, due to the effect of the boundaries ∂B, is found in the mean value given by all BCs for small volume sizes, the value being different from the effective one obtained for large specimens, so that the scaling law [2.24] must be applied with some caution. For small volumes, the average moduli obtained by simulations depend on the BCs: for instance, in the case of the dielectric permittivity, imposing E_0 on ∂B produces results close to the upper Wiener bound, while imposing D_0 on ∂B gives results close to the lower Wiener bound. This bias is well known (Huet 1990; Sab 1992; Ostoja-Starzewski 1998), and corresponds to the case of the two homogeneous trial fields from which these bounds are derived. It must be taken into account for the definition of the RVE. The result is that the mean value computed on too small specimens cannot represent the effective response for the composite material, even using the periodic BCs and a sufficient number of realizations. From simulations given in Kanit et al. (2003) and summarized in section 2.4.1, it appears also that the mean value obtained with the periodic BC is unbiased for smaller sizes of B than for the other BC, but it leads in general to higher variances than for the two other conditions, requiring a larger number of simulations to get a given precision for the effective property. The scaling with volume V of the error resulting from the bias is briefly introduced in section 2.3.4.

2.3.4. *Scaling of the bias, or systematic error*

The BCs applied to the domain B in $\mathbb{R}^3$ with volume V and surface area S may introduce a bias in the estimated apparent properties. It is due to an edge effect produced by a boundary layer with thickness t where the applied fields differ from the fields acting inside B: instead of $Z(x)$ the numerical simulation provides in each point x the value $Z(x) + Y(x)$, $Y(x)$ being the local systematic error. This boundary layer is mentioned in early papers on homogenization of random media for the case of the electric field in electrostatics (Beran 1965) and for elastic fields in micromechanics (Beran and McCoy 1970; Kröner 1972, p. 108; Willis 1982, among others). In usual homogenization of random media, domains B with infinite extension are considered, so that the impact of BCs is eliminated. In the case of finite domains B, boundaries must be accounted for. It is easy to work out a scaling law of the bias when assuming that the systematic error $Y(x)$ is localized in a small domain δB inside B (with average $\overline{Y}$ in δB), close to the boundary ∂B, so that $Y(x) = 0$ when $x \in B \cap \delta B^c$. When estimating $E(Z)$ by the volume average of $Z(x)$, the overall error of the estimation is given by:

$$
\begin{aligned}
\langle Z + Y \rangle - E(Z) &= \frac{1}{V}\int_B (Z(x) + Y(x))dx - E(Z) \\
&= \langle Z \rangle - E(Z) + \frac{1}{V}\int_{\delta B} Y(x)dx \\
&= \langle Z \rangle - E(Z) + \frac{1}{V}\overline{Y}tS
\end{aligned}
$$

From the above expression of $\langle Z + Y \rangle - E(Z)$ the variance of the overall error is given by:

$$
\begin{aligned}
&E\{(\langle Z + Y \rangle - E(Z))^2\} \\
&= D^2(Z) + \frac{1}{V^2}E\{(\overline{Y}tS)^2\} + \frac{2}{V^2}E\{(\langle Z \rangle - E(Z))\frac{1}{V}\int_{\delta B} Y(x)dx\} \\
&= D^2(Z) + \frac{S^2}{V^2}t^2E\{(\overline{Y})^2\}
\end{aligned}
$$

neglecting the correlation between fields $Z(x)$ and $Y(x)$ inside δB; as far as we know, this is implicitly assumed in published papers about the scaling of the systematic error.

The variance of the overall error is the sum of two terms, namely the usual variance of the statistical error $D^2(Z)$ and the second-order moment of the bias (error due to the boundary layer). Therefore, the contribution of the bias to the variance scales as $V^{-\frac{2}{3}}$ (or L^{-2}) if L is a measure of the size of B. As an immediate result, the contribution of the bias to the standard deviation scales as $V^{-\frac{1}{3}}$ (or L^{-1}). For homogenization in $\mathbb{R}^2$, the same L^{-2} scaling is expected after introduction of a boundary layer close to the perimeter of B. The pre-factor $t^2 E\{(\overline{Y})^2\}$ due to the bias depends on the BCs and on the scale of the RF $Z(x)$. It is difficult or even impossible to obtain its value in a closed form, in contrast with the pre-factor of the variance of the random error given in equation [2.24]. It is easy to deduce a similar scaling law for the absolute systematic error defined by (see, for instance, (Schneider et al. 2022)):

$$
\begin{aligned}
|E\{\langle Z+Y\rangle\} - E(Z)| = |E\{\langle Y\rangle\}| &= \frac{1}{V} E\{\left|\int_{\delta B} Y(x)dx\right|\} \\
&= \frac{1}{V} tSE\{|\overline{Y}|\}
\end{aligned}
$$

and the same scaling as for the standard deviation of the error is recovered, with the same difficulty to recover the pre-factor $tE\{|\overline{Y}|\}$ in general.

Considerable efforts have been recently made in the so-called "stochastic homogenization" community to provide theoretical results on the scaling of the variance $D^2(Z)$ and the absolute systematic error. In Egloffe et al. (2015), the L^{-1} scaling of the absolute systematic error is conjectured with the same boundary layer argument. It is successfully tested by numerical simulations in the discrete case of independent random conductivities in $\mathbb{Z}^2$ and $\mathbb{Z}^3$ and with Dirichlet BCs. Additional theoretical results are obtained in the discrete periodic case with independent random conductivities (Gloria et al. 2015; Duerinckx et al. 2020), where the following scaling of the absolute systematic error is given in $\mathbb{Z}^d$ by:

$$
|E\{\langle Z+Y\rangle\} - E(Z)| \lesssim L^{-d} \ln^d L \qquad [2.30]
$$

From this scaling of the error, a higher efficiency of the periodic version of random media is expected for homogenization. However, the degree of generality of this result, in particular concerning the structure of random media for which it can be applied, is presently unknown. Periodic simulations with independent random conductivities in $\mathbb{Z}^2$ and a low contrast equal to 4 (Egloffe et al. 2015) do follow the scaling law given by equation [2.30], but the standard L^{-1} scaling is observed for periodic media with a non-zero correlation length. In all cases, the average absolute systematic error was observed to be much smaller than the random error.

Simulations of non-overlapping objects (spheres and fibers in $\mathbb{R}^3$ and discs in $\mathbb{R}^2$) in Schneider et al. (2022) are used to estimate the thermal conductivity of random media with a low contrast equal to 6. For spheres and discs, the average absolute systematic error (estimated by considering the estimation of simulations of the largest sample as the true effective conductivity) seems to scale as $V^{-\frac{1}{3}}$ (or L^{-1}) for the periodic and the standard geometry, as seen from Figures 7 and 9 of Schneider et al. (2022). For short fibers, the L^{-1} scaling is observed in the standard case, while a possible L^{-2} scaling (dropping the logarithmic term in equation [2.30]) is obtained for the periodic case in Figure 2.12, but this last conclusion is derived from three sizes of simulations only, which is not very reliable. The data given in this paper are obtained on a very small range of scales (a factor limited to 16 for spheres, 32 for discs and 2 for fibers), so that the observations should be confirmed on a much wider range of scales, but apparently the L^{-1} scaling seems to be dominant for these more realistic microstructures than a discrete set of independent conductivities. In the simulations of Schneider et al. (2022), the average absolute systematic error is lower by an order of magnitude for the periodic case, and is lower than the random error (with an order of magnitude for spheres and for discs); for fibers, the systematic error and the random error are similar, showing a higher sensitivity to artifacts introduced by edge effects. From the available results, for large enough simulations, the bias generated by boundaries can be neglected, as done in most available statistical approaches of the RVE.

2.3.5. *Determination of the size of the RVE in the nonlinear case*

In the nonlinear case, it is wise to start from the apparent energy defined in section 2.2.3 for which the same approach as for the linear case can be followed: in micromechanics, for each set of applied conditions $\langle \sigma \rangle$ or $\langle e \rangle$, the variance of the average $\langle W(x, e) \rangle$ or $\langle U(x, \sigma) \rangle$ can be estimated, from which an integral range and the size of the RVE can be obtained. Consider for instance the potential energy $W(x, e)$ and its expectation $W^*(E\{e\})$:

$$2W(x, e) = \sigma(x)e(x)$$

From the Hill–Mandel lemma,

$$2\langle W(x, e) \rangle = \langle \sigma e \rangle = \langle \sigma \rangle \langle e \rangle$$

so that for a given $\langle e \rangle$ the variance of $\langle W(x, e) \rangle$ is proportional to the variance of $\langle \sigma \rangle$. Therefore, the determination of the size of the RVE in the nonlinear case is directly derived from the size of the RVE of the field $\sigma(x)$. This should be preferred to any statistical analysis of the parameters of the constitutive law obtained on a small scale, for which the point variance, and as a result the integral range, is not properly defined by equation [2.27].

2.3.6. *Remarks on the reduction of variance*

When using simulations of a random microstructure, some authors of recent publications try to get rid of fluctuations by fixing some parameters for each realization, for instance, the volume fraction of components, or even other morphological properties like set covariances or higher order moments, defining a so-called "representative unit cell" (Matouš et al. 2017). Sometimes the volume fraction is constrained to a unique value, whatever the scale of observation. This naive approach is proposed to reduce the variance of estimation of properties and homogenized coefficients in order to limit the number of simulations. Needless to say that such a procedure may be the source of uncontrolled biases, the effective behavior depending nonlinearly on such parameters as volume fractions or higher order morphological properties.

Hoping to be able to make a correct macroscopic prediction with a single small representative realization satisfying some constraints is a methodology very close to the traditional approaches considering a single RVE to represent the microstructure. Unless ergodicity conditions are reached by means of a single, large enough simulation (which can be checked through the estimation of integral ranges), there is no warranty on the quality of the prediction, due to some uncontrolled features, or due to the generation of bias by the effect of some BC, as illustrated by some of the studies presented later. Moreover, such an approach is non-physical and generates mathematical inconsistencies when working on random media: on the one hand, fixing the volume fraction of a composite on samples B with volume V is equivalent to constrain $A_3 = 0$ from equations [2.26] and [2.24]. On the other hand, considering random subvolumes of B, fluctuations of local volume fractions are automatically generated, from which a non-zero integral range can be estimated. It is difficult to predict the impact of such a contradiction on the predicted effective properties, more specifically for a large contrast of properties of various constituents. In this context, it is preferable to avoid any attempt of reduction of the variance.

2.4. Examples of application

In this section, the methodology developed to estimate the size of the statistical RVE is illustrated in various situations covering a wide range of applications. An exhaustive review of applications of the statistical RVE is not possible, since the first published work on this tool (Kanit et al. 2003) is currently mentioned in more than 2,300 publications. That is why we will limit our purpose to some significant applications. More details for each example are given in the references and Jeulin (2021, Chapter 19).

2.4.1. *Linear elastic properties and thermal conductivity*

Two-components media on the Voronoi mosaic

In a pioneering work (Kanit et al. 2003), followed by many applications published these last 20 years, the fluctuations of apparent elastic properties and thermal conductivity are studied from FE simulations. A two-component model is obtained from a random Voronoi mosaic (Figure 2.1): starting from a random Voronoi tessellation of the 3D space, every cell of the tessellation is randomly and independently affected to component 1 with the probability p and to component 2 with the probability $1 - p$. By convention, cells C have here a unit average volume $E\{V(C)\}$, so that the volume of each simulation is of the order of the number of cells that it contains. Simulations are made for two values of p (and consequently of the volume fraction): $p = 0.5$ and $p = 0.7$. For any random mosaic medium in $\mathbb{R}^3$, the integral range A_3 does not depend on p and is given by:

$$A_3 = \frac{E\{(V(C))^2\}}{E\{V(C)\}} \qquad [2.31]$$

Simulations of periodic random media are used from the periodic version of the Voronoi tessellations (Decker and Jeulin 2000). Most calculations of apparent properties are made for a contrast $c = 100$ of the Young's moduli and the thermal conductivities, the larger property being chosen with the probability p. Three types of BCs are investigated: KUBC, SUBC and periodic for elastic moduli; uniform temperature gradient UGT (or Dirichlet), uniform heat flux UHF (or Neumann) and periodic for the thermal conductivity. The main results are summarized as follows:

– The size effect given by equation [2.23] for the variance $D^2_Z(V)$ is recovered for the local volume fraction, while the size effect given by equation [2.28] is observed for apparent elastic moduli (bulk and shear moduli) and thermal conductivity, with a coefficient $\gamma \simeq 1$. The estimated value of the integral range from empirical variances obtained for $p = 0.5$ and $p = 0.7$ are equal to 1.178 and 1.111, in agreement with its theoretical value (here $A_3 = 1.179$), which validates the simulations and the procedure of estimation of A_3 from the data $D^2_Z(V)$.

– It is observed that the integral range of apparent properties (elastic moduli and thermal conductivity), deduced from a point variance calculated from equation [2.27], depends on the property of interest and on its contrast over the two components. It is of the order of twice the integral range of the volume fraction for the Voronoi mosaic, but this result cannot be generalized to any microstructure or physical property.

– Small sizes of realizations introduce a bias of apparent properties, as illustrated by Figures 2.2 and 2.3 in the case of the apparent bulk and shear moduli K and G. This bias is much higher for KUBC and SUBC (it disappears for volumes containing at least 1000 Voronoi cells in the case of elastic moduli) than for periodic BCs (the bias disappears for volumes containing at least 15 cells). Similarly, the bias is higher for UGT and UHF, where at least 100 cells are required for unbiased simulations, than for periodic BCs, requiring at least 15 cells.

– The variance of apparent properties is always larger for periodic BCs, requiring more realizations for a given statistical precision of the apparent properties, but smaller volumes can be used in simulations. Typically, about 1,700 realizations of domains containing on average 15 cells were shown to be used to estimate the elastic moduli with a relative precision $\epsilon_{rela} = 5\%$ in the case of $p = 0.7$ and $c = 100$.

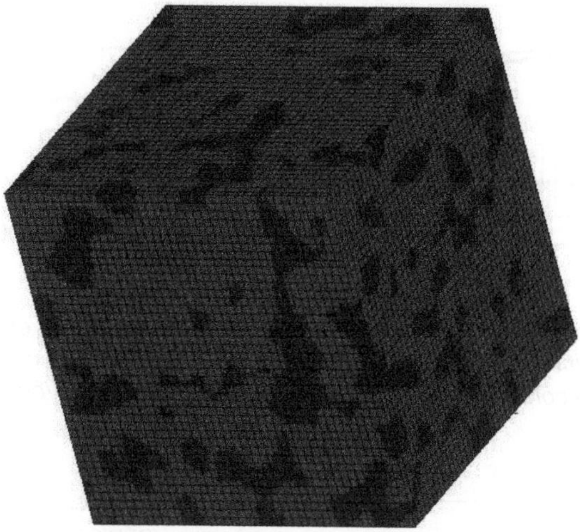

Figure 2.1. *Realization of a two-components Voronoi mosaic, with the FE mesh (Kanit et al. 2003). For a color version of this figure, see www.iste.co.uk/bernacki/digital.zip*

Materials from food industry

The previous methodology was applied for the first time to real 3D images of two materials from the food industry, namely, ice creams, obtained by optical sectioning with a confocal microscope (Kanit et al. 2006). The purpose was to explain from numerical calculations the different apparent properties observed for the two materials in the presence of two components with a high contrast ($1,000$ for Young's moduli and 100 for thermal conductivities), so that standard bounds are far apart and therefore inefficient to give estimates of the effective properties. The harder phase covers 70% of the volume, and a factor two is observed for the experimental Young's moduli of the two materials.

Numerical simulations of elastic and thermal fields are made by FEs with the three types of BCs: KUBC, SUBC and periodic for elasticity, and UGT, UHF and periodic for thermal properties.

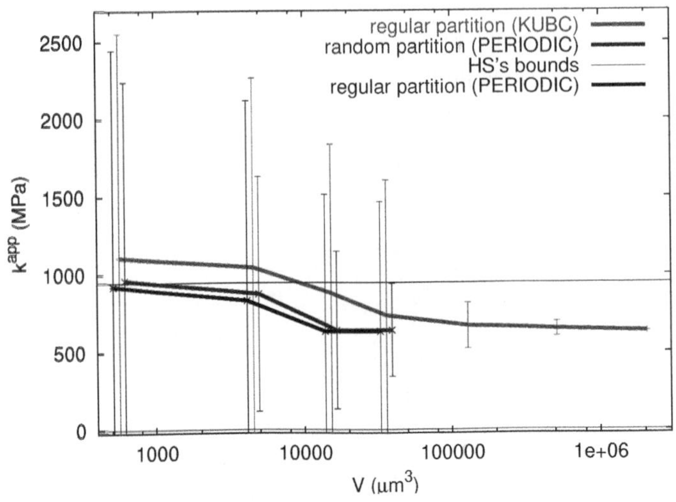

Figure 2.2. *Apparent bulk modulus K and interval of variation estimated from FE calculation with different BC and volumes V for the Voronoi mosaic with $p = 0.7$ (Kanit et al. 2003). For a color version of this figure, see www.iste.co.uk/bernacki/digital.zip*

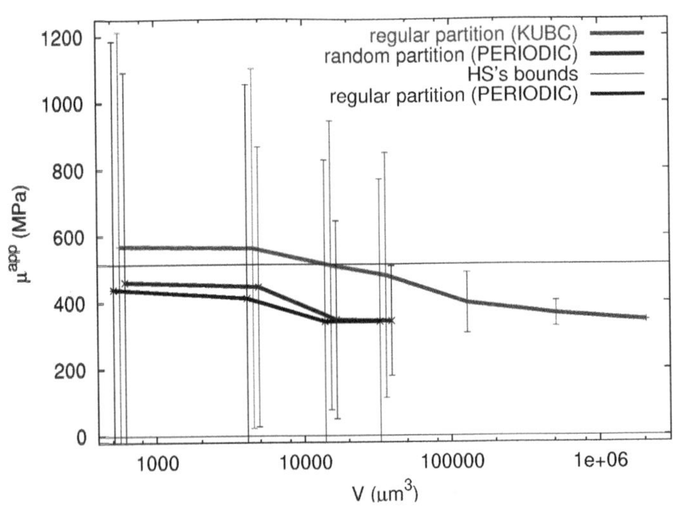

Figure 2.3. *Apparent shear modulus G and interval of variation estimated from FE calculation with different BC and volumes V for the Voronoi mosaic with $p = 0.7$ (Kanit et al. 2003). For a color version of this figure, see www.iste.co.uk/bernacki/digital.zip*

Having a single volume image for each material, the only way to use a statistical approach of the RVE is to cut subvolumes of various sizes, either by partitioning the image, or taking random subvolumes. It illustrates the efficiency of the statistical RVE approach, which can be operated for a single large realization of the microstructure. Here are the main results:

– To avoid bias in the estimation of apparent properties, volumes larger than $(24\,\mu m)^3 = 13824\,\mu m^3$ are required for periodic BCs. Periodic BCs are applied even if the microstructure is not periodic.

– The integral ranges of the properties are larger for the material with a coarser microstructure, as expected.

– The largest integral range of the elastic moduli is found for the shear modulus, resulting from the fact that the contrast in shear modulus between the two components is higher than for the bulk modulus.

– The integral range corresponding to the thermal conductivity is much higher (factor 2-4) than for elastic fields, resulting in this case in larger sizes of RVE.

– The investigated volume of the material with the finer microstructure is sufficient to provide apparent Young's modulus with a correct precision ($\epsilon_{rela} \simeq 15\%$ and 6%), but three equivalent volumes should be used to reach $\epsilon_{rela} \simeq 5\%$, requiring the input of more images.

Unidirectional fiber composite by finite elements

A unidirectional composite with a PA6 matrix and long parallel glass fibers is studied in (Oumarou et al. 2011). Image analysis and FE element calculations of thermal and elastic fields are made on transverse sections, where fibers are displayed as non-overlapping discs with heterogeneous locations, mainly due to the presence of plies in the manufacturing process. The average volume fraction is 0.4, and the average fiber diameter is16 μm. The RVE is estimated for all properties of the composite. The most sensitive property to fluctuations of the microstructure is the transverse shear modulus, for which the RVE size is the largest: for a square image with $854\,\mu m$ edge, containing about 1,310 fibers, its relative precision is 18.7% for a single image, and therefore 1.87% for 100 images.

Random fiber composites by iterations of FFT

In Altendorf et al. (2014), microtomographic images and 3D simulations of various random networks of finite fibers having a length of the order of the size of the samples, various distributions of fiber orientation and radius, are generated with the volume fraction $V_V \simeq 0.15$ to predict the elastic properties and the conductivity by iterations of FFT.

A glass-fiber reinforced polymer is examined by microtomography with a voxel sampling of $3.5\,\mu m$. After application of a segmentation procedure to separate fibers, length-weighted radius and orientation distributions are estimated from 3D images.

Estimated curvature parameters are introduced as input in a random-walk-based model of three-dimensional fiber systems to generate virtual fibrous media with controlled bending of fibers (Altendorf and Jeulin 2011).

Various types of generated periodic microstructures range from isotropic to transversely isotropic, and to orthotropic media, representing plausible virtual fibrous materials for fibers with average length of 1.19mm (or equivalently 340 voxels). The anisotropy of the macroscopic responses and the size of the corresponding RVE are examined numerically. Field calculations are made by FFT on simulations with typical size 600^3 voxels. The elastic and thermal properties of the system studied in Oumarou et al. (2011) are used for comparison.

It is found that the variance of the properties on a volume V scales as a power law $\sim 1/V^\gamma$ with $\gamma < 1$ and in the range $0.71 - -0.87$, in agreement with equation [2.28]. This is a consequence of the long-range correlations in the microstructure, in comparison to the size of simulations, resulting in a slower decrease in the variance of local averages as usual, but faster than for the Poisson fibers where $\gamma = 2/3$, as seen in section 2.4.1. Concerning the RVE size, a relative precision in the range $0.29 - -1.86\%$ is reached with 10 realizations in cubic domains with 400^3 voxels for the full set of parameters of interest: fiber volume fraction V_V, elastic moduli and thermal conductivity for all considered fibrous systems. Lower size volumes can be used from flat specimens accounting for the main orientations of the fibrous system.

Examples of random media with infinite integral range: Poisson fibers and gigantic RVE

A family of random media with infinite integral range is constructed as follows: by dilation of Poisson varieties V_k of dimension $k < n$ (Matheron 1975), various Boolean varieties are generated in the space with dimension n, $\mathbb{R}^n$ (Jeulin 1991a, 1991b, 2016). Let us consider a convex domain K in $\mathbb{R}^n$, with Lebesgue measure (volume in 3D) $\mu_n(K)$. The scaling law of the variance of the volume fraction given by equation [2.28] is theoretically obtained for the Boolean model built on isotropic Poisson varieties V_k (Jeulin 2016): in $\mathbb{R}^n$, the theoretical expression of the variance $D^2_Z(K)$ of the local fraction $Z = \frac{\mu_n(A \cap K)}{\mu_n(K)}$ of a Boolean model built on isotropic Poisson varieties of dimension k ($k = 0, 1, ..., n-1$) V_k is expressed as:

$$D^2_Z(K) = p(1-p)\left(\frac{A_k}{\mu_n(K)}\right)^{\frac{n-k}{n}} \qquad [2.32]$$

the scaling exponent being $\gamma = \frac{n-k}{n}$. As particular cases, Poisson points ($k = 0$) give the standard Boolean model with a finite integral range and $\gamma = 1$, Poisson lines ($k = 1$) generate Poisson fibers with $\gamma = \frac{n-1}{n}$ and Poisson hyperplanes ($k = n-1$) provide Poisson strata with $\gamma = \frac{1}{n}$. It turns out that the most penalizing situation with respect to the scaling of the variance is the case of Poisson strata, with a very slow decrease in the variance with the volume of the sample K, where $\gamma = \frac{1}{n}$. For isotropic Boolean fibers in 2D, the scaling exponent is $\gamma = \frac{1}{2}$ (working in 2D, dilated Poisson lines were used to model a fiber network (Delisée et al. 2001)). For isotropic Boolean fibers in 3D, the scaling exponent is $\gamma = \frac{2}{3}$. For isotropic Boolean strata in 3D, the scaling exponent is $\gamma = \frac{1}{3}$. This decrease in the variance with size is much slower than the case of a finite integral range. When considering random media with a nonlinear behavior, for instance, a viscoplastic material, a strong localization of strains resulting in shear bands is expected. This generates long range correlations of the fields that might be modeled by Boolean strata in 3D, and scaling laws similar to the dilated Poisson hyperplanes (with a scaling exponent close to $\frac{1}{3}$) might be recovered, so that a slow convergence toward the effective properties should be observed on numerical simulations with increasing sizes (Jeulin 2012). This point remains to be investigated by numerical experiments.

The theoretical and observed scaling laws of the variance for fibrous media made of long fibers at the scale of simulations make the statistical convergence of apparent properties to the effective properties difficult, which requires much larger RVE or much more realizations than the standard case. In addition, Poisson fibers (or Poisson strata) with infinite length cannot be periodized (any attempt to make them periodic would completely cover space), which makes numerical homogenization a challenge in these situations, despite their practical interest for some types of materials.

Simulations for the estimation by FEs (Dirrenberger et al. 2014) of the elastic properties and the conductivity of 3D random Poisson fibers (illustrated in Figure 2.4) surrounded by voids provide an empirical scaling law of the variance close to the theoretical one obtained for the volume fraction. This is expected, as a result of a high correlation between the elastic or thermal fields and the indicator function of the random set A.

Thermal conductivity λ and elastic properties are estimated over hundreds of realizations, with up to 800 fibers for the largest fields. The average measured fiber volume fraction is $p = 0.162 \pm 0.013$. The convergence of λ^{app} is reached for a volume $V = 10^3$ containing nearly 180 fibers. The relative discrepancy error on the mean value of λ^{app} between UTG and MTBC (mixed thermal BC satisfying the Hill–Mandel lemma, detailed in (Dirrenberger et al. 2014)) is about 16% for the largest sample. There is a strong discrepancy between the KUBC and MBC (mixed tri-axial loading detailed in (Dirrenberger et al. 2014)) estimates K^{app} and G^{app} for largest volumes, showing a lack of convergence of apparent elastic properties due to insufficient volumes. The variance of all measured parameters decreases slowly with the volume V, according to equation [2.28] with $\gamma = 2/3$, as theoretically predicted for the fiber volume fraction. This is illustrated for K^{app} in Figure 2.5. Calculation of the statistical relative precision obtained from the simulations show that they are in the range of 2.3–7.1%. Reaching a 5% relative precision with a single realization would require much larger volumes than used in the present numerical simulations, from 164^3 to 489^3, depending on the property and the BC.

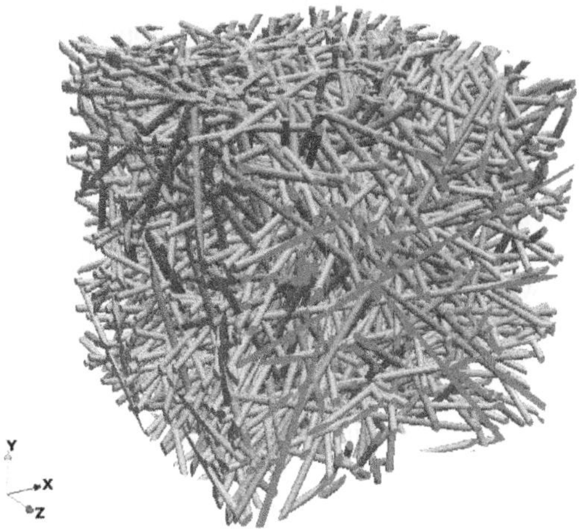

Figure 2.4. *3D simulation of Poisson fibers (radius 1 in a cube with edge 60) (Dirrenberger et al. 2014). For a color version of this figure, see www.iste.co.uk/bernacki/digital.zip*

Anisotropic elasticity in bulk polycrystals and thin films

The statistical approach to RVE size is applied to the effective elastic bulk properties of metallic polycrystals by El Houdaigui et al. (2007a). The example deals with a Voronoi tessellation of copper grains as shown in Figure 2.6. A uniform distribution of crystal lattice orientations is attributed to the grains of the volume element. Copper single crystals exhibit cubic anisotropy and three independent elastic moduli are:

$$C_{11} = 168,000\,\text{MPa}, \quad C_{12} = 121,400\,\text{MPa}, \quad C_{44} = 75,390\,\text{MPa}$$

at 300 K. The effective behavior of such a polycrystal is isotropic. It is characterized by its bulk modulus, which is identical to that of the cubic single crystal, and by its shear modulus which is estimated here by means of FE simulations. The statistical approach yields the results in Figure 2.6, which shows a rapid convergence of the apparent shear modulus, when using periodic BCs, toward a value close to the self-consistent estimate. It turns out that the effective shear modulus is estimated with a relative precision of 1% by means of 20 realizations of the polycrystals made of 200 grains in average.

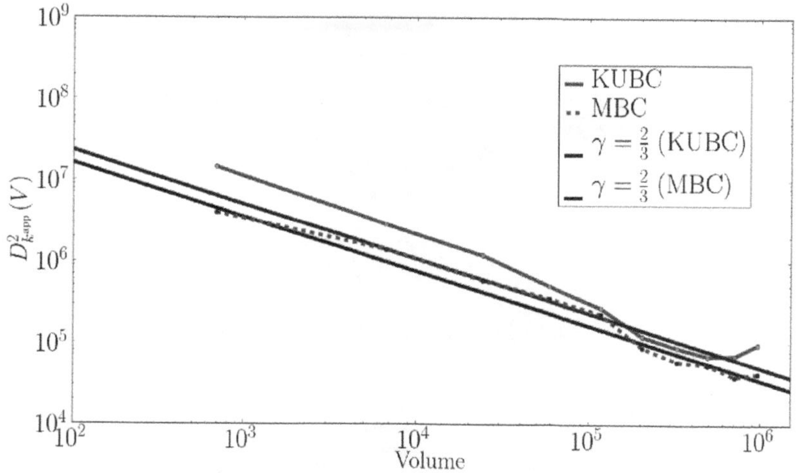

Figure 2.5. *Variance scaling of the apparent bulk modulus K^{app} with the volume of simulations (logarithmic scales) (Dirrenberger et al. 2014). For a color version of this figure, see www.iste.co.uk/bernacki/digital.zip*

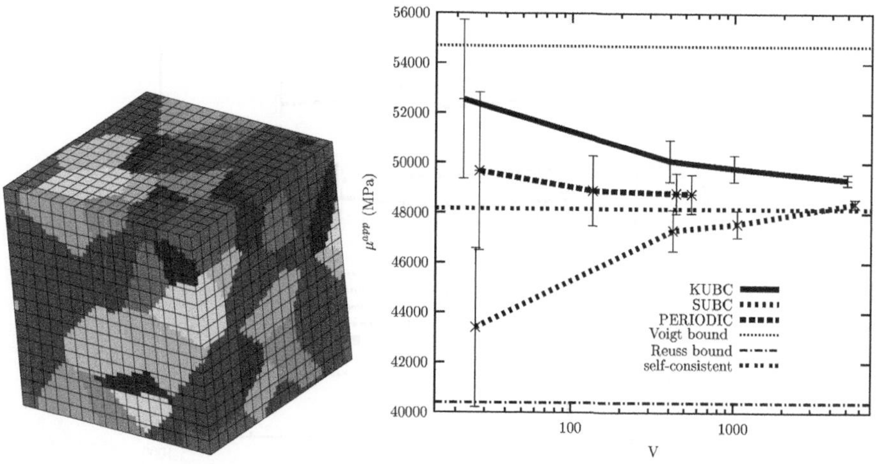

Figure 2.6. *Apparent shear modulus for isotropic bulk polycrystalline copper as a function of volume size (after El Houdaigui et al. (2007a)). For a color version of this figure, see www.iste.co.uk/bernacki/digital.zip*

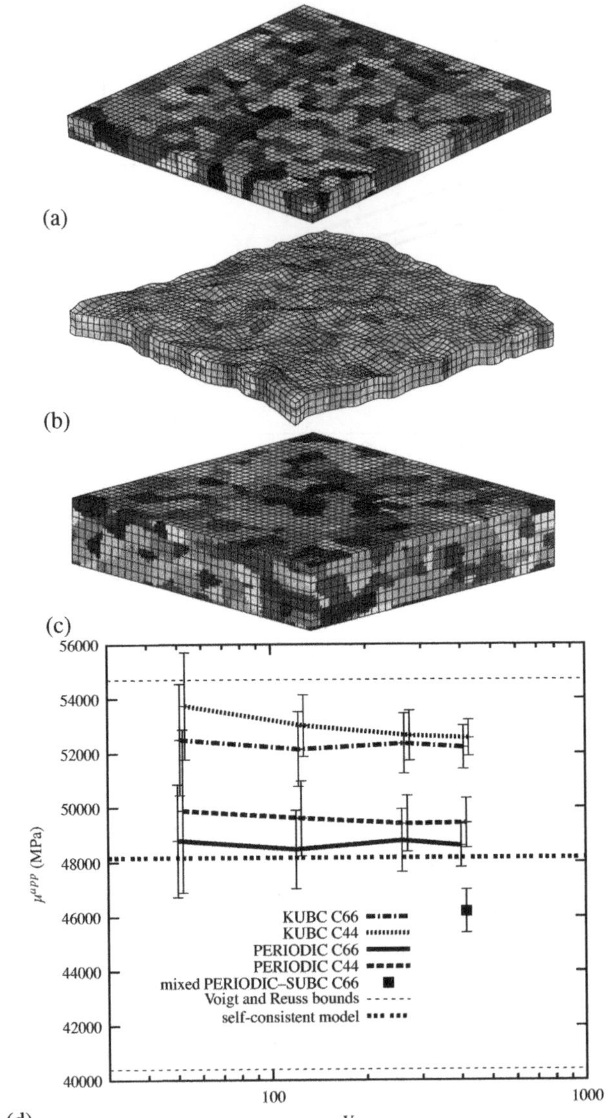

Figure 2.7. *Apparent shear modulus of polycrystalline thin films: (a) example of film with one grain within the thickness; and (b) its deformed state after in-plane shear with stress concentration contour (center); (c) example of a thin film morphology with three grains within the thickness; (d) apparent shear moduli of the one-grain thick film depending on system size and type of boundary conditions. For a color version of this figure, see www.iste.co.uk/bernacki/digital.zip*

The question of RVE size can be extended to thin polycrystalline films that possess only a few grains within the thickness, as illustrated in Figure 2.7. Thin films are produced from Voronoi tessellation by random slices of the bulk volumes with fixed thickness. The upper and lower surfaces are free of forces, whereas periodic BCs are applied in the film plane. This corresponds to a mixed type of BCs. Again, isotropic distributions of lattice orientation is considered. For the statistical analysis, the in-plane dimensions of the plate are increased for a constant grain size, whereas the film thickness is unchanged. Several questions arise. First, how many grains within the thickness are needed to retrieve the elastic properties of the bulk materials? This question can be answered using the statistical approach and FE simulations carried out by El Houdaigui et al. (2007b) show that three to five grains are sufficient in the thickness to recover the bulk properties, depending on the wanted precision.

A second question is: how much does the in-plane C_{66} and out-of-plane C_{44} shear moduli differ for a thin film with one grain within the thickness in average? It is found that the effective properties of the film differ from that of bulk copper, as shown in Figure 2.7. The in-plane shear modulus is found to be closer to the bulk one than the out-of-plane effective shear modulus. These results have implications for the properties of metallic films and coatings used in industry.

Granular media studied by microtomography

The two examples introduced now are based on 3D FFT computation of fields in granular materials from images obtained by microtomography: the first one concerns a propellant material made of spheroidal grains embedded in a polymer matrix (Willot et al. 2013). The second material is a mortar (Escoda et al. 2011). After a morphological segmentation of images, FFT computations provide thermal and elastic fields. In addition to the estimation of the apparent properties and their statistical RVE, a morphological analysis of the fields provides instructive information on local field concentrations, not detailed here.

Granular material

A (926 $\times$ 926 $\times$ 463) image is extracted from a microtomographic image obtained on a cylinder of a propellant material. Two segmentations provide either disconnected grains, or a connected microstructure with slightly overlapping grains. The thermal response of the image is obtained by the FFT method, with the following conductivities, obtained from experimental measurements: $\lambda_1 = 0.15\text{J}/\text{mKs}$ for the matrix and $\lambda_2 = 0.609\text{J}/\text{mKs}$ for grains. A macroscopic temperature gradient with averages $\langle \partial_i T \rangle = \Delta T_0 \delta_{ij}$ is applied, with periodic BC. The elastic response is computed for isotropic elastic moduli of components with two sets of parameters: for the matrix, (i) $G = 1/2$, $K = 1/3$; (ii) $G = 1/2$, $K = 2/3$; elastic properties of grains are the same up to a factor 2. Two macroscopic conditions are imposed: hydrostatic strain loading and shear strain loading. Full field calculation is made on the two segmentations. The apparent conductivity is isotropic with, for the connected microstructure, $\lambda^{app} \simeq 0.387$, close to the lower Hashin–Shtrikman bound HS$^-$ (0.368) (Willis 1982), as expected for a granular medium (Jeulin 2021, Chapter 18). The integral range of the heat flux is $A_3 = 13^3$ and its point variance $D_q^2 = 0.23$. The statistical relative precision is very good, due to the low contrast: $\epsilon_{rela} = 0.4\%$. The apparent elastic properties in case (i) are $G^{app} = 0.6534$ and $K^{app} = 4455$, very close to HS$^-$. With this very low contrast, a recourse to numerical homogenization seems superfluous. However, the examination of full fields with a relatively high space resolution gives an interesting insight into the local response of the medium: local hot spots of the heat flux appear in the matrix at the interface between close grains, on parts of grain boundaries orthogonal to the macroscopic temperature gradient. Similar local maxima are seen for the component of the temperature gradient parallel to the macroscopic gradient. These hotspots slightly increase the point variance of the fields, and might alter the sustainability of this material if submitted to a thermal shock during its storage.

Mortar microstructure

A 3D gray scale image ($1{,}000 \times 1{,}200 \times 1{,}000$ voxels, 25μm per voxel) of a mortar sample is obtained by microtomography (Escoda et al. 2011). Three components are present: voids, fine aggregates and the surrounding matrix cement paste. In the final segmented image by a morphological procedure detailed in Escoda et al. (2011) and Chapter 19 in Jeulin (2021), a three-phase microstructure is obtained with aggregates ($V_V = 0.334$), cement matrix ($V_V = 0.64$) and voids ($V_V = 0.0235$). Elastic fields are computed by iterations of FFT for two types of loading conditions and with periodic BC, as before: hydrostatic strain loading and shear strain loading. To deal with the presence of voids, the augmented Lagrangian algorithm (Michel et al. 2000) is used on images with 500^3 or 735^3 voxels. The contrast is monitored by the ratio of Young's moduli of aggregates and of the matrix, $c = E(a)/E(m) = 10^{-8}$, 3, 100, 1,000, 10,000 with a fixed value for $E(m)$. The contrast $c = 3$ corresponds to the real medium. Other ratios enhance the contrast and allow us to anticipate the situations of damage (10^{-8}) or of creep ($c \geq 100$). When $E(m) = 20$MPa and $E(a) = 60$MPa, the apparent Young's modulus is $E^{app} = 27.7$GPa, consistent with the experimental value (26.4GPa), which is expected for a low contrast ($c = 3$).

Concerning the RVE for elastic fields, $\epsilon_{rela} = 5\%$ is reached at low contrasts ($c \leq 100$) for volumes of the order of $V_0 = 500^3$ voxels. In the rigid case, the precision of the results remains between 5.4% and 11.4% for the given volumes. The morphological RVE for $\epsilon_{rela} = 5\%$ is much larger than the specimen size for pore volume fraction (1676^3), due to its low value and to a heterogeneous distribution of voids in the available image.

In a complementary study (Escoda et al. 2016), the same approach is applied to virtual concrete materials generated by multiscale simulation of three families of Poisson polyhedra to reproduce the size distribution of aggregates. Grains are randomly located by a sequential algorithm, without overlaps. Elastic fields are computed by FFT on large 3D simulations (1,600^3 with 0.09mm/voxel) with three scales of polyhedra. For comparison, a separation of scales is assumed to introduce in a smaller size computation (in terms of voxels: 1,024^3 with 0.24mm/voxel), a two components medium containing large gravels and a composite matrix made up of cement and smaller gravels with elastic properties estimated from a former homogenization. This simplified approach lowers the computational cost. The predicted moduli are very close: $K^{app}/K_m = 9.0$ for three scales, 10.5 for scale separation; $G^{app}/G_m = 10.3$ for three scales, 11.5 for scale separation. The relative precision of estimates is very high, since the RVE for $\epsilon_{rela} = 5\%$ are in the range 266^3–308^3, depending on the moduli and on the type of calculation.

Particulate random composites

Two-component random media made up of non-overlapping particles with a single volume v (spheres or ellipsoids) are simulated in El Moumen et al. (2015) for low volume fractions ($V_V = 0.13$ and 0.23). The elastic properties K^{app} and G^{app} are estimated by FEs with periodic BCs. The integral range deduced from equation [2.24] with a point variance estimated by equation [2.27], and fitted from simulated moduli, is compared to a proposed expression of the integral range (V_V/N_V), N_V being the number of particles contained in the volume V. The integral range in that case does not depend on the shape of particles. The two estimates of A_3 are in good agreement. We have to point out that two implicit approximations are used by the authors: first, for a vanishing volume fraction of non-overlapping particles with a large separation, the integral range of the mosaic model given by equation [2.31] is recovered, so that in the present case $A_3 \simeq v = V_V/N_V$. A lower value of A_3 is expected for higher volume fractions, since for a hard-core process generated by the particles, the set covariance $\overline{C}(h)$ presents negative correlations for some specific distances h. The second approximation concerns the use of equation [2.27] in the presence of a low contrast of elastic moduli.

In the remaining part of the paper, the authors look at the change of the volume of the RVE with particulate volume fraction. Apparently using the expression V_V/N_V for A_3 and ignoring the variations of N_V, inconsistent with the change of V_V, they predict a maximum of the volume of the RVE for $V_V = \frac{2}{3}$, which is incorrect: using the correct approximation of the integral range $A_3 \simeq v = v$ and equation [2.27], a maximum of the volume of the RVE is expected for $V_V = \frac{1}{2}$.

It would be interesting to extend this work to a population of particles with different volumes. In that case, equation [2.31] should be used for a first approximation of A_3.

As reminded in section 2.4.1, for a medium of this type with reinforcing non-overlapping particles, a good estimate of the effective properties is provided by the lower Hashin–Shtrikman bound HS$^-$. Therefore, the point variance deduced from this model (see Bornert and Suquet 2001, p. 71; Jeulin 2021, pp. 688, 690) would be more appropriate than equation [2.27] used in this chapter.

Boolean models by FFT

The effective bulk modulus of the Boolean model of spheres (with diameter 18 voxels), illustrated by Figure 2.8, is estimated by iterations of FFT from 10 realizations with size 256^3 covering the full range of volume fractions (Willot and Jeulin 2009). Two extreme situations are explored: the porous case (for spheres with zero moduli) and the rigid case (in fact for a very high contrast, with $K_1 = G_1 = 10^3$ in spheres and $K_2 = G_2 = 1$ in the matrix). Elastic fields computation is made by means of the augmented Lagrangian version of the FFT algorithm, well adapted to infinite contrast (Michel et al. 2000), applying a macroscopic deformation equivalent to a hydrostatic loading ($\langle e_m \rangle = 1$). The effective bulk moduli K^* is subject to percolation effects: K^* vanishes for a 0.95 pore volume fraction (the solid phase being then disconnected) and starts to increase in the rigid case for a sphere volume fraction close to 0.29 for which spheres percolate.

Figure 2.8. *3D simulation of a Boolean model of spheres. For a color version of this figure, see www.iste.co.uk/bernacki/digital.zip*

Integral ranges are estimated from the statistics of field averages at different scales, and are plotted in Figure 2.9 as functions of the sphere volume fraction, after normalization for a sphere with unit diameter. The integral range of the microstructure, obtained by direct calculation from the theoretical expression of the covariance of the Boolean model of spheres, is lower than the integral range of the mean stress component σ_m, except for very high pore volume fraction, or for sphere volume fraction lower than the percolation threshold in the rigid case.

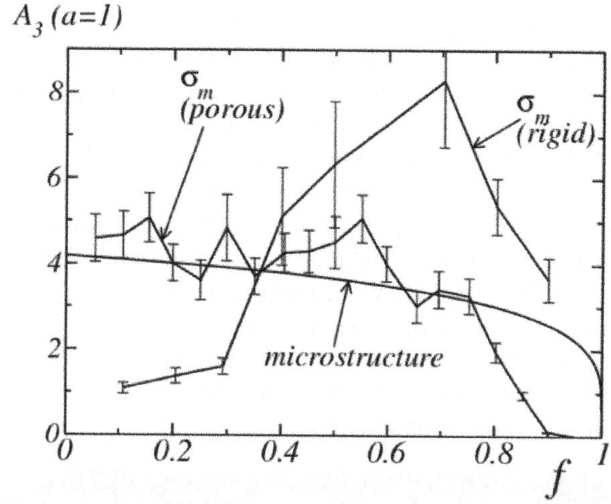

Figure 2.9. *Integral range of σ_m, for porous and rigid Boolean models of spheres as a function of sphere volume fraction f. A_3 is normalized for a sphere of diameter 1 (Willot and Jeulin 2009)*

Multiscale media by FFT

Two-scale and three scale random media generated by Cox Boolean models can show a lowering of the percolation threshold, as compared to the Boolean model (section 2.4.1). The impact of this change of morphology on the effective properties of composites is studied by means of numerical simulations and iterative computations of fields by FFT for a high contrast of properties of various Cox Boolean models (Willot and Jeulin 2011). The two-scale model is a three-parameter model, obtained by keeping Poisson germs falling in a first Boolean model of spheres (with large radius R_1), the two scales having the same volume fraction $V_V^{1/2}$. Small spheres of the second scale and final model have a radius $R_2 \ll R_1$. An example of simulation of a two-scale 2D Cox Boolean model is given in Figure 2.10. The three scale model is similar, with an additional intermediary scale and a final radius $R_3 \ll R_2 \ll R_1$, each scale having the volume fraction $V_V^{1/3}$. For each type of microstructure, four realizations are generated in domains B with 512^3 or 750^3 voxels.

The statistical relative precision of estimated apparent properties is in the range 1–5% for almost all simulations. The effective conductivity, estimated for a 10^4 contrast, is much more sensitive to percolation effects than the elastic properties.

Since the largest RVE sizes for the one-scale model correspond to rigidly reinforced media with $V_v \approx 0.7$ (see Figure 2.9), the variances $D^2_Z(V)$ and D^2_Z, with $Z = \sigma_m$ (1/3 of the trace of the stress tensor), are estimated for the two-scale Cox Boolean model when $V_{v1} = V_{v2} = 0.7$ for a contrast higher than 10,000 of the bulk and shear moduli, and a scale ratio equal to 10. It is found that the integral range of the two-scale medium is increased by a factor close to 5.8 times, whereas the point variance increases by a factor 32. A 10% precision (i.e. $\epsilon_{rela} = 0.1$) is achieved when $V \approx 550^3$, which corresponds to an increase of the RVE by a factor $5.8 \times 32 = 5.7^3$. The dramatic increase in the point variance in the case of the two-scale random medium underlies very large local stresses that could induce more damage in the matrix if it contains a defect, as compared to the one-scale model for the same volume fraction. Larger RVE, combined to the difficulty of observation of multiscale microstructures in single images, makes the experimental study of such materials and their simulation much more challenging than their theoretical approach, for which some approximations of physical properties are obtained based on the assumption of a large separation of scales.

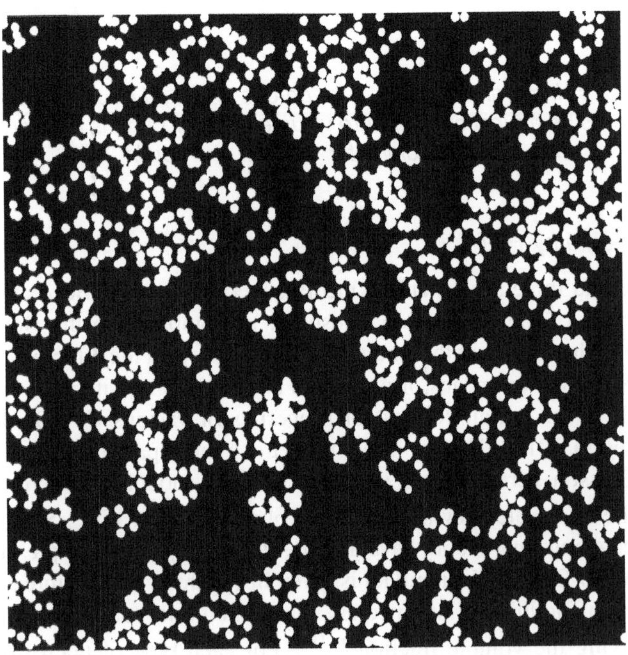

Figure 2.10. *Example of simulation* (800×800) *of a 2D Cox Boolean model of discs with radius* 5 *pixels with centers inside a Boolean model of discs with radius* 32 *pixels*

2.4.2. *Elastic and viscoelastic properties of nanocomposites*

The prediction of properties of rubber reinforced by carbon black fillers (CB) is of major importance to improve the quality of tires in industry. Elastic and viscoelastic properties of rubber reinforced by CB were estimated from simulations of the microstructure. In Jean et al. (2011a) and Chapter 6 of Jeulin (2021), a three-step Cox Boolean model is identified from TEM micrographs to locate CB nanoparticles inside spheroidal aggregates and outside of exclusion zones. In Figliuzzi et al. (2016), a variant of this model replaces the exclusion zones made up of a Boolean model of spheres by a Johnson Mehl mosaic, and carbon black aggregates are located in a second Boolean model of spheres.

Elastic properties

The prediction of elastic properties of carbon black rubber composites is made by FE and FFT in Jean et al. (2011b). For FE calculation, priority is given to generate a FE mesh that follows the microstructure interfaces, according to the conforming meshing technique. FFT calculations are made with the augmented Lagrangian algorithm (Michel et al. 2000) on 3D images with different volumes: 200^3, 350^3 and 500^3 for a voxel size of 3.2nm. It is applied to two versions of simulations, images used as input for FE, which are not periodic, and a slightly modified version to get periodic images.

The shear modulus estimated by FE from SUBC and KUBC converges for the largest volumes (from $(800\text{nm})^3$). The elastic response computed by FFT for large volumes is in good agreement with FE computations. The obtained shear modulus is $G^{app} = 2.4\text{MPa}$, close to the lower Hashin Shtrikman bound, while the measured one is $G^{ex} = 1.81\text{MPa}$. This overestimation may be due to the presence of a layer of elastomer with weaker properties around carbon black particles, or to the morphological parameters identification from TEM images.

Integral ranges of the field estimated by application of equation [2.24] are $A_3^{SUBC} = 3 \times 10^7\text{nm}^3$ and $A_3^{PBC} = 1.06 \times 10^7\text{nm}^3$, insuring a correct statistical estimation of G^{app} for the largest simulated volumes (cube with edge equal to 1500nm and volume $V = 337.5 \times 10^7\text{nm}^3$).

Prediction of viscoelastic properties by FFT

To predict the viscoelastic properties from the microstructure of carbon black reinforced rubber studied from TEM images (Figliuzzi et al. 2016), simulations of two-scale Cox Boolean models are generated by implantation of random spheres in a Johnson–Mehl mosaic. Viscoelastic moduli are estimated from FFT computations on simulated images. Two components are considered: the filler with an elastic behavior, and the matrix with a viscoelastic behavior obtained from complex elastic moduli, viscous effect being accounted for by their imaginary part. The same approach as for elastostatics is followed using now complex properties and fields. The discrete scheme of Willot and Pellegrini (2008) is used to compute the fields by iterations of FFT. The local properties are set to 78,400 and 30,000 MPa for the bulk and shear moduli of the filler. For the polymer, a viscoelastic law with Prony series given in Laiarinandrasana et al. (2012) is used for 23 frequencies. Domains with volume $(0.8\,\mu\text{m})^3$ digitized on 400^3 voxels are simulated. Hydrostatic strain loading and shear strain loading are applied.

The relative precision of the estimated moduli is calculated from the variance of the fields at different scales. At a frequency $\omega = 117$Hz, the relative precision for the real part and the imaginary part of K^{app} is equal to 0.2% and 0.3%. For the real part and the imaginary parts of G^{app}, it is equal to 17% and 3.8%, respectively.

This study was followed by intensive simulations on 4,000 realizations of the model with different input parameters on 512^3 images to cover a wide range of apparent properties (Koishi et al. 2017). The simulations made on the largest supercomputer TSUBAME in Tokyo University provided a large set of results to a data mining procedure to apply a multi-objective optimization process in order to design new filled rubber nanostructures with improved usage properties.

2.4.3. *Optical properties of nanocomposites*

Several studies are oriented toward the estimation of maps of the electrostatic fields $E(x)$ and $D(x)$ for validation of the approach, and to get indications on fields fluctuations and on the statistical RVE (Moreaud 2006; Jeulin and Moreaud 2008). In the context of a quasistatic approximation valid for light wavelengths λ larger than the scale of the nanostructure, the effective dielectric permittivity $\epsilon^*(\lambda)$ is then used to predict the change of optical properties of a random medium with the wavelength λ. This is illustrated for two examples of applications to paints.

2D Dielectric random media

Simulations of 2D models of autodual random sets (where the morphological properties of the two components are identical, and therefore with surface fraction 0.5) provide interesting examples of digital materials for testing numerical homogenization tools. Indeed, the exact effective permittivity ϵ^* of 2D autodual random sets and the fields variance are known, as recalled in Jeulin (2021, Chapter 18). For an autodual random set in $\mathbb{R}^2$, the effective permittivity is given by the geometric average (Matheron 1967):

$$\epsilon^* = \sqrt{\epsilon_1 \epsilon_2} \qquad [2.33]$$

The theoretical point variance of the i component of the dielectric displacement $D(x)$, obtained from an applied average electric field E, is given by Jeulin (2021, p. 688):

$$E\left\{(D_i(x) - \epsilon^* E_i)^2\right\} = \frac{1}{2} E_i^2 \sqrt{\epsilon_1 \epsilon_2} \left(\sqrt{\epsilon_1} - \sqrt{\epsilon_2}\right)^2 \qquad [2.34]$$

For comparison, three models of autodual random sets (RS) are simulated, each realization being periodic. The integral range A_2 of their set covariance is estimated as usually, from the scaling law given by equation [2.24] applied to the variance of local area fractions. The simulated models are:

– a symmetric colour Dead Leaves of discs (see Jeulin 2021, Chapter 11) (Figure 2.11) with a 10 pixels diameter (theoretical $A_2 = 54.234$ pixels^2, experimental $A_2 = 54.239$ pixels^2);

– two excursion sets of Gaussian RF: one obtained asymptotically from dilution RF (Jeulin 2021, Chapter 15) with discs with a 10 pixels diameter (Dil) ($A_2 = 4.64$ pixels^2), and one with a Gaussian covariance (Jeulin 2021, Chapter 4) ($A_2 = 5.23$ pixels^2).

These models show very different morphologies, but own the same effective properties given by equation [2.33]. The dielectric displacement field $D(x)$ is computed from simulations by FFT, applying an average electric field E_0 in the horizontal direction Ox and using the Eyre–Milton algorithm (Eyre and Milton 1999). Sizes of images are in the range 32^2–$2,048^2$. For the white phase, $\epsilon_1 = 1$. The dielectric contrast c is monitored through $\epsilon_2 = 2$, 10, 100, 1,000, 10,000.

The values of the apparent dielectric permittivity ϵ^* and the point variance of D_x, for $E_x = 1$, are in excellent agreement with the theoretical values given by equation [2.33] ($\sqrt{\epsilon_1 \epsilon_2}$) and by equation [2.34]. It is important to notice that the variance of $\epsilon(x)$, obtained from equation [2.27] ($0.25(\epsilon_2 - \epsilon_1)^2$), is much larger than the point variance of D_x. Its use to estimate the integral range would underestimate it by a factor equal to 2.74 for $c = 10$ and to 509 for $c = 10^4$! It is therefore wise to use the correct point variance to get a correct estimation of the scale carried by a physical field. Otherwise, some erroneous conclusions can be drawn, like an apparent decrease in the integral range of the field with c (Jeulin 2005). However, sampling calculations to evaluate the RVE size, as shown in previous studies, are correct when using as input the experimental variance $D_Z^2(V)$.

The integral range of the component D_x is always larger than the corresponding value A_2 for the area fraction and increases with the contrast c. The 5% relative precision RVE for a single realization depends on the random model: it is in the range 64^2–256^2 when $c = 10$ and from 256^2 to $1,024^2$ when $c = 1,000$.

Figure 2.11. *Simulation of an autodual RS: Symmetric dead leaves of discs (non-periodic* 512×512 *image)*

3D Boolean model of spheres

Boolean models of spheres (with diameter 10 voxels) are simulated for three nominal volume fractions p (0.2, 0.3 and 0.5) (Moreaud 2006; Jeulin and Moreaud 2008). The matrix has a permittivity $\epsilon_1 = 1$ and spheres a larger permittivity with increasing contrast $\epsilon_2 = 10, 100, 1{,}000$. Fields $D(x)$ are estimated by iterations of FFT with application of an horizontal field E_x, with $E_y = E_z = 0$. The effective permittivity, obtained from 10 realizations on 256^3 periodic images, is obtained as a function of p and of the contrast c. Results are close to the lower third-order Beran bound (Beran 1965) for $c < 100$ and $p < 0.3$, and to the upper bound for $p = 0.5$. Overall, bounds and the self-consistent model provide a poor prediction for this model.

The RVE V_{RVE} to estimate ϵ^* with a relative precision better than 5% is given by $V_{RVE} = 32^3, 64^3, 128^3$ for a contrast $c = 10, 100, 1{,}000$ and $p = 0.3$. Point variance and integral ranges are estimated from 256^3 simulations. The microstructure integral range A_3, theoretically calculated from the covariance, is underestimated by $10\%–15\%$ in simulations. Except for a low contrast and a low volume fraction, the integral range of D_x is larger than A_3 and strongly increases with contrast, like the point variance.

Optical properties of paints

TiO_2 paint

In Azzimonti et al. (2013), the prediction of optical properties of simulated deposit models for paints is investigated. The coating is a mixture of various components with different permittivity. The optical properties (and their change with the light wavelength λ) of the layer are governed by the effective permittivity of the mixture, since the optical index is given by $n = (\epsilon^*)^{0.5}$. Paint coatings are simulated by a random deposition process of nanoparticles with different shapes: spheres or rhombus-shaped inclusions. The typical size of the nanoparticles being much smaller than the light wavelengths, electrostatics can be used to estimate ϵ^*. This is made from electrostatic fields computed by iterations of FFT on simulated paint layers (in $1{,}200^3$ voxels volumes) using the complex dielectric function $\epsilon(\lambda)$ of the TiO^2 pigment and of the matrix glass. Since the deposit shows transverse isotropy, this property appears in the permittivity tensor with the following diagonal components: $\epsilon^*_{xx} = \epsilon^*_{yy}$ and ϵ^*_{zz}. Hot spots (similar to the thermal case in section 2.4.1) are generated on the field maps, around corners of very close rhombus nanoparticles, or between two nearly touching spherical inclusions, when the zone of contact is orthogonal to the applied macroscopic field E. For these simulations, this effect is stronger for the imaginary part of the displacement field, which generates color on the macroscopic scale.

The size of the RVE of $D(x)$ is much larger than the pigment size, in particular for its imaginary part as a result of hot spots. The relative precision of the estimation of the average real part of D ($\mathrm{Re}(D)$) is 0.82%. It is not so good, but acceptable for the average of its imaginary part $\mathrm{Im}(D)$ (8.84%). A 1% relative precision would require the simulation of a volume made of 204^3 voxels for the volume fraction V_V, $1{,}050^3$ for $\mathrm{Re}(D_x)$ and $5{,}130^3$ for $\mathrm{Im}(D_x)$.

Hematite coating

In a second study, a calibrated paint with an epoxy resin containing a dispersion of hematite nanocubes with 0.1 volume fraction is modeled by a two-scale random set after identification of its parameters from quantitative SEM image analysis on $30\,\mu\mathrm{m} \times 30\,\mu\mathrm{m}$ images (Couka et al. 2015b). On the scale of a few micrometers, aggregates of hematite nano-particles (similar to dice with round edges and corners, and with diagonal close to 300nm) are organized along thin curved channels. A two-scale random model was developed to simulate the dispersion of nanoparticles: cubic nanoparticles with uniform orientations are located according to a random sequential addition RSA hard core model of spheres outside of exclusion zones, almost free from hematite particles. Nanoparticles are packed by translation until intersection. For real SEM images and for simulations, the 2D integral range of the area fraction obtained on sections is $A_2 = 3.52 \times 10^{-2} \mu m^2$, so that the area fraction of a single 2D image is obtained with a 11.7% relative precision.

Ellipsometry measurements of the complex refractive index as a function of the light wavelength (in the range 381–881 nm) are used to estimate the anisotropic permittivity of hematite particles, resin and coating (Couka et al. 2015a). The apparent permittivity is estimated from FFT full-field computations of the complex dielectric displacement D on a realization of the 3D random model containing 99,886 nanoparticles, digitized on a grid of 500^3 voxels with a resolution of $31\text{nm}/\text{voxel}$, so that each nanoparticle is digitized as a cube of about 5^3 voxels. The visible spectrum is digitized in 24 equispaced values for $\lambda = 381$–781 nm.

For illustration, the relative precision obtained from statistical RVE computation by means of the integral range derived from equation [2.24] fitted from subvolumes of a large simulation is given for three wavelengths and a single 3D image in the below table. When $\lambda = 441\text{nm}$, a volume $V = 4{,}285^3$ voxels corresponding to $132.8\mu\text{m}^3$ is required to reach a precision better than 0.1% for $\text{Im}(\varepsilon^*)$. The largest RVE corresponds to the lowest wavelength ($\lambda = 441\text{nm}$).

$\lambda(\text{nm})$	441	621	800
Relative precision $\text{Re}(\varepsilon^*)$	1.29%	1.27%	0.92%
Relative precision $\text{Im}(\varepsilon^*)$	2.51%	0.21%	0.36%

The predicted apparent permittivity from FFT computations on simulated microstructures is very close to the measurements on a sample of hematite coating, except for the imaginary part on shortest wavelengths, for which the quasi-static approximation reaches its limits.

2.4.4. *Fluid flow and permeability*

The emergence of Darcy's law on a macroscopic level is a direct outcome of the linear character of the Stokes equation [2.3] obtained from Navier–Stokes equations by linearization while neglecting inertial effects for a fluid with viscosity μ (Matheron 1967), combined to the continuity equation for incompressible fluids:

$$\partial_i u_i = 0 \qquad [2.35]$$

and with the following BC (no slip) applied on pore boundary:

$$\partial A : u_i(x) = 0 \text{ for } x \in \partial A$$

On a macroscopic scale, the fluid flow satisfies Darcy's law relating the fluid flux Q^i to the pressure gradient $\partial_i P$:

$$\langle u_i \rangle = Q^i = -\frac{1}{\mu} K^{ij} \partial_j P$$

The permeability K has the dimension of a surface (length^2) and is usually expressed in cm^2. The BC acting on the pore boundary with a complex geometry differ from the conditions operating in electrostatics, so that generally the permeability K cannot be deduced in a simple way from the permittivity or the resistivity of the porous medium (Matheron 1967).

The Darcy permeability of random porous media with a full range of pore volume fractions is studied by solving the Stokes equations by FFT on periodic simulations of Boolean models of spheres (Abdallah et al. 2015) and cylinders (Willot et al. 2016). A numerical method of resolution was developed by Wiegmann (2007). It relies on the resolution of four Poisson equations (one for pressure p and one for each component of the velocity u_i), the inversion of the Laplacian being obtained by FFT. The no-slip condition is constrained by forces introduced along ∂A. In Abdallah et al. (2015), FFT solutions are obtained for a Boolean model of spheres with diameter $d = 15$ voxels generating two porous media: solid phase is the union of spheres (case A, with permeability k_A), or alternatively pores are the union of spheres (case B, with permeability k_B). Simulations in cubic volumes 100^3, 256^3 and 512^3 are generated to cover the full range of pore volume fractions.

For a large range of pore volume fractions of the Boolean model of spheres and the two cases A and B, the size of the statistical RVE, calculated from the integral range and the point variance of the velocity field is much larger than the microstructural RVE or the RVE for elastic moduli or thermal conductivity. It strongly increases when reaching the percolation threshold. In Bignonnet (2020), a different scheme is introduced to compute velocity fields on images by FFT. It makes use of a variational principle and trial force fields to produce an upper bound of the permeability. Therefore, results obtained by this method should be compared to other calculations. It is used to estimate the permeability of the Voronoi mosaic for the full range of pore volume fractions on simulated images with size 512^3. The integral range of the velocity is much larger than the integral range of volume fraction. Again, the RVE size shows similar variations with p as obtained for the Boolean model of spheres in Abdallah et al. (2015), and is the largest close to the percolation threshold (0.1453 in the present case for site percolation).

It would be interesting to estimate the Darcy permeability of multi-scale models such as Cox Boolean models. However, this task may be insurmountable, since gigantic RVE are expected (as for Poisson fibers in section 2.4.1).

2.4.5. *Acoustic properties and wave propagation in fibrous media*

For engineering purposes, predicting acoustic properties of random media from their microstructure is a first step to designing new materials with improved properties, like a better sound attenuation. One of the challenging problems to solve concerns the fluctuations of the microstructure and the local physical fields involved in the homogenization of the acoustic behavior (Peyrega and Jeulin 2013). The results outlined in this subsection concern the estimation of acoustic properties of random fibers by numerical homogenization, and the study of corresponding RVE.

On the microscopic scale, viscous and thermal dissipations of energy occur in the air saturating voids of a porous medium. To predict the acoustic behavior of random porous media, use is made of a thermoacoustic formalism on realizations. Properties like harmonic acoustic velocity and temperature are homogenized by averaging after a numerical computation of the harmonic fields (flow velocity, fluid temperature and pressure) by FEs, using the FE software Comsol Multiphysics, for frequencies in the range 0 – 4,032 Hz. More details are given in Peyrega and Jeulin (2013) and Jeulin (2021, Chapter 19). Differently from other examples presented in this chapter, here we have to deal with the solutions of a dynamic problem of wave propagation.

Twenty realizations are generated for a model of random fibrous media of infinite parallel fibers, with cross-sections made of three Boolean models of discs with radius $R_F = 42\,\mu m$, $100\,\mu m$ and $300\,\mu m$, and with average pore volume fraction 0.64 corresponding to real materials with wooden fibers used for acoustic attenuation. Square fields have the following edge lengths: 2.52, 6 and 18mm. Acoustic fields are not homogeneous and local fluctuations appear, resulting in variations of the acoustic properties between realizations. For each radius, the average absorption coefficient is estimated from permeabilities averaged over the $n = 20$ independent periodic microstructures. Concerning the radius $42\,\mu m$, the prediction of the coefficient of absorption $\alpha(\omega)$ fits the experimental results, as for a simplified cell made of a single fiber building a regular medium. For larger fibers ($R_F = 100\ \mu m$ and $300\ \mu m$), the frequency evolutions are globally similar with almost identical amplitudes, but a frequency shift is observed between regular and random microstructures.

The size of the statistical RVE is estimated for properties from results on the dispersion. Concerning the area fraction, the integral range of sections is directly calculated from the theoretical expression of the covariance of the 2D Boolean model of discs (0.0049, 0.0277 and 0.2495mm^2) in agreement with estimations from simulations. The relative precision for the pore volume fraction on a single realization is $\epsilon_{rela} = 3.7\%$.

Harmonic acoustic velocity and temperature being complex variables, the estimation of the RVE operates on their real and imaginary parts. Detailed results are available in Peyrega and Jeulin (2013). Integral ranges A_2, ϵ_{rela} and the RVE size of the four average scalar fields decrease with the frequency for $R_F = 42\ \mu m$ and $R_F = 100\ \mu m$. This can be explained by the frequency dependence of the thickness of the boundary layer of the interface pore-solid, involving a decrease in correlation lengths at high frequency. The relative precision of the four scalar fields is much better than 10%. The RVE size does not change with frequency for $R_F = 300\ \mu m$.

Integral ranges of fields are much larger than the morphological integral ranges. Concerning the acoustic absorption $\alpha(\omega)$, its relative precision for 20 realizations is better than 5% for $R_F = 42\ \mu m$ and $R_F = 100\ \mu m$.

Finally, in the present case, thermal properties fluctuate more and develop longer correlation lengths than velocity fields. The thermal properties are therefore the limiting factor to define a RVE in this application.

2.4.6. *Nonlinear properties*

Viscoplastic composite

The same approach was applied to a nonlinear behavior, namely, for 2D/3D viscoplastic composite materials (Madi et al. 2006), the creep following a simple power law with exponents 3.5 in the hard phase and 4 in the soft phase. Shear tests are simulated by FEs under a macroscopic shear stress, and the statistical analysis is performed on the shear strain rate field. In that case, it is found that for the same statistical precision, the linear elastic RVE is larger than the nonlinear RVE. The degree of generality of this result is not known. It is expected that in the case of a strong localization with a nonlinear behavior, long range correlations of the fields might appear and scaling laws similar to the dilated Poisson hyperplanes (with a scaling exponent close to $\frac{1}{3}$) might be recovered, so that a slow convergence toward the effective properties might be observed on numerical simulations; however, this should be confirmed by further simulations.

Elastoplasticity of porous media

The yield surface for random porous elastoplastic materials was determined for the first time by Fritzen et al. (2012) following a statistical approach. Previous studies performed since the early 1980s to analyze void growth and ductile fracture mechanisms were limited to periodic unit cells containing a single void. The use of a random model is closer to the physical reality of engineering materials and allows for more general interactions between voids. The proposed hard core model of non-intersecting spherical voids of single size is illustrated in Figure 2.12. The elastoplastic matrix is assumed to obey to J_2-plasticity in the absence of hardening. The yield stress is taken as $\sigma_0 = 100$ MPa. Periodic BCs are applied controlling the applied axial deformation E_{11}, E_{22}, E_{33} of the cubic volume:

$$\begin{aligned}
&E_{11}(t) = t\dot{\varepsilon}_0(\alpha + \beta),\\
&E_{22}(t) = t\dot{\varepsilon}_0(-\alpha + \beta),\\
&E_{33}(t) = t\dot{\varepsilon}_0\beta,\\
&\Sigma_{12}(t) = \Sigma_{13}(t) = \Sigma_{23}(t) = 0
\end{aligned} \qquad [2.36]$$

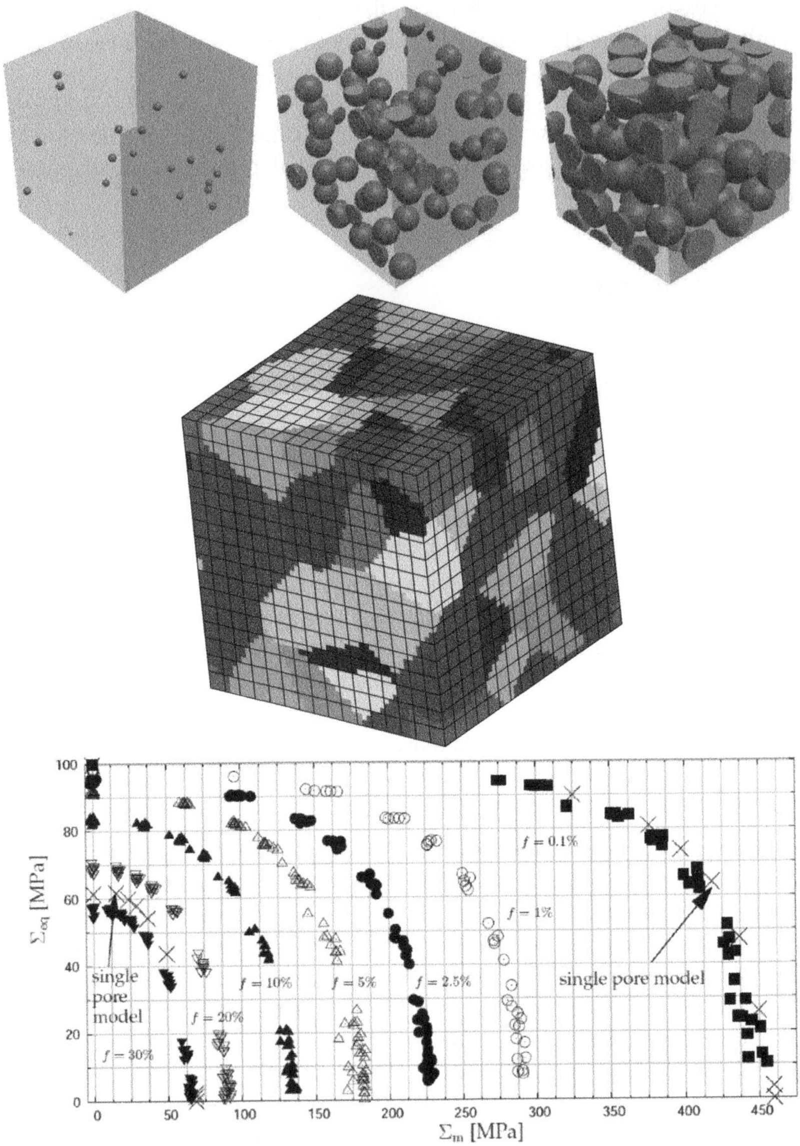

Figure 2.12. *Hard core model of non-intersecting spherical holes in a matrix for three volume fractions ($f = 0.001, 0.1, 0.3$ from the left to the right) and associated finite element meshes ($f = 0.001, 0.01, 0.1, 0.3$). The bottom plot shows the determined limit analysis surface for various void volume fractions (after Fritzen et al. (2012)). For a color version of this figure, see www.iste.co.uk/bernacki/digital.zip*

Time t is here a loading parameter, since rate-independent plasticity is considered. It is increased until an overall limit stress state is reached for which the von Mises equivalent stress Σ_{eq} and the mean stress Σ_m (opposite to the hydrostratic pressure) are recorded for each value of the loading parameters α and β. The latter parameters explore the deviatoric and hydrostatic parts of the strain space. The values $\alpha = 0; 1/2; 1$ and $0 \leq \beta \leq 1$ are considered. The results are shown in Figure 2.12 (bottom) and they show the yield surfaces for various values of the porosity f. They are compared to computations based on periodic unit cells (single pore model). The unit cell provides an excellent estimate of the yield locus for $f = 0.001$ but a poor prediction for high porosity $f = 0.3$. Figure 2.13 shows the results of the statistical approach in the form of the dependence of the asymptotic von Mises equivalent and hydrostatic stress responses as functions of the number of voids in the volume element. In the case $N = 50$, 10 realizations were considered to evaluate the average which is compared well with estimated for larger N values, at least for periodic BCs. In contrast, uniform Dirichlet conditions only provide upper bounds.

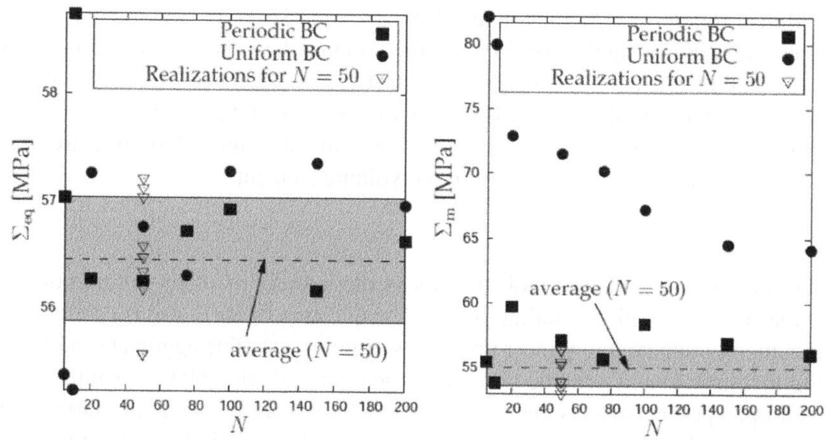

Figure 2.13. *Comparison of the limit asymptotic von Mises equivalent stress (left) and hydrostatic stress (right) for various numbers of pores and types of boundary conditions (interval of confidence in gray), $f = 0.2, \alpha = 1, \beta = 0.25$ (after Fritzen et al. (2012))*

The obtained yield functions are used by Fritzen et al. (2012) to calibrate an effective Gurson–Tvergaard–Needleman potential with parameters taken as explicit functions of porosity f :

$$\phi = \frac{\Sigma_{eq}^2}{\sigma_F^2} + 2q_1 f \cosh\left(q_2 \frac{3\Sigma_m}{2\sigma_F}\right) - 1 - q_3 f^2 \qquad [2.37]$$

$$\text{with} \quad q_1(f) = \theta_0 - \theta_1 f, \quad q_2(f) = \theta_2, \quad q_3(f) = (q_1(f))^2 \qquad [2.38]$$

The involved coefficients are identified as: $\theta_0 = 0.169, \quad \theta_1 = 1, \quad \theta_2 = 0.92$. This meta-model can be used for many applications within the considered porosity range.

This statistical approach was extended by Fritzen et al. (2013) to double porosity media with separation of scales, meaning that holes are embedded in an homogenized elastoplastic matrix. The latter is endowed with a compressible plasticity model involving an elliptic yield function with respect to von Mises equivalent stress and hydrostatic stress. Beyond limit analysis, the statistical method can be applied to the occurrence of localization phenomena and failure analysis for a Boolean model of holes, as proposed by Cadet et al. (2021). This requires large deformation elastoplasticity formulation and detection of void coalescence phenomena. This can be performed by means of Rice's strain localization criterion applied to the effective incremental tangent behavior of the unit cell, as done by Cadet et al. (2022). Surface responses for the critical strain at overall localization can be predicted in that way considering various realizations of the random microstructure. In the presence of instabilities, the anisotropy bias induced by the cube shape of the volume element becomes particularly significant. In simplified approaches considering a single pore in a cube volume element, this bias is very strong. It is alleviated by consideration of a random model but still exists. That is why rotation of the loading was applied systematically by Cadet et al. (2022) to determine the interval of fracture strain induced by spurious anisotropy effect of the volume element.

Elastoviscoplastic polycrystals

The plasticity of metallic polycrystals is the subject of many contributions in micromechanics of bulk crystalline solids. The question of RVE size is central in the computation of apparent nonlinear properties of polycrystalline aggregates addressed for the first time in a systematic way by Barbe et al. (2001b, 2003). A simple model for isotropic polycrystals is the Voronoi tessellation as depicted in Barbe et al. (2001a). It can be extended to account for more realistic and elaborate grain shape distributions, as done in Quey and Renversade (2018). The Voronoi model is sufficient to highlight the tremendous heterogeneity of plastic deformation in polycrystals when comparing the overall stress–strain curve to mean response of individual grains or individual material points. The description of intragranular heterogeneities goes beyond estimations by micromechanical models and is one of the greatest advances of full field polycrystal plasticity simulations. The latter are based on the appropriate choice of continuum crystal plasticity constitutive models that were developed since the early 1970s. The number of grains required in the polycrystalline aggregates to estimate with a given precision the overall response of the polycrystal is still an open question. Heuristic results are presented in the cited references but they are limited to particular loading conditions like tensile curves, cyclic loading or rolling conditions in metal forming. A systematic procedure cannot easily be defined due to the path-dependent character of elastoplasticity laws. Strong non-linearities arise from the selection of activated slip systems and interaction between slip systems in the hardening functions. Large deformations and crystal lattice rotation are other sources of nonlinearity of the problem.

An illustrative example is given here for polycrystalline copper films as explored by Šiška et al. (2007a, 2007b). Columnar grains are defined by appropriate two-dimensional Voronoi tessellation. The average in-plane grain size d is taken equal to the film thickness h. The heterogeneity of polycrystals stems from the crystal lattice orientation distribution among the grains. Thin metallic films usually possess preferential grain orientations associated with a non uniform crystallographic texture. Copper thin films discussed in Šiška et al. (2007a, 2007b) are characterized by the pole figure shown in Figure 2.15 (top left), with 90% of $\{111\}$ grains, 6% of $\{001\}$ grains and 4% random grains. The $\{111\}$ grains, for instance, have one $\{111\}$ plane parallel to the film but the orthogonal in-plane orientation is random. In-plane periodicity is enforced and the remaining lateral surfaces are free of forces in free-standing films or one face remains flat due to the presence of a rigid substrate. Results are provided here in the latter case. The film is subjected to cyclic uniaxial in-plane tension-compression with prescribed overall strain $E_{22} = \pm 0.5\%$. The corresponding overall stress-strain response is shown in Figure 2.14 in the case of a 250-grain film. This estimation of the effective tension-compression behavior of the film is compared to the ensemble average of 10 realizations of 50-grains aggregates. The cyclic loops exhibit some scatter represented by error-bars in the figure. The difference between both curves remains small, indicating that the 250-grain film can be regarded as a RVE for the considered loading conditions.

Cyclic loading leads to an evolution of the stress–strain loops when increasing the number of applied cycles. Convergence of the loop is found after 100 cycles according to Šiška et al. (2007b). The first and the stabilized loops are shown in Figure 2.15 for the various components of the crystallographic textures. The results are given for five realizations of 50-grains films. Almost no scatter is observed for the dominant $\{111\}$ texture component due to the sufficiently large number of grains sharing this orientation. In contrast, significant scatter is observed for the $\{100\}$ and random crystallographic texture components. It means that the obtained results may not be representative for this population of grains and number of grains larger than 50 should be considered.

The cyclic elastoplasticity behavior of bulk FCC polycrystals was studied by Farooq et al. (2019). The authors provide a thorough statistical analysis of the strain distribution inside the polycrystalline aggregates for complex non-symmetric stress-based or strain-based cyclic loading conditions. The results show the ability of full-field crystal plasticity simulations to account for complex ratcheting and cyclic stress relaxation behavior, which is classically observed experimentally in engineering alloys like nickel-based superalloys for aeronautic applications.

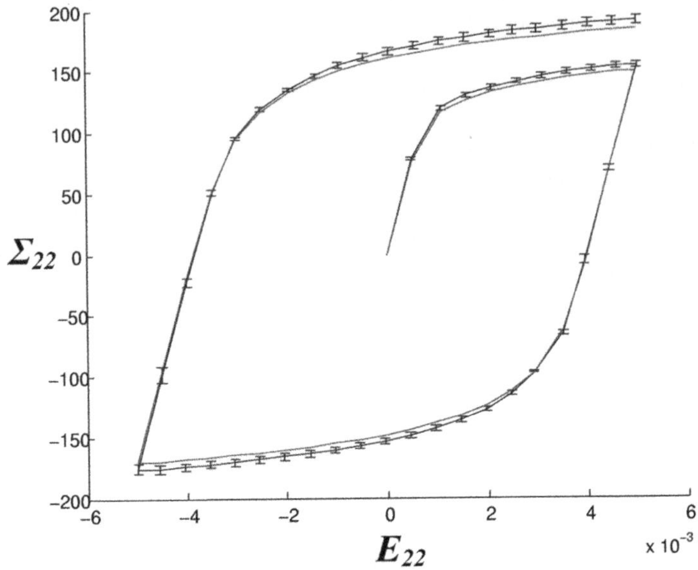

Figure 2.14. *First cyclic overall stress–strain loop for textured polycrystalline copper thin films made of 225 grains (red curve) and ensemble average of 10 simulations with 50 grains films (blue curve) with corresponding standard deviation indicated by error bars (after Šiška et al. (2007b)). For a color version of this figure, see www.iste.co.uk/bernacki/digital.zip*

The free surface behavior of polycrystals is important not only for films but also for comparison with experimental results showing strain and lattice rotation fields at surfaces by means of standard optical or electronic microscopy methods. The major problem lies in the fact that, generally, the 3D grain morphology of underlying grain layers are not accessible to these measurements; also, they contribute to the surface response. Ensemble averaging methods were employed to analyze the surface behavior of polycrystalline copper polycrystals in Zeghadi et al. (2007a, 2007b) For that purpose, random grain microstructures were simulated with the constraint of prescribed morphology of the intersection of the grains with the free surface and prescribed crystal orientation of surface grains. The results highlight the strong impact of 3D grain morphology on the strain and stress response at the free surface. In particular, it is demonstrated for the considered FCC material with isotropic orientation distributions that two to three grain layers are significantly affecting the surface response. This was confirmed in the case of cyclic behavior of tantalum polycrystals by Colas et al. (2019, 2021).

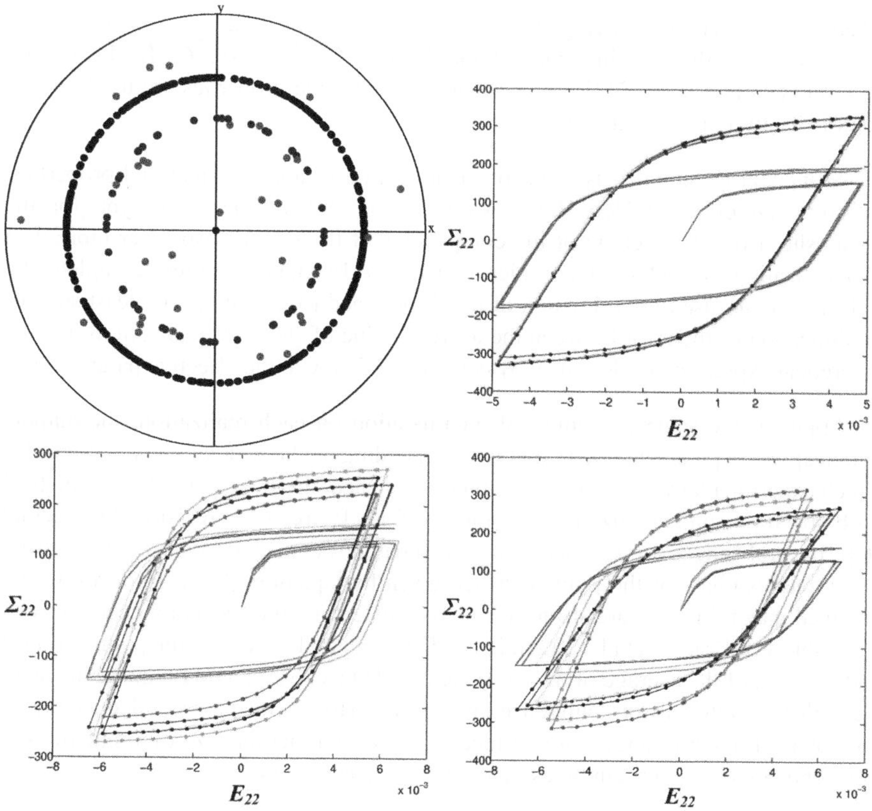

Figure 2.15. *Crystallographic texture of the copper thin films (top left pole figure); first and stabilized cyclic loops for the average stress–strain in {111} grains (top right), in {001} grains (bottom left) and random grains (bottom right) (after Šiška et al. (2007b)). For a color version of this figure, see www.iste.co.uk/bernacki/digital.zip*

2D hyperelastic Boolean random media

In Brändel et al. (2022), the macroscopic hyperelastic parameters of 2D Boolean models with circular and rectangular grains stiffer than the matrix are estimated by homogenization, the two components following a Neo–Hooke law so that they are both hyperelastic with different coefficients. In this detailed and complete work (which is very briefly summarized here) are a full range of area fractions $A_A = 0.1$–0.9, finite strains in finite elasticity are applied on realizations of microstructures (512^2 images) by FEs calculations with periodic BC. The Neo–Hooke strain energy Ψ is expressed as a function of the Lamé constants λ and μ by:

$$\Psi = \frac{\mu}{2}(I_1 - 3) + \frac{\lambda}{2}\ln^2 J - \mu \ln J \qquad [2.39]$$

where $J = \det(F)$, F being the deformation gradient, and $I_1 = trace(C) = trace(F^T F)$ is the first invariant of the Cauchy–Green tensor C. In the matrix, $\lambda = 100\text{GPa}$ and $\mu = 75\text{GPa}$. For the grains, the same properties are used up to a contrast factor c in the range of 2–100.

A uniaxial tension test is used for identification of apparent material properties, assuming an effective Neo–Hooke strain energy Ψ^* of the same type for the composite, from the average of the components of the stress tensor over more than 200 realizations of each model of microstructure. A biaxial tension test, completed by a shear test, are used for validation of the fit obtained for λ^* and μ^* by least squares minimization of the error between the average value of $\langle\Psi\rangle$ over realizations and the theoretical expression of Ψ^* given by equation [2.39] with the effective parameters.

Applying the same procedure of identification on each realization, the authors obtain apparent parameters λ^{app} and μ^{app} depending on the scale of computations, for which experimental variances are computed. For each 2D Boolean model, a statistical RVE size for a single realization is computed for a 1% relative precision. In a second step, integral ranges are estimated for the apparent parameters, unfortunately using the wrong estimate of the point variance given by equation [2.27]. As a result, the obtained integral ranges are strongly underestimated and unexpectedly decrease with c in Table 4 of Brändel et al. (2022). Instead, use should be made of the point variance and variance of the average of the stress field to deduce the integral range of this field. Nevertheless, intervals of confidence given in the paper, which are not deduced from the obtained integral ranges, are correct. It is checked that the size of simulations is large enough to get an accurate estimation of the parameters.

In the case of Boolean models of discs with a fixed radius, the size of images to reach the asymptotic estimations and the RVE size are the largest for $A_A \simeq 0.6$. We have to notice that it corresponds to the percolation threshold of discs in 2D. When increasing the anisotropy of the primary grain (from discs to more or less elongated rectangles), the convergence of apparent properties with the size of simulations is accelerated, as a result of a better percolation of the grains. More elongated grains provide stiffer media, as already shown in elasticity, but percolation effects are amplified in the 2D case. Finally, populations of discs with a lognormal distribution of their radii are simulated. To maintain a good statistical precision and a low systematic error, larger RVE and larger number of realizations have to be used when increasing the coefficient of variation c_v of the distribution. With the given parameters, a relative precision of the order of 3% is reached when $c_v = 1$, for which the 2D medium is 20% softer than in the case of discs with a single radius.

2.5. Conclusion

The determination by a statistical approach of the size of the RVE to estimate effective properties by numerical simulations of digital materials with a given precision is now well established, after nearly two decades of development and applications. From available data on simulations, the estimation of effective properties (linear and nonlinear elastic moduli, linear and nonlinear conductivity, flow permeability, etc.) of the common Boolean model of spheres, its multiscale versions and other models are now well documented. Intervals of confidence of the obtained apparent properties are given as a function of the scale of computations for a correct use of predictions from images obtained on real or on simulated microstructures. This is mandatory in order to master small-scale fluctuations. The best strategy to minimize size and edge effects in numerical simulations is to operate with periodic BCs, and when possible on periodic generated realizations of microstructures from some probabilistic model.

For a nonlinear constitutive behavior, in micromechanics, it is wise to estimate the macroscopic stress–strain relationships with their intervals of confidence and their own statistical RVE sizes, before estimating the parameters of the underlying constitutive equations.

Further use of this statistical approach is expected in various areas: numerical homogenization of nonlinear physical properties, damage, wave propagation and fracture toughness as suggested in preliminary studies (see Jeulin (2021, Chapters 20 and 21)).

2.6. References

Abdallah, B., Willot, F., Jeulin, D. (2015). Stokes flow through a Boolean model of spheres: Representative volume element. *Transport in Porous Media*, 109, 711–726.

Altendorf, H. and Jeulin, D. (2011). Random-walk-based stochastic modeling of three-dimensional fiber systems. *Physical Review E*, 83(4), 041804.

Altendorf, H., Jeulin, D., Willot, F. (2014). Influence of the fiber geometry on the macroscopic elastic and thermal properties. *International Journal of Solids and Structures*, 51(23–24), 3807–3822.

Armstrong, S., Kuusi, T., Mourrat, J.-C. (2019). *Quantitative Stochastic Homogenization and Large-scale Regularity*. Springer, Berlin.

Azzimonti, D., Willot, F., Jeulin, D. (2013). Optical properties of deposit models for paints: Full-fields FFT computations and representative volume element. *Journal of Modern Optics*, 60(7), 519–528.

Barbe, F., Decker, L., Jeulin, D., Cailletaud, G. (2001a). Intergranular and intragranular behavior of polycrystalline aggregates. Part 1: Fe model. *International Journal of Plasticity*, 17(4), 513–536.

Barbe, F., Forest, S., Cailletaud, G. (2001b). Intergranular and intragranular behavior of polycrystalline aggregates. Part 2: Results. *International Journal of Plasticity*, 17(4), 537–563.

Barbe, F., Forest, S., Quilici, S., Cailletaud, G. (2003). Numerical study of crystalline plasticity: Measurements of the heterogeneities due to grain boundaries under small strains. *La Revue de Métallurgie*, 101, 815–823.

Beran, M. (1965). Use of the variational approach to determine bounds for the effective permittivity in random media. *Il Nuovo Cimento (1955-1965)*, 38(2), 771–782.

Beran, M. and McCoy, J. (1970). Mean field variations in a statistical sample of heterogeneous linearly elastic solids. *International Journal of Solids and Structures*, 6(8), 1035–1054.

Bignonnet, F. (2020). Efficient FFT-based upscaling of the permeability of porous media discretized on uniform grids with estimation of RVE size. *Computer Methods in Applied Mechanics and Engineering*, 369, 113237.

Bobeth, M. and Diener, G. (1986). Field fluctuations in multicomponent mixtures. *Journal of the Mechanics and Physics of Solids*, 34(1), 1–17.

Bornert, M. and Suquet, P. (2001). Propriétés non linéaires des composites : approches par les potentiels. In *Homogénéisation en mécanique des matériaux*, volume 2, Bornert, M., Bretheau, T., Gilormini, P. (eds). Hermès, Paris.

Brändel, M., Brands, D., Maike, S., Rheinbach, O., Schröder, J., Schwarz, A., Stoyan, D. (2022). Effective hyperelastic material parameters from microstructures constructed using the planar Boolean model. *Computational Mechanics*, 69(6), 1295–1321.

Buryachenko, V.A. (2007). *Micromechanics of Heterogeneous Materials*. Springer, Berlin.

Buryachenko, V.A. and Kreher, W.S. (1995). Internal residual stresses in heterogeneous solids–A statistical theory for particulate composites. *Journal of the Mechanics and Physics of Solids*, 43(7), 1105–1125.

Cadet, C., Besson, J., Flouriot, S., Forest, S., Kerfriden, P., Rancourt, V.D. (2021). Ductile fracture of materials with randomly distributed defects. *International Journal of Fracture*, 230, 193–223.

Cadet, C., Besson, J., Flouriot, S., Forest, S., Kerfriden, P., Lacourt, L., de Rancourt, V. (2022). Strain localization analysis in materials containing randomly distributed voids: Competition between extension and shear failure modes. *Journal of the Mechanics and Physics of Solids*, 166, 104933.

Cailletaud, G., Jeulin, D., Rolland, P. (1994). Size effect on elastic properties of random composites. *Engineering Computations*, 11, 99–110.

Colas, D., Finot, E., Flouriot, S., Forest, S., Mazière, M., Paris, T. (2019). Local ratcheting phenomena in the cyclic behavior of polycrystalline tantalum. *JOM Journal of the Minerals, Metals & Materials Society*, 71, 2586–2599.

Colas, D., Finot, E., Flouriot, S., Forest, S., Mazière, M., Paris, T. (2021). Experimental and computational approach to fatigue behavior of polycrystalline tantalum. *Metals*, 11, 416.

Couka, E., Willot, F., Callet, P., Jeulin, D. (2015a). Optical response of a hematite coating: Ellipsometry data versus Fourier-based computations. *Advanced Science, Engineering and Medicine*, 7, 925–931.

Couka, E., Willot, F., Jeulin, D., Ben Achour, M., Chesnaud, A., Thorel, A. (2015b). Modeling of the multiscale dispersion of nanoparticles in a hematite coating. *Journal of Nanoscience and Nanotechnology*, 15, 3515–3521.

Decker, L. and Jeulin, D. (2000). Simulation 3D de matériaux aléatoires polycristallins. *Revue de Métallurgie*, 97(2), 271–175.

Delisée, C., Jeulin, D., Michaud, F. (2001). Caractérisation morphologique et porosité en 3d de matériaux fibreux cellulosiques. *Comptes Rendus de l'Académie des Sciences-Series IIB-Mechanics*, 329(3), 179–185.

Dirrenberger, J., Forest, S., Jeulin, D. (2014). Towards gigantic RVE sizes for 3D stochastic fibrous networks. *International Journal of Solids and Structures*, 51, 359–376.

Duerinckx, M., Gloria, A., Otto, F. (2020). The structure of fluctuations in stochastic homogenization. *Communications in Mathematical Physics*, 377(1), 259–306.

Egloffe, A.-C., Gloria, A., Mourrat, J.-C., Nguyen, T.N. (2015). Random walk in random environment, corrector equation and homogenized coefficients: From theory to numerics, back and forth. *IMA Journal of Numerical Analysis*, 35(2), 499–545.

El Houdaigui, F., Forest, S., Gourgues, A.F., Jeulin, D. (2007a). On the size of the representative volume element for isotropic elastic polycrystalline copper. In *IUTAM Symposium on Mechanical Behavior and Micro-Mechanics of Nanostructured Materials*. Springer, Berlin.

El Houdaigui, F., Forest, S., Gourgues, A.-F., Jeulin, D. (2007b). Representative volume element sizes for copper bulk polycrystals and thin layer. In *Colloque 3M Matériaux, Mécanique, Microstructures, sur le thème Interfaces : de l'atome au polycristal*. CEA Saclay/INSTN, Paris.

El Moumen, A., Kanit, T., Imad, A., El Minor, H. (2015). Effect of reinforcement shape on physical properties and representative volume element of particles-reinforced composites: Statistical and numerical approaches, *Mechanics of Materials*, 83, 1–16.

Escoda, J., Willot, F., Jeulin, D., Sanahuja, J., Toulemonde, C. (2011). Estimation of local stresses and elastic properties of a mortar sample by FFT computation of fields on a 3D image. *Cement and Concrete Research*, 41(5), 542–556.

Escoda, J., Willot, F., Jeulin, D., Sanahuja, J., Toulemonde, C. (2016). Influence of the multiscale distribution of particles on elastic properties of concrete. *International Journal of Engineering Science*, 98, 60–71.

Eyre, D.J. and Milton, G.W. (1999). A fast numerical scheme for computing the response of composites using grid refinement. *The European Physical Journal-Applied Physics*, 6(1), 41–47.

Farooq, H., Cailletaud, G., Forest, S., Ryckelynck, D. (2019). Crystal plasticity modeling of the cyclic behavior of polycrystalline aggregates under non-symmetric uniaxial loading: Global and local analyses. *International Journal of Plasticity*, 126, 102619.

Figliuzzi, B., Jeulin, D., Faessel, M., Willot, F., Koishi, M., Kowatari, N. (2016). Modelling the microstructure and the viscoelastic behaviour of carbon black filled rubber materials from 3D simulations. *Technische Mechanik*, 32, 22–46.

Fritzen, F., Forest, S., Böhlke, T., Kondo, D., Kanit, T. (2012). Computational homogenization of elasto-plastic porous metals. *International Journal of Plasticity*, 29, 102–119.

Fritzen, F., Forest, S., Kondo, D., Böhlke, T. (2013). Computational homogenization of porous materials of Green type. *Computational Mechanics*, 52, 121–134.

Gloria, A., Neukamm, S., Otto, F. (2015). Quantification of ergodicity in stochastic homogenization: Optimal bounds via spectral gap on Glauber dynamics. *Inventiones Mathematicae*, 199(2), 455–515.

Hersant, T. and Jeulin, D. (1976). L'échantillonnage dans les analyses quantitatives d'images. Exemples d'application aux mesures des teneurs de phases dans les agglomérés et des inclusions dans les aciers. *Mémoires et Etudes Scientifiques de la Revue de Métallurgie*, 73, 503.

Hill, R. (1963). Elastic properties of reinforced solids: Some theoretical principles. *Journal of the Mechanics and Physics of Solids*, 11(5), 357–372.

Huet, C. (1990). Application of variational concepts to size effects in elastic heterogeneous bodies. *Journal of the Mechanics and Physics of Solids*, 38(6), 813–841.

Jean, A., Jeulin, D., Forest, S., Cantournet, S., NGuyen, F. (2011a). A multiscale microstructure model of carbon black distribution in rubber. *Journal of Microscopy*, 241(3), 243–260.

Jean, A., Willot, F., Cantournet, S., Forest, S., Jeulin, D. (2011b). Large-scale computations of effective elastic properties of rubber with carbon black fillers. *International Journal for Multiscale Computational Engineering*, 9(3), 272–303.

Jeulin, D. (1991a). Modèles de fonctions aléatoires multivariables. *Sci. Terre*, 30, 225–256.

Jeulin, D. (1991b). Modèles morphologiques de structures aléatoires et de changement d'echelle. PhD Thesis, Université de Caen, France.

Jeulin, D. (2001). Random structure models for homogenization and fracture statistics. In *Mechanics of Random and Multiscale Microstructures*, Jeulin, D. and Ostoja-Starzewski, M. (eds). Springer, Berlin.

Jeulin, D. (2005). Random structures in physics. In *Space, Structure and Randomness: Contributions in Honor of Georges Matheron in the Field of Geostatistics, Random Sets and Mathematical Morphology*, Bilodeau, M., Meyer, F., Schmit, M. (eds). Springer, New York.

Jeulin, D. (2012). Morphology and effective properties of multi-scale random sets: A review. *C. R. Mécanique*, 340, 219–229.

Jeulin, D. (2016). Power laws variance scaling of Boolean random varieties. *Methodology and Computing in Applied Probability*, 18(4), 1065–1079.

Jeulin, D. (2021). *Morphological Models of Random Structures*. Springer, Berlin.

Jeulin, D. and Moreaud, M. (2008). Statistical representative volume element for predicting the dielectric permittivity of random media. In *Proceedings CMDS 11*. Les Presses de l'Ecole des Mines de Paris, Paris.

Jikov, V.V., Kozlov, S.M., Oleinik, O.A. (1994). *Homogenization of Differential Operators and Integral Functionals*. Springer Science, Berlin.

Kanit, T., Forest, S., Galliet, I., Mounoury, V., Jeulin, D. (2003). Determination of the size of the representative volume element for random composites: Statistical and numerical approach. *International Journal of Solids and Structures*, 40, 3647–3679.

Kanit, T., Nguyen, F., Forest, S., Jeulin, D., Reed, M., Singleton, S. (2006). Apparent and effective physical properties of heterogeneous materials: Representativity of samples of two materials from food industry. *Computer Methods in Applied Mechanics and Engineering*, 195, 3960–3982.

Kanit, T., Forest, S., Jeulin, D., N'Guyen, F., Singleton, S. (2011). Virtual improvement of ice cream properties by computational homogenization of microstructures. *Mechanics Research Communications*, 38(2), 136–140.

Koishi, M., Kowatari, N., Figliuzzi, B., Faessel, M., Willot, F., Jeulin, D. (2017). Computational material design of filled rubbers using multi-objective design exploration. In *Constitutive Models for Rubber X*, Lion, A. and Johlitz, M. (eds). CRC Press, Boca Raton.

Kreher, W. (1990). Residual stresses and stored elastic energy of composites and polycrystals. *Journal of the Mechanics and Physics of Solids*, 38(1), 115–128.

Kreher, W. and Pompe, W. (1985). Field fluctuations in a heterogeneous elastic material – An information theory approach. *Journal of the Mechanics and Physics of Solids*, 33(5), 419–445.

Kröner, E. (1972). *Statistical Continuum Mechanics*. Springer, Berlin.

Laiarinandrasana, L., Jean, A., Jeulin, D., Forest, S. (2012). Modelling the effects of various contents of fillers on the relaxation rate of elastomers. *Materials & Design*, 33, 75–82.

Lantuejoul, C. (1991). Ergodicity and integral range. *Journal of Microscopy*, 161(3), 387–403.

Madi, K., Forest, S., Jeulin, D., Boussuge, M. (2006). Estimating RVE sizes for 2D/3D viscoplastic composite materials. In *Matériaux 2006*, Dijon.

Matheron, G. (1965). *Les variables régionalisées et leur estimation : une application de la théorie de fonctions aléatoires aux sciences de la nature*. Masson, Paris.

Matheron, G. (1967). *Eléments pour une théorie des milieux poreux*. Masson, Paris.

Matheron, G. (1968). Composition des perméabilités en milieu poreux hétérogène : critique de la règle de pondération géométrique. *Rev. IFP*, 23, 201–218.

Matheron, G. (1971). *The Theory of Regionalized Variables and Its Applications*. École National Supérieure des Mines de Paris, Paris.

Matheron, G. (1975). *Random Sets and Integral Geometry*. Wiley, New York.

Matheron, G. (1989). *Estimating and Choosing: An Essay on Probability in Practice*. Springer, Berlin.

Matouš, K., Geers, M.G.D., Kouznetsova, V.G., Gillman, A. (2017). A review of predictive nonlinear theories for multiscale modeling of heterogeneous materials. *Journal of Computational Physics*, 330, 192–220.

Michel, J.C., Moulinec, H., Suquet, P. (2000). A computational method based on augmented lagrangians and fast Fourier transforms for composites with high contrast. *CMES (Computer Modelling in Engineering & Sciences)*, 1(2), 79–88.

Moreaud, M. (2006). Propriétés morphologiques multi-échelles et prévision du comportement diélectrique de nanocomposites. PhD Thesis, École Nationale Supérieure des Mines de Paris.

Moulinec, H. and Suquet, P. (1994). A fast numerical method for computing the linear and nonlinear mechanical properties of composites. *Comptes rendus de l'Académie des sciences. Série 2, Mécanique, physique, chimie, astronomie*, 1417–1423.

Ostoja-Starzewski, M. (1998). Random field models of heterogeneous materials. *International Journal of Solids and Structures*, 35(19), 2429–2455.

Oumarou, M., Jeulin, D., Renard, J. (2011). Etude statistique multi-échelle du comportement élastique et thermique d'un composite thermoplastique. *Revue des composites et des matériaux avancés*, 21, 221–254.

Peyrega, C. and Jeulin, D. (2013). Estimation of acoustic properties and of the representative volume element of random fibrous media. *Journal of Applied Physics*, 113, 104901.

Ponte Castaneda, P. and Willis, J.R. (1988). On the overall properties of nonlinearly viscous composites. *Proceedings of the Royal Society of London. A. Mathematical and Physical Sciences*, 416(1850), 217–244.

Quey, R. and Renversade, L. (2018). Optimal polyhedral description of 3D polycrystals: Method and application to statistical and synchrotron X-ray diffraction data. *Computer Methods in Applied Mechanics and Engineering*, 330, 308–333.

Sab, K. (1992). On the homogenization and the simulation of random materials. *Eur. J. Mech., A/Solids*, 11, 585–607.

Sanchez-Palencia, E. and Zaoui, A. (1987). *Homogenization Techniques for Composite Media*. Springer, Berlin.

Schneider, M., Josien, M., Otto, F. (2022). Representative volume elements for matrix-inclusion composites – A computational study on the effects of an improper treatment of particles intersecting the boundary and the benefits of periodizing the ensemble. *Journal of the Mechanics and Physics of Solids*, 158, 104652.

Šiška, F., Forest, S., Gumbsch, P. (2007a). Simulation of stress–strain heterogeneities in copper thin films: Texture and substrate effects. *Computational Materials Science*, 39, 137–141.

Šiška, F., Forest, S., Gumbsch, P., Weygand, D. (2007b). Finite element simulations of the cyclic elastoplastic behavior of copper thin films. *Modelling and Simulation in Materials Science and Engineering*, 15, S217–S238.

Suquet, P. and Ponte Castaneda, P. (1993). Small-contrast perturbation expansions for the effective properties of nonlinear composites. *Comptes rendus de l'Académie des sciences. Série 2, Mécanique, physique, chimie, sciences de l'univers, sciences de la terre*, 317, 1515–1522.

Talbot, D.R.S. and Willis, J. (1985). Variational principles for inhomogeneous non-linear media. *IMA Journal of Applied Mathematics*, 35(1), 39–54.

Wiegmann, A. (2007). Computation of the permeability of porous materials from their microstructure by FFF-Stokes. *ITWM Bericht*, 129.

Willis, J.R. (1982). Elasticity theory of composites. In *Mechanics of Solids*, Hill, R., Hopkins, H.G., Sewell, M.J. (eds). Pergamon, Press, Oxford, 653–686.

Willis, J.R. (1991). On methods for bounding the overall properties of nonlinear composites. *Journal of the Mechanics and Physics of Solids*, 39(1), 73–86.

Willis, J.R. (2001). Lectures on mechanics of random media. In *Mechanics of Random and Multiscale Microstructures*, Jeulin, D. and Ostoja-Starzewski, M. (eds). Springer, Berlin.

Willot, F. and Jeulin, D. (2009). Elastic behavior of composites containing Boolean random sets of inhomogeneities. *International Journal of Engineering Sciences*, 47, 313–324.

Willot, F. and Jeulin, D. (2011). Elastic and electrical behavior of some random multiscale highly-contrasted composites. *International Journal of Multiscale Computational Engineering*, 9, 305–326.

Willot, F. and Pellegrini, Y.P. (2008). Fast Fourier transform computations and build-up of plastic deformation in 2D, elastic-perfectly plastic, pixelwise disordered porous media. In *Proceedings CMDS 11*. Les Presses de l'Ecole des Mines de Paris, Paris.

Willot, F., Gillibert, L., Jeulin, D. (2013). Microstructure-induced hotspots in the thermal and elastic responses of granular media. *International Journal of Solids and Structures*, 50(10), 1699–1709.

Willot, F., Abdallah, B., Jeulin, D. (2016). The permeability of Boolean sets of cylinders. *Oil & Gas Science and Technology–Revue d'IFP Energies nouvelles*, 71, 52.

Zeghadi, A., Forest, S., Gourgues, A.-F., Bouaziz, O. (2007a). Ensemble averaging stress–strain fields in polycrystalline aggregates with a constrained surface microstructure–Part 2: Crystal plasticity. *Philosophical Magazine*, 87, 1425–1446.

Zeghadi, A., Nguyen, F., Forest, S., Gourgues, A.-F., Bouaziz, O. (2007b). Ensemble averaging stress–strain fields in polycrystalline aggregates with a constrained surface microstructure–Part 1: Anisotropic elastic behaviour. *Philosophical Magazine*, 87, 1401–1424.

3

Analytical Micromechanical Methods for Elasto-Viscoplastic Composites and Polycrystals

Stéphane BERBENNI[1] and Samuel FOREST[2]

[1] *LEM3, CNRS, Arts et Métiers ParisTech, Université de Lorraine, Metz, France*
[2] *CNRS UMR 7633, Centre des Matériaux, MINES Paris, PSL Research University, Evry, France*

3.1. Introduction

Materials like composites or metallic alloys exhibit time-dependent behaviors, which are elasto-viscoplastic (or nonlinear viscoelastic) and involve different orders of time differentiation. This leads to a "long-term memory effect" due to spatial–temporal interactions between the constituents of these composite materials. As an illustration of this effect, the effective (overall) behavior of an aggregate of Maxwellian constituents is no more Maxwellian (Suquet 1987). In addition, such materials exhibit nonlinear behaviors, which make the prediction accuracy of mean field homogenization models in comparison with full-field computations more difficult. Generally, mean field homogenization models first require a linearization step, which needs to be optimized and which is crucial to derive accurate estimates of the effective behavior and phase responses. Two main homogenization strategies were reported in the literature for elasto-viscoplastic composites and polycrystals.

For a color version of all figures in this chapter, see www.iste.co.uk/bernacki/digital.zip.

Digital Materials,
coordinated by Marc BERNACKI and Samuel FOREST.

The first is based on Riemann–Stieltjes integral-based "hereditary" homogenization approaches for non-ageing linear viscoelastic materials using Laplace–Carson (LC) transforms and the correspondence principle (Mandel 1966). Different homogenization schemes were developed using this principle, (see, for example, Hashin 1969; Laws and McLaughlin 1978; Rougier et al. 1994; Li and Weng 1997; Masson and Zaoui 1999; Pierard and Doghri 2006). However, LC inversion is needed at each resolution step in these homogenization methods, which severely limits their practical applications. The second more efficient way is based on time-incremental "internal variables approaches", whereby the stress history in phases is recorded through internal variables. Besides, complex loading paths in time domains and ageing behaviors can be treated. Different formulations were recently developed in this direction. Among them, variational approaches were developed based on an implicit Euler time discretization scheme for composite materials (Lahellec and Suquet 2007, 2013; Brassard et al. 2012; Agoras et al. 2016; Cotelo et al. 2020) and polycrystals (Das and Ponte Castañeda 2021). Direct internal variables approaches using the LC inversions and the collocation method were developed by Ricaud and Masson (2009) for two-phase composites, and for linear anisotropic polycrystals with a self-consistent (SC) procedure in Masson et al. (2012). More recently, this approach was also extended to nonlinear elasto-viscoplastic particulate composites with ageing effects in Masson et al. (2020), using a "modified secant" formulation as linearization for the viscoplastic strain rate in the matrix phase. As other direct internal variables approaches avoid LC inversions, the "additive interaction" (AI) law for the Maxwellian Eshelby inclusion problem was derived by Molinari et al. (1997); Molinari (2002), and, more recently applied to the Mori–Tanaka (MT) and SC estimates, using a "tangent" linearization formulation in Mercier and Molinari (2009) and Mercier et al. (2012). The numerical results of the AI for nonlinear particulate elasto-viscoplastic composites were compared to full-field finite element (FEM) calculations in Czarnota et al. (2015). Another similar approach was proposed by Kowalczyk-Gajewska and Petryk (2011), in which the linearized sub-problems for an inhomogeneity/matrix system with elastic or viscoplastic interaction rules were solved sequentially within one incremental step. In the context of polycrystal plasticity, different extensions of the AI law were developed by Wang et al. (2010), and, more recently by Zecevic and Lebensohn (2020). Doghri et al. (2010) developed an incremental-homogenization model based on a first-order "affine" linearization and a fully implicit backward Euler scheme with an algorithmic tangent operator. Their numerically efficient method was applied to inclusion-reinforced elasto-viscoplastic composites for non-monotonic and non-proportional loadings and was compared to FEM calculations. The "translated field" (TF) method was developed for linear viscoelastic and elasto-viscoplastic composites in Paquin et al. (1999) and in Sabar et al. (2002) using a classic "secant" formulation. The TF method was adapted to treat linear viscoelastic composites with the MT scheme in Mercier et al. (2012), where it was successively compared to the AI law and full-field fast Fourier transform (FFT) calculations. The TF method was extended to nonlinear elasto-viscoplastic composites and polycrystals with a first-order "affine" formulation to improve the effective behavior estimates of composites (Berbenni and Capolungo 2015) and polycrystals (Mareau and Berbenni 2015; Lhadi et al. 2018). Another "internal variables" approach was introduced in Berbenni et al. (2015) and extended to elasto-viscoplastic particulate composites based on a modified secant formulation in Berbenni (2021); however, it will not be presented here. The TF method was developed for both MT and SC homogenization schemes and will be detailed in this chapter. Some applications of the TF method at small strains with different linearization schemes will be reported for both two-phase composites and polycrystals.

For practical applications, the micromechanical models can be used for different purposes: detection of damage initiation in individual phases, optimization of the microstructure under constraint of limited plasticity or damage and structural computations of engineering components. In the latter case, the micromechanical model must be used as the constitutive law evaluated at each integration point of the finite element (FE) mesh of the structure. In that respect, it competes with standard phenomenological models whose computational efficiency is ensured by proper implicit integration schemes. Computing consistent tangent operators is not an easy task for micromechanical models that involve a large number of internal variables, proportional to the number of phases. This means thousands of internal variables for polycrystal models, for instance Cailletaud and Pilvin (1994). That is why simplified or reduced order models have been developed based on original micromechanical schemes or full field simulations of representative volume elements. Among these models, the attention will be focused on the dynasty of so-called β-models created by Cailletaud (1992); Pilvin (1996) and recently reviewed and reworked in Cailletaud and Coudon (2016) and Coudon et al. (2019). This class of micromechanical models involves phenomenological tensor valued variables $\underset{\sim}{\beta}^{(r)}$ for each phase that account for nonlinear elastoviscoplastic accommodation of strain incompatibilities. The evolution equations for these variables contain parameters to be calibrated from other micromechanical model predictions or from full-field simulations. Numerical comparisons between the results given by the secant TF method of Sabar et al. (2002), the β model and crystal plasticity FE simulations were provided in Coudon et al. (2019). Applications of these β-models will be described and discussed, in particular for polycrystals at finite deformations.

In the whole chapter, the Einstein's summation convention is used for repeated indices and the notations for a given quantity X are the following:

- vector: $\underline{X}$;
- second-order tensor: $\underset{\sim}{X}$;
- fourth-order tensor: $\underset{\approx}{X}$.

Their indicial notations are:

– vector: X_i;

– second-order tensor: X_{ij};

– fourth-order tensor: X_{ijkl}.

The main symbols used are the following:

– the second-order unit tensor is $\underset{\sim}{\delta}$ where $\delta_{ij} = 1$ if $i = j$ and $\delta_{ij} = 0$ if $i \neq j$;

– the fourth-order unit tensor is $\underset{\approx}{I}$ where $I_{ijkl} = \frac{1}{2}\left(\delta_{ik}\delta_{jl} + \delta_{il}\delta_{kj}\right)$;

– the fourth-order projection tensor $\underset{\approx}{J}$ is given by $J_{ijkl} = \frac{1}{3}\delta_{ij}\delta_{kl}$;

– the fourth-order projection tensor is $\underset{\approx}{K}$ is given by $K_{ijkl} = I_{ijkl} - J_{ijkl} = \frac{1}{2}\left(\delta_{ik}\delta_{jl} + \delta_{il}\delta_{kj}\right) - \frac{1}{3}\delta_{ij}\delta_{kl}$;

– "." deals with the simple product;

– ":" deals with the contracted product;

– "$\otimes$" deals with the tensor product;

– the spatial convolution is denoted as $\star$;

– the gradient operator is denoted ∇ such that $(\nabla \underline{X})_{ji} = X_{i,j} = \partial X_i / \partial x_j$, where $\underline{x}$ is a spatial position vector. ∇^s denotes the symmetric part of the gradient;

– the divergence operator is denoted $\underline{div}$ such that $(\underline{div}\ \underset{\sim}{X})_i = X_{ij,j}$;

– the Eulerian temporal derivative of $\underset{\sim}{A}$ is denoted as $\dot{\underset{\sim}{A}}$.

3.2. Translated field method

3.2.1. *Field equations and integral equation*

Considering a heterogeneous medium V with Maxwellian constituents, local linear elastic moduli $\underset{\approx}{c}$ (elastic compliances $\underset{\approx}{s} = \underset{\approx}{c}^{-1}$) and viscoplastic strains $\dot{\underset{\sim}{\varepsilon}}^{vp}$, the problem is treated in the framework of quasi-static equilibrium with infinitesimal strains and no volume force.

Field equations of the elasto-viscoplastic heterogeneous problem are given by:

$$
\begin{aligned}
&\dot{\underset{\sim}{\varepsilon}} = \dot{\underset{\sim}{\varepsilon}}^{e} + \underset{\sim}{g}(\underset{\sim}{\sigma}) \\
&\underline{div}\,\dot{\underset{\sim}{\sigma}} = 0 \\
&\underline{div}\,\underset{\sim}{\sigma} = 0 \\
&\dot{\underset{\sim}{\varepsilon}} = \nabla^{s}\underline{\dot{u}} \\
&\underline{\dot{u}}^{d} = \dot{\underset{\sim}{E}} \cdot \underline{x}
\end{aligned}
\qquad [3.1]
$$

where $\dot{\underset{\sim}{\varepsilon}}$ is the total strain rate, $\underset{\sim}{\sigma}$ is the Cauchy stress , and$\dot{\underset{\sim}{\varepsilon}}^{e} = \underset{\approx}{s} : \dot{\underset{\sim}{\sigma}}$ (respectively, $\dot{\underset{\sim}{\varepsilon}}^{vp} = \underset{\sim}{g}(\underset{\sim}{\sigma})$, this function can be complemented by internal variables) is the elastic (respectively, viscoplastic) strain rate. The balance equation for unknown stress rate $\dot{\underset{\sim}{\sigma}}$ is incrementally solved in the time step dt (i.e. between t and $t + dt$). At time t, the balance equation for the known stress $\underset{\sim}{\sigma}$ is verified. Here, a homogeneous velocity vector $\underline{\dot{u}}^{d}$ is imposed on the boundary ∂V where $\dot{\underset{\sim}{E}}$ is prescribed. In this time incremental scheme and in contrast with integral hereditary approaches, the viscoplastic strains are considered as internal variables containing the whole history of the material.

The local viscoplastic strain rate $\dot{\underset{\sim}{\varepsilon}}^{vp} = \underset{\sim}{g}(\underset{\sim}{\sigma})$ is a nonlinear function dependent on the stress tensor $\underset{\sim}{\sigma}$. The solution consists of determining the strain and stress rates satisfying all these field equations. In the following, in order to be able to apply analytical homogenization methods, this nonlinear behavior is approximated by a thermoelastic like linear comparison material (Ponte Castañeda 1991). Therefore, the first step is to solve a linear viscoelastic problem with a linearized viscoplastic strain rate:

$$
\dot{\underset{\sim}{\varepsilon}}^{vp} = \underset{\approx}{m} : \underset{\sim}{\sigma} + \dot{\underset{\sim}{\varepsilon}}^{v0} \qquad [3.2]
$$

where $\underset{\approx}{m}$ is a linearized viscoplastic compliance and $\dot{\underset{\sim}{\varepsilon}}^{v0}$ is the “back-extrapolated” strain rate (i.e. the viscoplastic strain rate at zero stress). In this chapter, three different linearization schemes will be adopted, namely, the classical “secant” procedure, the classical “affine” procedure and the “modified secant” procedure (see section 3.2.3). Let us introduce an infinite homogeneous reference medium with linear elastic compliances compliances $\underset{\approx}{S}$, linear viscoplastic compliances $\underset{\approx}{M}$ and with homogeneous reference “back-extrapolated” strain $\dot{\underset{\sim}{E}}^{v0}$. The corresponding heterogeneous linear elastic and linear viscoplastic compliances $\underset{\approx}{s}$, $\underset{\approx}{m}$ and "back-extrapolated" strain rate $\dot{\underset{\sim}{\varepsilon}}^{v0}$ can be written in terms of spatial fluctuations $\delta \underset{\approx}{s}$, $\delta \underset{\approx}{b}$ and $\delta \dot{\underset{\sim}{\varepsilon}}^{v0}$ with respect to the infinite homogeneous reference medium as follows: $\underset{\approx}{s} = \underset{\approx}{S} + \delta \underset{\approx}{s}$, $\underset{\approx}{m} = \underset{\approx}{M} + \delta \underset{\approx}{m}$ and $\dot{\underset{\sim}{\varepsilon}}^{v0} = \dot{\underset{\sim}{E}}^{v0} + \delta \dot{\underset{\sim}{\varepsilon}}^{v0}$ with $\underset{\approx}{S} = (\underset{\approx}{C})^{-1}$ and $\underset{\approx}{M} = (\underset{\approx}{B})^{-1}$. Using Green's function technique on the statically admissible fields $\underset{\sim}{\sigma}$ and $\dot{\underset{\sim}{\sigma}}$ and on the compatible field $\dot{\underset{\sim}{\varepsilon}}$, the following integral equation is obtained (Berbenni and Capolungo 2015; Mareau and Berbenni 2015):

$$\dot{\underset{\sim}{\varepsilon}} = \dot{\underset{\sim}{E}} - \underset{\approx}{\Gamma}^C \star \left(\delta \underset{\approx}{c} : \dot{\underset{\sim}{\varepsilon}}^e\right) - \underset{\approx}{\Gamma}^B \star \left(\delta \underset{\approx}{b} : \dot{\underset{\sim}{\varepsilon}}^{vp} - \underset{\approx}{b} : \dot{\underset{\sim}{\varepsilon}}^{v0}\right) + \left(\underset{\approx}{\Gamma}^C : \underset{\approx}{C} - \underset{\approx}{\Gamma}^B : \underset{\approx}{B}\right) \star \dot{\underset{\sim}{\varepsilon}}^{vp} \quad [3.3]$$

where $\underset{\approx}{\Gamma}^C$ and $\underset{\approx}{\Gamma}^B$ are the modified Green tensors associated with $\underset{\approx}{C}$ and $\underset{\approx}{B}$ (Kröner 1990).

3.2.2. *TF-based method: theoretical foundations*

This section describes the TF method in the case of heterogeneous linear viscoelastic (Maxwellian) materials within a small strain assumption. Here, the viscoplastic strain rates are supposed to be linear following equation [3.2]. The TF method is a mean field homogenization (MFH) method based on the following additive decomposition for the total strain rate field (Paquin et al. 1999, 2001; Berbenni and Capolungo 2015; Mareau and Berbenni 2015):

$$\dot{\underset{\sim}{\varepsilon}} = \dot{\underset{\sim}{\varepsilon}}^e + \dot{\underset{\sim}{\varepsilon}}^{vp} = \dot{\underset{\sim}{e}} + \dot{\tilde{\underset{\sim}{\varepsilon}}} = \dot{\underset{\sim}{e}}^e + \dot{\underset{\sim}{e}}^{vp} + \dot{\tilde{\underset{\sim}{\varepsilon}}}^e + \dot{\tilde{\underset{\sim}{\varepsilon}}}^{vp} \quad [3.4]$$

where $\dot{\underset{\sim}{e}}^e$, $\dot{\underset{\sim}{e}}^{vp}$ are pure elastic (respectively, pure viscoplastic) compatible fields subjected to the same stress state as the elasto-viscoplastic heterogeneous material. Furthermore, $\dot{\tilde{\underset{\sim}{\varepsilon}}}^e$ and $\dot{\tilde{\underset{\sim}{\varepsilon}}}^{vp}$ are considered as "residual" elastic (respectively, viscoplastic) strain rate fields defined as $\dot{\tilde{\underset{\sim}{\varepsilon}}}^e = \dot{\underset{\sim}{\varepsilon}}^e - \dot{\underset{\sim}{e}}^e$ and $\dot{\tilde{\underset{\sim}{\varepsilon}}}^{vp} = \dot{\underset{\sim}{\varepsilon}}^{vp} - \dot{\underset{\sim}{e}}^{vp}$, respectively. The translated field decomposition can be written in terms of stress rate decomposition for elasticity and stress decomposition for viscoplasticity:

$$\begin{aligned} \dot{\underset{\sim}{\sigma}} &= \underset{\approx}{c} : \dot{\underset{\sim}{\varepsilon}}^e = \underset{\approx}{c} : \dot{\underset{\sim}{e}}^e + \dot{\tilde{\underset{\sim}{\sigma}}}^* \\ \underset{\sim}{\sigma} &= \underset{\approx}{b} : \left(\dot{\underset{\sim}{\varepsilon}}^{vp} - \dot{\underset{\sim}{\varepsilon}}^{v0}\right) = \underset{\approx}{b} : \left(\dot{\underset{\sim}{e}}^{vp} - \dot{\underset{\sim}{\varepsilon}}^{v0}\right) + \tilde{\underset{\sim}{\sigma}}'' \end{aligned} \quad [3.5]$$

where $\dot{\tilde{\underset{\sim}{\sigma}}}^* = \underset{\approx}{c} : \dot{\tilde{\underset{\sim}{\varepsilon}}}^e$ and $\tilde{\underset{\sim}{\sigma}}'' = \underset{\approx}{b} : \dot{\tilde{\underset{\sim}{\varepsilon}}}^{vp}$ are, respectively, residual stress rate and stress associated with $\dot{\tilde{\underset{\sim}{\varepsilon}}}^e$ and $\dot{\tilde{\underset{\sim}{\varepsilon}}}^{vp}$ with $\dot{\tilde{\underset{\sim}{\sigma}}}^* \neq \dot{\tilde{\underset{\sim}{\sigma}}}''$. The compatible fields $\dot{\underset{\sim}{e}}^e$ and $\dot{\underset{\sim}{e}}^{vp}$ in equation [3.4] are solutions of the purely (linear) elastic and (nonlinear) viscoplastic problems:

$$\begin{aligned} \dot{\underset{\sim}{e}}^{e} &= \overline{\dot{\underset{\sim}{e}}^{e}} - \underset{\approx}{\Gamma}^{C} \star \left(\delta \underset{\approx}{c} : \dot{\underset{\sim}{e}}^{e}\right) = \underset{\approx}{A}^{C} : \overline{\dot{\underset{\sim}{e}}^{e}} \\ \dot{\underset{\sim}{e}}^{vp} &= \overline{\dot{\underset{\sim}{e}}^{vp}} - \underset{\approx}{\Gamma}^{B} \star \left(\delta \underset{\approx}{b} : \dot{\underset{\sim}{e}}^{vp} - \underset{\approx}{b} : \dot{\underset{\sim}{\varepsilon}}^{v0}\right) = \underset{\approx}{A}^{B} : \overline{\dot{\underset{\sim}{e}}^{vp}} + \underset{\sim}{a} \end{aligned} \quad [3.6]$$

where $\underset{\approx}{A}^C$ is an elastic strain rate concentration tensor for the purely elastic problem and $\underset{\approx}{A}^B$ and $\underset{\sim}{a}$ are strain rate concentration tensors associated with the purely viscoplastic problem. Using the linear comparison material, the latter reduces to a classic linear thermoelastic problem. These concentration tensors depend on the homogenization schemes introduced in the next sections. Macroscopic averages of equations [3.4] and [3.5] yield:

$$\begin{aligned} &\overline{\dot{\tilde{\underset{\sim}{\varepsilon}}}^{e}} = 0\,;\, \overline{\dot{\tilde{\underset{\sim}{\varepsilon}}}^{vp}} = 0\,;\, \overline{\dot{\tilde{\underset{\sim}{\varepsilon}}}} = 0 \\ &\overline{\dot{\tilde{\underset{\sim}{\sigma}}}'} = 0\,;\, \overline{\dot{\tilde{\underset{\sim}{\sigma}}}''} = 0 \end{aligned} \quad [3.7]$$

Due to the compatibility of $\dot{\underset{\sim}{e}}$, the total translated strain rate field $\dot{\tilde{\underset{\sim}{\varepsilon}}}$ is also compatible from equation [3.4]. Furthermore, from equation [3.5], the residual stress fields $\dot{\tilde{\underset{\sim}{\sigma}}}^{*}$ and $\dot{\tilde{\underset{\sim}{\sigma}}}''$ are balanced. Thus, to solve the field equations (equation [3.1]) for the translated problem, the following integral equation of the translated problem is obtained:

$$\dot{\tilde{\underset{\sim}{\varepsilon}}} = \overline{\dot{\tilde{\underset{\sim}{\varepsilon}}}} - \underset{\approx}{\Gamma}^{C} \star \left(\delta \underset{\approx}{c} : \dot{\tilde{\underset{\sim}{\varepsilon}}}^{e}\right) - \underset{\approx}{\Gamma}^{B} \star \left(\delta \underset{\approx}{b} : \dot{\tilde{\underset{\sim}{\varepsilon}}}^{vp}\right) + \left(\underset{\approx}{\Gamma}^{C} : \underset{\approx}{C} - \underset{\approx}{\Gamma}^{B} : \underset{\approx}{B}\right) \star \dot{\tilde{\underset{\sim}{\varepsilon}}}^{vp} \quad [3.8]$$

Considering only the local parts of Green's operators, we will see in the next applications with the MT scheme for two-phase composites and the SC scheme for polycrystals, that once the translated problem (equation [3.8]) is solved, the solution of equation [3.3) can be reached using equations [3.4] and [3.6].

3.2.3. *TF method for elasto-viscoplastic two-phase composite materials*

In the case of two-phase composite materials, the RVE is constituted of one inhomogeneity phase I with volume fraction $f^{(I)} = f$ embedded in a matrix phase M with volume fraction $f^{(M)} = 1 - f$. The MT scheme (Mori and Tanaka 1973; Weng 1984; Benveniste 1987) is applied using the strain rate concentration equation of the TF method for each inhomogeneity phase I. It is well suited to estimate the effective properties of such materials, provided the volume fraction of inhomogeneities is relatively low. The RVE is subjected at its boundary ∂V to the homogeneous macroscopic strain rate denoted as $\dot{\underset{\sim}{E}}$ satisfying:

$$f\dot{\underset{\sim}{\varepsilon}}^{(I)} + (1-f)\dot{\underset{\sim}{\varepsilon}}^{(M)} = \dot{\underset{\sim}{E}} \qquad [3.9]$$

The MT procedure assumes the matrix phase M as the homogeneous reference medium and reduces the elastic and viscoplastic Green's operators of the TF method to their local parts to make use of the elastic and viscoplastic Eshelby tensors (Eshelby and Peierls 1957). Here, the inhomogeneity phase I and the matrix phase M are both Maxwellian with respective elastic $\underset{\approx}{c}^{(I)}$, $\underset{\approx}{C}$ and linearized viscoplastic $\underset{\approx}{b}^{(I)}$, $\underset{\approx}{B}$ moduli (and conversely respective elastic $\underset{\approx}{s}^{(I)}$, $\underset{\approx}{S}$ and linearized viscoplastic $\underset{\approx}{m}^{(I)}$, $\underset{\approx}{M}$ compliances), back-extrapolated strains $\dot{\underset{\sim}{\varepsilon}}^{v0(I)}$, $\dot{\underset{\sim}{\varepsilon}}^{v0(M)}$. Therefore, the spatial fluctuations of elastic and nonlinear viscoplastic properties are defined by $\delta\underset{\approx}{c}^{(I)} = \underset{\approx}{c}^{(I)} - \underset{\approx}{C}$, $\delta\underset{\approx}{s}^{(I)} = \underset{\approx}{s}^{(I)} - \underset{\approx}{S}$, $\delta\underset{\approx}{b}^{(I)} = \underset{\approx}{b}^{(I)} - \underset{\approx}{B}$, $\delta\underset{\approx}{m}^{(I)} = \underset{\approx}{m}^{(I)} - \underset{\approx}{M}$, $\delta\dot{\underset{\sim}{\varepsilon}}^{v0^{(I)}} = \dot{\underset{\sim}{\varepsilon}}^{v0(I)} - \dot{\underset{\sim}{\varepsilon}}^{v0(M)}$.

Following the TF method (see section 3.2.2), we obtain: $\tilde{\dot{\underset{\sim}{\varepsilon}}}^{(I)} = \tilde{\dot{\underset{\sim}{\varepsilon}}}^{e(I)} + \tilde{\dot{\underset{\sim}{\varepsilon}}}^{vp(I)}$ and the total strain rate in the inhomogeneity phase I constituted of a random distribution of ellipsoidal inhomogeneities reads:

$$\dot{\underset{\sim}{\varepsilon}}^{(I)} = \underset{\approx}{A}^{C} : \dot{\underset{\sim}{\varepsilon}}^{e(M)} + \underset{\approx}{A}^{B} : \dot{\underset{\sim}{\varepsilon}}^{vp(M)} + \underset{\sim}{a}^{(I)} + \tilde{\dot{\underset{\sim}{\varepsilon}}}^{(I)} \qquad [3.10]$$

where:

$$\begin{aligned}
&\underset{\approx}{A}^{C} = \left(\underset{\approx}{I} + \underset{\approx}{\Gamma_l}^{C} : \delta\underset{\approx}{c}^{(I)}\right)^{-1} \\
&\underset{\approx}{A}^{B} = \left(\underset{\approx}{I} + \underset{\approx}{\Gamma_l}^{B} : \delta\underset{\approx}{b}^{(I)}\right)^{-1} \\
&\underset{\sim}{a}^{(I)} = \left(\underset{\approx}{I} - \underset{\approx}{A}^{B}\right) : \left(\delta\underset{\approx}{b}^{(I)}\right)^{-1} : \left(\underset{\approx}{b}^{(I)} : \dot{\underset{\sim}{\varepsilon}}^{v0(I)} - \underset{\approx}{B} : \dot{\underset{\sim}{\varepsilon}}^{v0(M)}\right) \\
&\tilde{\dot{\underset{\sim}{\varepsilon}}}^{(I)} = \tilde{\dot{\underset{\sim}{\varepsilon}}}^{(M)} - \underset{\approx}{\Gamma_l}^{C} : \delta\underset{\approx}{c}^{(I)} : \tilde{\dot{\underset{\sim}{\varepsilon}}}^{e(I)} - \underset{\approx}{\Gamma_l}^{B} : \delta\underset{\approx}{b}^{(I)} : \tilde{\dot{\underset{\sim}{\varepsilon}}}^{vp(I)} \\
&\quad + \left(\underset{\approx}{\Gamma_l}^{C} : \underset{\approx}{C} - \underset{\approx}{\Gamma_l}^{B} : \underset{\approx}{B}\right) : \tilde{\dot{\underset{\sim}{\varepsilon}}}^{vp(I)}
\end{aligned} \qquad [3.11]$$

with the elastic (respectively, viscoplastic) translated strain rates given by:

$$\begin{aligned} \tilde{\dot{\underset{\sim}{\varepsilon}}}^{e(I)} &= \dot{\underset{\sim}{\varepsilon}}^{e(I)} - \underset{\approx}{A}^C : \dot{\underset{\sim}{\varepsilon}}^{e(M)} \\ \tilde{\dot{\underset{\sim}{\varepsilon}}}^{vp(I)} &= \dot{\underset{\sim}{\varepsilon}}^{vp(I)} - \underset{\approx}{A}^B : \dot{\underset{\sim}{\varepsilon}}^{vp(M)} - \underset{\sim}{a}^{(I)} \end{aligned} \qquad [3.12]$$

Hence, from the TF decomposition, $\tilde{\dot{\underset{\sim}{\varepsilon}}}^{(M)} = 0$. The tensors $\underset{\approx}{\Gamma_l}^*$ with $* = C, B$ represent the local parts of Green's operators. They are numerically computed using the Gauss–Legendre integration method from the following integrals over the volume V_I of the inhomogeneity:

$$\begin{aligned} \underset{\approx}{\Gamma_l}^C &= \int_{V_I} \underset{\approx}{\Gamma}^C (\underline{x} - \underline{x}')\, dV' \\ \underset{\approx}{\Gamma_l}^B &= \int_{V_I} \underset{\approx}{\Gamma}^B (\underline{x} - \underline{x}')\, dV' \end{aligned} \qquad [3.13]$$

After straightforward algebraic manipulations using equations [3.10]–[3.12], the strain rate concentration equation for the phase I is given by:

$$\begin{aligned} \dot{\underset{\sim}{\varepsilon}}^{(I)} &= \underset{\approx}{A}^C : \dot{\underset{\sim}{\varepsilon}}^{(M)} + \dot{\underset{\sim}{\varepsilon}}^{vp(I)} - \underset{\approx}{A}^C : \left(\underset{\approx}{A}^B\right)^{-1} : \left(\dot{\underset{\sim}{\varepsilon}}^{vp(I)} - \underset{\sim}{a}^{(I)}\right) \\ &+ \underset{\approx}{A}^C : \left(\underset{\approx}{\Gamma_l}^C : \underset{\approx}{C} - \underset{\approx}{\Gamma_l}^B : \underset{\approx}{B}\right) : \left(\dot{\underset{\sim}{\varepsilon}}^{vp(I)} - \underset{\approx}{A}^B : \dot{\underset{\sim}{\varepsilon}}^{vp(M)} - \underset{\sim}{a}^{(I)}\right) \end{aligned} \qquad [3.14]$$

Thus, the strain rate concentration equation for the matrix phase M is easily obtained from equation [3.14] together with equation [3.9].

The stress rate interaction law between both phases is determined from equation [3.14] together with the linear elastic Hooke's law:

$$\begin{aligned} \dot{\underset{\sim}{\sigma}}^{(I)} - \dot{\underset{\sim}{\sigma}}^{(M)} &= \left(\underset{\approx}{C} - \left(\underset{\approx}{\Gamma_l}^C\right)^{-1}\right) : \left(\dot{\underset{\sim}{\varepsilon}}^{(I)} - \dot{\underset{\sim}{\varepsilon}}^{(M)}\right) \\ &- \underset{\approx}{C} : \left(\dot{\underset{\sim}{\varepsilon}}^{vp(I)} - \dot{\underset{\sim}{\varepsilon}}^{vp(M)}\right) \\ &- \left(\underset{\approx}{\Gamma_l}^C\right)^{-1} : \underset{\approx}{\Gamma_l}^B : \underset{\approx}{\delta b}^{(I)} : \dot{\underset{\sim}{\varepsilon}}^{vp(I)} + \left(\underset{\approx}{\Gamma_l}^C\right)^{-1} : \left(\underset{\approx}{A}^B\right)^{-1} : \underset{\sim}{a}^{(I)} \\ &+ \left(\underset{\approx}{C} - \left(\underset{\approx}{\Gamma_l}^C\right)^{-1} : \underset{\approx}{\Gamma_l}^B : \underset{\approx}{B}\right) : \left(\dot{\underset{\sim}{\varepsilon}}^{vp(I)} - \underset{\approx}{A}^B : \dot{\underset{\sim}{\varepsilon}}^{vp(M)} - \underset{\sim}{a}^{(I)}\right) \end{aligned} \qquad [3.15]$$

The first term contains the classic elastic Hill's constraint tensor (Hill 1965) $\underset{\approx}{L}^* = \left(\underset{\approx}{\Gamma_l}^C\right)^{-1} - \underset{\approx}{C}$, which solves the purely elastic interaction problem between both phases. The other terms represent the elastic-viscoplastic interactions between inclusion I and matrix M at the origin of the long memory effect. This equation was reported in Berbenni and Capolungo (2015) for nonlinear viscoelastic composites. In linear viscoelasticity and in the particular case of isotropic incompressible viscoelastic matrix and inhomogeneity phases, it was reported in Mercier et al. (2012) that the interaction law given by equation [3.15] provides the same result as the exact solution of Hashin (1969). For compressible linear viscoelasticity, some comparisons with the "additive" interaction law (Molinari et al. 1997; Molinari 2002; Mercier and Molinari 2009; Czarnota et al. 2015) were also provided in Mercier et al. (2012).

The interaction law (equation [3.15]) contains two asymptotic states. For small physical times, that is, when viscoplastic strains are negligible ($\underset{\sim}{\dot{\varepsilon}}^{vp(I)} = \underset{\sim}{\dot{\varepsilon}}^{vp(M)} \to 0$), equation [3.14] reduces to the classical Hill's elastic stress interaction rule (Hill 1965):

$$\underset{\sim}{\dot{\sigma}}^{(I)} - \underset{\sim}{\dot{\sigma}}^{(M)} = -\underset{\approx}{L}^* : \left(\underset{\sim}{\dot{\varepsilon}}^{(I)} - \underset{\sim}{\dot{\varepsilon}}^{(M)}\right) \qquad [3.16]$$

At large physical times, equation [3.15] can be simplified by considering the steady state given by $\underset{\sim}{\dot{\sigma}}^{(I)} = \underset{\sim}{\dot{\sigma}}^{(M)} \to \underset{\sim}{0}$, $\underset{\sim}{\dot{\varepsilon}}^{(I)} \to \underset{\sim}{\dot{\varepsilon}}^{vp(I)}$, $\underset{\sim}{\dot{\varepsilon}}^{(M)} \to \underset{\sim}{\dot{\varepsilon}}^{vp(M)}$ and $\underset{\sim}{\dot{\varepsilon}}^{vp(I)} \to \underset{\approx}{A}^B : \underset{\sim}{\dot{\varepsilon}}^{vp(M)} + \underset{\sim}{a}^{(I)}$. After simplifications, the asymptotic nonlinear viscoplastic interaction law reads:

$$\underset{\sim}{\sigma}^{(I)} - \underset{\sim}{\sigma}^{(M)} = \left(\underset{\approx}{B} - \left(\underset{\approx}{\Gamma_l}^B\right)^{-1}\right) : \left(\underset{\sim}{\dot{\varepsilon}}^{vp(I)} - \underset{\sim}{\dot{\varepsilon}}^{vp(M)}\right) \qquad [3.17]$$

which is the same as the one reported in Molinari et al. (1987) for the "tangent" formulation.

To study the effect of linear versus nonlinear viscosity on the overall composite responses, the present MT scheme based on the TF method is extended to nonlinear viscoelasticity using the linearized relationship described in equation [3.2]. Here, we assume that both phases (I, M) are supposed to be elasto-viscoplastic with nonlinear incompressible viscoplasticity given by the dissipation potential Φ^{vp}:

$$\underset{\sim}{\dot{\varepsilon}} = \underset{\approx}{s} : \underset{\sim}{\dot{\sigma}} + \underset{\sim}{\dot{\varepsilon}}^{vp} = \underset{\approx}{s} : \underset{\sim}{\dot{\sigma}} + \frac{\partial \Phi^{vp}}{\partial \underset{\sim}{\sigma}'} \qquad [3.18]$$

where $\underset{\sim}{\sigma}'$ is the deviatoric stress tensor, and Φ^{vp} is given for each phase (I, M) by a the power law (Norton's model):

$$\Phi^{vp}\left(\underset{\sim}{\sigma}'\right) = \frac{\dot{\varepsilon}_0 \sigma_0}{n+1}\left(\frac{\sigma_{eq}}{\sigma_0}\right)^{n+1} \qquad [3.19]$$

where $m = 1/n$ is the strain-rate sensitivity, which is assumed to be the same for both phases, σ_{eq} is the equivalent Von Mises stress, $\dot{\varepsilon}_0$ is a reference strain rate and σ_0 is the material's flow stress. The value $m = 1$ corresponds to linear viscoelasticity. For $m \neq 1$, the behavior is nonlinear, which is the case for metallic alloys.

The elastic compliance tensor $\underset{\approx}{s}$ is isotropic for both phases I and M, which are elastically compressible:

$$\underset{\approx}{s} = \frac{1}{3k}\underset{\approx}{J} + \frac{1}{2\mu}\underset{\approx}{K} \qquad [3.20]$$

where k ad μ are the elastic bulk and shear moduli, respectively. Equation [3.18] is also written as:

$$\underset{\sim}{\dot{\varepsilon}} = \underset{\approx}{s} : \underset{\sim}{\dot{\sigma}} + \frac{1}{2\eta}\underset{\sim}{\sigma}' \qquad [3.21]$$

where η is the shear viscous modulus given by:

$$\eta = \frac{\sigma_0}{3\dot{\varepsilon}_0}\left(\frac{\sigma_0}{\sigma_{eq}}\right)^{n-1} \qquad [3.22]$$

Using equation [3.2], three different linearization schemes for the linear comparison material are considered:

– the classical "secant" linearization (denoted as "SEC");

– the classical "affine" linearization (denoted as "AFF");

– the "modified secant" linearization (denoted as "SEC MODIF").

It is noteworthy that the last one is considered for the first time within the TF method inspired by several recent studies on homogenization-based methods considering second moments of stresses in two-phase Maxwellian composites (Li and Weng 1997; Badulescu et al. 2015; Masson et al. 2020; Berbenni 2021). The classical SEC and AFF linearization procedures only consider the first-order stress moments (i.e. the phase average stresses), whereas the SEC MODIF linearization procedure includes second-order stress moments (Qiu and Weng 1992; Suquet 1995;

Hu 1996; Ponte Castañeda and Suquet 2002). The SEC MODIF linearization procedure was developed to better account for intra-phase heterogeneities in nonlinear composites and makes use of an isotropic linear comparison composite. Since the MT scheme is used, the stress in the inhomogeneity phase I is here supposed to be homogeneous in the linear comparison composite. Therefore, for all the investigated nonlinear extensions and in contrast with the matrix phase M, the "effective stress" in the inhomogeneity phase σ_{eq}^{I} is computed from the averaged deviatoric stress over phase I denoted $\bar{\underset{\sim}{\sigma}}'^{I}$:

$$\sigma_{eq}^{I} = \overline{\sigma_{eq}}^{I} = \left(\frac{3}{2}\bar{\underset{\sim}{\sigma}}'^{I} : \bar{\underset{\sim}{\sigma}}'^{I}\right)^{1/2} \quad [3.23]$$

In the case of the classical SEC linearization, the linearized viscoplastic compliance and the "back-extrapolated" strain rate in equation [3.2] become for both phases:

$$\begin{aligned} \underset{\approx}{m} &= \frac{1}{2\eta}\underset{\approx}{K} \\ \dot{\underset{\sim}{\varepsilon}}^{v0} &= \underset{\sim}{0} \end{aligned} \quad [3.24]$$

where $\underset{\approx}{m}$ is the classic secant compliance tensor. In the case of the AFF linearization, we have:

$$\begin{aligned} \underset{\approx}{m} &= \frac{1}{2\eta}\left(\underset{\approx}{K} + \frac{3}{2}(n-1)\frac{\underset{\sim}{\sigma}'}{\sigma_{eq}} \otimes \frac{\underset{\sim}{\sigma}'}{\sigma_{eq}}\right) \\ \dot{\underset{\sim}{\varepsilon}}^{v0} &= \frac{\partial \Phi^{vp}}{\partial \underset{\sim}{\sigma}'} - \underset{\approx}{m} : \underset{\sim}{\sigma} \end{aligned} \quad [3.25]$$

where $\underset{\approx}{m}$ is the tangent compliance tensor (see, for example, Gilormini 1995). For these two methods, $\underset{\approx}{m}$ depends on the first-order moment of deviatoric stress field with an effective stress given for both phases by $\overline{\sigma_{eq}}$ as defined in equation [3.23]. In contrast, the modified SEC MODIF linearization is based on the estimate of the "effective stress" in the matrix phase σ_{eq}^{M} based on the second-order moment of the deviatoric stress field:

$$\sigma_{eq}^{M} = \overline{\overline{\sigma_{eq}}}^{M} = \left(\frac{3}{2}\bar{\bar{S}}^{M}\right)^{1/2} \quad [3.26]$$

where $\bar{\bar{S}}^{M} = \overline{\underset{\sim}{\sigma}' : \underset{\sim}{\sigma}'}^{M}$. In order to evaluate the time evolution of $\overline{\overline{\sigma_{eq}}}^{M}$, a direct method is used (Li and Weng 1997; Masson et al. 2020). The method is based on the application of the Hill's lemma (Hill 1967), $\overline{\underset{\sim}{\sigma} : \underset{\sim}{\dot{\varepsilon}}} = \overline{\underset{\sim}{\sigma}} : \overline{\underset{\sim}{\dot{\varepsilon}}}$. Disregarding the second-order moment of the hydrostatic stress field in the matrix phase (Li and Weng 1997; Masson et al. 2020), straightforward algebraic manipulations lead to the following time evolution equation for $\bar{\bar{S}}^M$:

$$\dot{\bar{\bar{S}}}^M + \frac{2}{\tau^{(M)}} \bar{\bar{S}}^M = 4\mu^{(M)} \frac{X}{1-f} \quad [3.27]$$

where $\tau^{(M)} = \frac{\eta^{(M)}}{\mu^{(M)}}$ is the relaxation time of the matrix phase, and X is given by:

$$\begin{aligned} X = \overline{\underset{\sim}{\sigma}} : \overline{\underset{\sim}{\dot{\varepsilon}}} - f\left(\frac{1}{2\mu^{(I)}} \dot{\overline{\underset{\sim}{\sigma}'}}^{I} : \overline{\underset{\sim}{\sigma}'}^{(I)} + \frac{1}{2\eta^{(I)}} \overline{\underset{\sim}{\sigma}'}^{(I)} : \overline{\underset{\sim}{\sigma}'}^{(I)}\right) \\ - \sum_{r=I}^{r=M} f^{(r)} \frac{1}{9k^{(r)}} \overline{\dot{\sigma}_{kk}}^{(r)} \overline{\sigma_{kk}}^{(r)} \end{aligned} \quad [3.28]$$

with σ_{kk} being the trace of $\underset{\sim}{\sigma}$. At t = 0, it is assumed that $\underset{\sim}{\sigma}'^{(I)} = \underset{\approx}{0}$ and $\bar{\bar{S}}^M = 0$. Then, $\bar{\bar{S}}^M$ can be calculated incrementally using equation [3.27] together with equation [3.28] to derive the modified secant compliance tensor of the matrix phase $\underset{\approx}{m}(\overline{\overline{\sigma_{eq}}}^M)$ as well as the "effective secant viscosity" $\eta(\overline{\overline{\sigma_{eq}}}^M)$ in equations [3.24] and [3.22], respectively.

3.2.4. *TF method for elasto-viscoplastic polycrystals using the SC scheme*

The TF method is applied to the SC scheme for polycrystals. The detailed mathematical derivations of the method with the SC scheme can be found in Paquin et al. (1999) and Mareau and Berbenni (2015). In Paquin et al. (1999), a classical "secant" linearization procedure ("TF-SEC") was first developed, whereas an "affine" linearization procedure ("TF-AFF") was developed in Mareau and Berbenni (2015). Many applications of the TF method to polycrystalline alloys (steels, Ti alloys, etc.) were reported in the two last decades (Paquin et al. 2001; Sabar et al. 2002; Berbenni et al. 2004, 2007; Mareau et al. 2009; Nicaise et al. 2011; Mareau and Berbenni 2015; Robert and Mareau 2015; Terentyev et al. 2015; Amouzou et al. 2016; Lhadi et al. 2018). In the case of slip rates on slip systems following a power law in face-centered cubic (f.c.c.) metals, the "secant" and "affine" linearization procedures are reported in section 3.4.1. Following the SC scheme based on the "TF-AFF" method (labeled as "SC-TF-AFF"), the strain rate equation for each "grain" (i.e. each crystal orientation) of the polycrystal is given by Mareau and Berbenni (2015) and Lhadi et al. (2018):

$$\begin{aligned}
\dot{\underset{\sim}{\varepsilon}} &= \underset{\approx}{A}^{C^e} : \dot{\underset{\sim}{E}} + \underset{\approx}{A}^{C^e} : \underset{\approx}{\Gamma_l}^{B^e} : \left(\underset{\approx}{b} : \dot{\underset{\sim}{\varepsilon}}^{v0} - \underset{\approx}{B}^e : \dot{\underset{\sim}{E}}^{v0e} \right) \\
&+ \underset{\approx}{A}^{C^e} : \underset{\approx}{\Gamma_l}^{C^e} : \left(\underset{\approx}{\delta c} : \dot{\underset{\sim}{\varepsilon}}^{vp} + \underset{\approx}{c} : \underset{\approx}{A}^{C^e} : \left(\overline{\dot{\underset{\sim}{\varepsilon}}^{vp}} - \dot{\underset{\sim}{E}}^{vpe} \right) \right) \\
&- \underset{\approx}{A}^{C^e} : \underset{\approx}{\Gamma_l}^{B^e} : \left(\underset{\approx}{\delta b} : \dot{\underset{\sim}{\varepsilon}}^{vp} - \underset{\approx}{b} : \underset{\approx}{A}^{B^e} : \left(\underset{\approx}{M}^e : \underset{\sim}{\Sigma} + \dot{\underset{\sim}{E}}^{v0e} - \overline{\dot{\underset{\sim}{\varepsilon}}^{vp}} \right) \right) \\
&+ \underset{\approx}{A}^{C^e} : \left(\underset{\approx}{\Gamma_l}^{C^e} : \underset{\approx}{C}^e - \underset{\approx}{\Gamma_l}^{B^e} : \underset{\approx}{B}^e \right) : \left(\dot{\underset{\sim}{\varepsilon}}^{vp} - \underset{\approx}{A}^{B^e} : \overline{\dot{\underset{\sim}{\varepsilon}}^{vp}} \right) \\
&- \underset{\approx}{A}^{C^e} : \left(\underset{\approx}{\Gamma_l}^{C^e} : \underset{\approx}{C}^e - \underset{\approx}{\Gamma_l}^{B^e} : \underset{\approx}{B}^e \right) : \left(\underset{\approx}{A}^{B^e} : \underset{\approx}{\Gamma_l}^{B^e} : \left(\underset{\approx}{b} : \dot{\underset{\sim}{\varepsilon}}^{v0} - \underset{\approx}{B}^e : \dot{\underset{\sim}{E}}^{v0e} \right) \right)
\end{aligned} \qquad [3.29]$$

In this equation, $\dot{\underset{\sim}{E}}^{vpe}$ is given by Mareau and Berbenni (2015) and (Lhadi et al. 2018):

$$\begin{aligned}
\dot{\underset{\sim}{E}}^{vpe} &= \underset{\approx}{Z} : \overline{\underset{\approx}{c} : \dot{\underset{\sim}{\varepsilon}}^{vp}} - \underset{\approx}{Z} : \overline{\underset{\approx}{c} : \underset{\approx}{A}^{C^e} : \underset{\approx}{\Gamma_l}^{B^e} : \left(\underset{\approx}{b} : \dot{\underset{\sim}{\varepsilon}}^{v0} - \underset{\approx}{B}^e : \dot{\underset{\sim}{E}}^{v0e} \right)} \\
&- \underset{\approx}{Z} : \overline{\underset{\approx}{c} : \underset{\approx}{A}^{C^e} : \underset{\approx}{\Gamma_l}^{C^e} : \left(\underset{\approx}{\delta c} : \dot{\underset{\sim}{\varepsilon}}^{vp} + \underset{\approx}{c} : \underset{\approx}{A}^{C^e} : \overline{\dot{\underset{\sim}{\varepsilon}}^{vp}} \right)} \\
&+ \underset{\approx}{Z} : \overline{\underset{\approx}{c} : \underset{\approx}{A}^{C^e} : \underset{\approx}{\Gamma_l}^{B^e} : \left(\underset{\approx}{\delta b} : \dot{\underset{\sim}{\varepsilon}}^{vp} - \underset{\approx}{b} : \underset{\approx}{A}^{B^e} : \left(\underset{\approx}{M}^e : \underset{\sim}{\Sigma} + \dot{\underset{\sim}{E}}^{v0e} - \overline{\dot{\underset{\sim}{\varepsilon}}^{vp}} \right) \right)} \\
&- \underset{\approx}{Z} : \overline{\underset{\approx}{c} : \underset{\approx}{A}^{C^e} : \left(\underset{\approx}{\Gamma_l}^{C^e} : \underset{\approx}{C}^e - \underset{\approx}{\Gamma_l}^{B^e} : \underset{\approx}{B}^e \right) : \left(\dot{\underset{\sim}{\varepsilon}}^{vp} - \underset{\approx}{A}^{B^e} : \overline{\dot{\underset{\sim}{\varepsilon}}^{vp}} \right)} \\
&+ \underset{\approx}{Z} : \overline{\underset{\approx}{c} : \underset{\approx}{A}^{C^e} : \left(\underset{\approx}{\Gamma_l}^{C^e} : \underset{\approx}{C}^e - \underset{\approx}{\Gamma_l}^{B^e} : \underset{\approx}{B}^e \right) : \left(\underset{\approx}{A}^{B^e} : \underset{\approx}{\Gamma_l}^{B^e} : \left(\underset{\approx}{b} : \dot{\underset{\sim}{\varepsilon}}^{v0} - \underset{\approx}{B}^e : \dot{\underset{\sim}{E}}^{v0e} \right) \right)}
\end{aligned} \qquad [3.30]$$

with $\underset{\approx}{Z}$ given by:

$$\underset{\approx}{Z} = \left(\underset{\approx}{I} - \underset{\approx}{S}^e : \overline{\underset{\approx}{c} : \underset{\approx}{A}^{C^e} : \underset{\approx}{\Gamma_l}^{C^e} : \underset{\approx}{c} : \underset{\approx}{A}^{C^e}} \right)^{-1} : \underset{\approx}{S}^e \qquad [3.31]$$

In both "SC-TF-SEC" and "SC-TF-AFF" methods, the effective elastic (respectively, viscoplastic) moduli $\underset{\approx}{C}^e$ (respectively, $\underset{\approx}{B}^e$) are considered as homogeneous properties to compute the purely elastic (respectively, viscoplastic) strain rate concentration tensors $\underset{\approx}{A}^{C^e}$ (respectively, $\underset{\approx}{A}^{B^e}$) and the tensors $\underset{\approx}{\Gamma_l}^{C^e}$ (respectively, $\underset{\approx}{\Gamma_l}^{B^e}$), using equations [3.11] and [3.13]. The elastic (respectively, viscoplastic) Eshelby tensor is given by $\underset{\approx}{S}^{Esh^{C^e}} = \underset{\approx}{\Gamma_l}^{C^e} : \underset{\approx}{C}^e$ (respectively, $\underset{\approx}{S}^{Esh^{B^e}} = \underset{\approx}{\Gamma_l}^{B^e} : \underset{\approx}{B}^e$). An iterative scheme using a fixed-point algorithm is used to compute $\underset{\approx}{C}^e$ (respectively, $\underset{\approx}{B}^e$) using:

$$\begin{aligned} \underset{\approx}{C}^e &= \overline{\underset{\approx}{c} : \underset{\approx}{A}^{C^e}} : \left(\overline{\underset{\approx}{A}^{C^e}}\right)^{-1} \\ \underset{\approx}{B}^e &= \overline{\underset{\approx}{b} : \underset{\approx}{A}^{B^e}} : \left(\overline{\underset{\approx}{A}^{B^e}}\right)^{-1} \end{aligned} \quad [3.32]$$

In the case of the "SC-TF-AFF" method, $\dot{\underset{\sim}{E}}^{v0e}$ in equations [3.29] and [3.30] is given by Mareau and Berbenni (2015) and Lhadi et al. (2018):

$$\dot{\underset{\sim}{E}}^{v0e} = \underset{\approx}{M}^e : \left(\overline{\underset{\approx}{b} : \underset{\approx}{A}^{B^e} : \underset{\approx}{\Gamma_l}^{B^e}} - \underset{\approx}{I}\right)^{-1} : \overline{\underset{\approx}{b} : \underset{\approx}{A}^{B^e} : \underset{\approx}{\Gamma_l}^{B^e} : \underset{\approx}{b} : \dot{\underset{\sim}{\varepsilon}}^{v0}} - \overline{\underset{\approx}{b} : \dot{\underset{\sim}{\varepsilon}}^{v0}} \quad [3.33]$$

3.3. The β-model dynasty for multiphase elastoviscoplastic polycrystals

This section is dedicated to a class of simplified homogenization models that aim at an efficient evaluation of the effective behavior of multiphase polycrystals with a view to application to structural computations. A shortcoming of homogenization models is the often prohibitive computation cost of their implementation at Gauss points of an FE mesh in a structural computation. The proposed efficient approach is based on the formulation of explicit scale transition rules providing the mean stress values in each phase from current values of a series of internal variables. Its origin goes back to the contributions by Cailletaud (1992) and Pilvin (1996) who proposed the following form of explicit scale transition rule for multiphase materials:

$$\underset{\sim}{\sigma}^{(r)} = \underset{\sim}{\overline{\sigma}} + \underset{\approx}{C}^{eff} : (\underset{\approx}{I} - \underset{\approx}{S}^{Esh}) : (\underset{\sim}{\beta} - \underset{\sim}{\beta}^{(r)}) \quad [3.34]$$

where $\underset{\sim}{\sigma}^{(r)}$ is the mean stress in phase number r, $\underset{\sim}{\overline{\sigma}}$ is the macroscopic stress, that is, the average over all phases and $\underset{\approx}{C}^{eff}$ is the effective tensor of elastic moduli. The Eshelby tensor $\underset{\approx}{S}^{Esh}$ is computed from the effective elastic moduli and the ellipsoidal inclusion shape retained to model each phase. The key variable is the $\underset{\sim}{\beta}^{(r)}$ tensor variable attributed to each phase, $\underset{\sim}{\beta}$ being the associated average. The β-variables are subjected to the following evolution equation:

$$\dot{\underset{\sim}{\beta}}^{(r)} = \dot{\underset{\sim}{\varepsilon}}^{p(r)} - \underset{\approx}{D} : \underset{\sim}{\beta}^{(r)} \|\dot{\underset{\sim}{\varepsilon}}^{p(r)}\| \quad [3.35]$$

This evolution equation is inspired from the concept of kinematic hardening well known in phenomenological equations of plasticity (Besson et al. 2009). It involves the plastic strain rate of the considered phase and its norm, for instance, chosen as the usual J_2 norm. It contains a fourth-order tensor $\underset{\approx}{D}$ of parameters to be determined depending on the homogenization scheme retained in the analysis. Note that if $\underset{\approx}{D}$ vanishes, the β-variables coincide with the plastic strain in each phase, thus retrieving the celebrated Kröner scale transition law. The relation [3.34] is valid in the case of homogeneous elastic behavior of the phases. It was initially applied to polycrystals with isotropic elasticity. Originally, a scalar parameter D was calibrated from the response of Berveiller and Zaoui's SC model suited for isotropic overall behavior (Berveiller and Zaoui 1979). The calibration can also be performed from more realistic predictions by means of FE (or FFT) analysis of representative volume elements like polycrystalline aggregates following (Hlilou et al. 2009). Similar comparisons between affine homogenization models and FE analyses were recently reported (Pilvin et al. 2017). The calibration is performed by comparing the overall stress–strain curves of the reference and β-models, but also the mean stress/mean strain curves for each considered crystal orientation.

The calibrated β-model can then be used to predict the polycrystal responses under complex loading conditions as done in Cailletaud and Sai (2008). In particular, this is an efficient computational tool for the prediction of macroscopic yield surface distorsion during various strain paths (Cailletaud 2009).

The β-models are computationally sufficiently efficient to be used at the integration point level in FE structural analyses. The reason lies in the explicit relation [3.34] that avoids any complex integration scheme in addition to the usual time-integration of internal variables in each phase. This possibility of performing multiscale simulations was demonstrated by Cailletaud and Pilvin (1994), where 2D simulations of a turbine disk and 3D simulations of a notched specimen were documented. Other authors have also applied polycrystal homogenization schemes to structural computations and simulations of forming processes such as Segurado et al. (2012) and Soho et al. (2014). Such attempts remain however seldom, even though these models make it possible to efficiently and realistically handle anisotropic behavior, complex distorsion of yield surfaces and sophisticated hardening rules. A reduced β-model is used by Rousselier and Leclercq (2006) to simulate anistropic plasticity and fracture of metal sheets by limiting to a minimum the number of considered crystallographic orientations (see also Luo and Rousselier 2014).

One of the most recent accounts of the genealogy of β-models can be found in the book chapter by Cailletaud and Coudon (2016). This strategy of model simplification for structural applications experiences today a renewal in the form of alternative strategies, such as reduced order models, and machine learning for the development of metamodels. The advantage of the class of β-models is that they already incorporate some knowledge from homogenization theory and from constitutive modeling of viscoplasticity. They should therefore be combined with modern reduced order modeling techniques and machine learning approaches, which require many evaluations of the constitutive model.

The application of β-models is not limited to polycrystals. The case of two-phase single crystal nickel-based superalloys was explored by Forest and Pilvin (1996). Extension to elastomeric composites was proposed by Parenteau et al. (2014).

An important application is the modeling of columnar grain microstructures, which are encountered in directionally solidified turbine blades for jet engines. All the grains share a common axial orientation but the transverse orientation is random. Simplified models are needed in engineering applications for the design of such turbine blades made up of nickel-based superalloys. A first attempt was put forward by Sai et al. (2006) and then further developed by Martin et al. (2014) and Coudon et al. (2019). Anisotropic thermo-elasto-visco-plasticity of individual grains is taken into account. Comparisons between the results of stochastic full-field simulations of the blades containing individual columnar grains and β-model predictions were provided. Also, Martin et al. (2014) computed a 3D notched specimen of DS material with the β-model. The authors then compared the model predictions of Hill's incremental SC model, TFs and β-models.

The paradigm of β-models is now illustrated more thoroughly in the case of elastoviscoplastic FCC polycrystals, following the work of Gu et al. (2019). Note that these authors also envisage the existence of residual stresses, which are not reported here. Let us start by setting the notations for the constitutive relations in each phase:

$$\underset{\sim}{\varepsilon}^{(r)} = \underset{\sim}{\varepsilon}^{e(r)} + \underset{\sim}{\varepsilon}^{p(r)}, \quad \underset{\sim}{\sigma}^{(r)} = \underset{\approx}{C}^{(r)} : \underset{\sim}{\varepsilon}^{e(r)} \qquad [3.36]$$

and at the macroscopic level:

$$\underset{\sim}{\overline{\sigma}} = \sum_r f_r \underset{\sim}{\sigma}^{(r)} = \underset{\approx}{C}^{eff} : \underset{\sim}{\varepsilon}^{e} \qquad [3.37]$$

with $\underset{\sim}{\overline{\varepsilon}} = \underset{\sim}{\varepsilon}^{e} + \underset{\sim}{\varepsilon}^{p}$. Note that the effective elastic and plastic strain tensors $\underset{\sim}{\varepsilon}^{e}$ and $\underset{\sim}{\varepsilon}^{p}$ are generally not equal to the averaged values of the local quantities. This is however the case when elasticity is homogeneous among the grains (isotropic case or epitaxy). In the case of inhomogeneous and anisotropic elasticity in the grains, the scale transition rule [3.34] must be replaced by

$$\underset{\sim}{\sigma}^{(r)} = \underset{\approx}{B}^{(r)} : \left(\underset{\sim}{\overline{\sigma}} + \underset{\approx}{C}^{eff} : (\underset{\approx}{I} - \underset{\approx}{S}^{Esh}) : (\underset{\sim}{\beta} - \underset{\sim}{\beta}^{(r)}) \right) \qquad [3.38]$$

where the concentration tensor is given by

$$\underset{\approx}{B}^{(r)} = \underset{\approx}{C}^{(r)} : \left(\underset{\approx}{I} + \underset{\approx}{S}^{\mathrm{Esh}} : \underset{\approx}{C}^{eff-1} : (\underset{\approx}{C}^{(r)} - \underset{\approx}{C}^{eff})\right)^{-1} : \underset{\approx}{C}^{eff-1} \qquad [3.39]$$

with:

$$< \underset{\approx}{B}^{(r)} >= \sum_r f^r \underset{\approx}{B}^{(c)} = \underset{\approx}{I}$$

The volume fraction of each crystallographic texture component r is denoted by f_r. Note that these tensors depend only on elasticity tensors and are therefore evaluated only once, thus saving a considerable amount of time. All nonlinear accommodation effects are carried by the nonlinear evolution of the β-variables. In this context, the definition of the overall variable $\underset{\sim}{\beta}$ in [3.38] must be amended as follows:

$$\underset{\sim}{\beta} = \sum_r f_r \underset{\approx}{L}^{*-1}{}_C : \underset{\approx}{B}^{(r)} : \underset{\approx}{L}^{*}{}_C : \underset{\sim}{\beta}^{(r)} \qquad [3.40]$$

with $\underset{\approx}{L}^{*}{}_C = \underset{\approx}{C}^{eff} : (\underset{\approx}{I} - \underset{\approx}{S}_{esh})$. The previous relation is obtained from the condition that the average value of the stress $\underset{\sim}{\sigma}^{(r)}$ must coincide with the macroscopic stress.

The macroscopic plastic strain $\underset{\sim}{\varepsilon}^p$ is obtained after substitution of the local and overall elastic laws into the scale transition rule:

$$\underset{\sim}{\varepsilon}^p = \overline{\underset{\sim}{\varepsilon}^p} + \sum_r f_r \underset{\approx}{C}^{(r)-1} : \underset{\approx}{B}^{(r)} : \underset{\approx}{L}^{*} : (\underset{\sim}{\beta} - \underset{\sim}{\beta}^{(r)}) \qquad [3.41]$$

See Gu et al. (2019) for the derivation. At this stage, any constitutive law is allowed to compute the plastic strain rate in each phase from the computed stress tensor in this grain. In particular, viscoplastic flow rules can be considered as extensions of the Schmid law involving the resolved shear stresses and suitable hardening variables. Dislocation densities can be used and their evolution rules can be chosen independently of the scale transition rules. Thermal and residual strains can be added in a similar way.

Validation of the β-models, regarded as reduced order models, must take place at two levels. First, the calibrated-scale transition rule must be verified after calibration by testing new loading conditions by performing the corresponding reference simulations on representative volume elements or using more sophisticated homogenization models like those presented in this chapter. Second, model predictions must be compared to experimental results at the macroscopic scale (overall stress–strain curves) but also at the phase level. For instance, Gu et al. (2019) performed neutron diffraction experiments to determine the average elastic strain in copper and niobium phases of two-phase polycrystals. These values were successfully compared with the β-model predictions, provided that residual stresses induced by the material processing are taken into account. Developments of the β-models could be oriented toward predictions strain heterogeneities inside the phases following homogenization schemes proposed for this purpose, like second-order approaches.

3.4. Applications of the analytical micromechanical methods

3.4.1. *Applications of the TF method to composites and polycrystals*

As numerical applications of the TF method for two-phase composites, the inhomogeneity phase I is constituted of long cylindrical fibers aligned in the (x_3)-direction and perpendicular to the (x_1, x_2) plane. Materials parameters are reported in Table 3.1. "Hard fibers" are simulated when $\sigma_0^{(I)} = 5\sigma_0^{(M)}$ and "soft fibers" are simulated when $\sigma_0^{(I)} = 0.2\sigma_0^{(M)}$. The elastic properties are homogeneous and given in Table 3.1 (E is elastic Young's modulus E and ν is the elastic Poisson ratio). In the present calculations, the inhomogeneities are described for the TF method by axisymmetric ellipsoidal inclusions elongated in the principal fiber direction with aspect ratio $c/a = 1/100$ (a is the semi-axis length of the fibers along the (x_3)-direction and c is their radius in the (x_1, x_2) plane). The applied strain rate $\dot{\underset{\sim}{E}}$ to the RVE is isochoric and corresponds to in-plane shear loading at 45° with respect to the (x_1) axis: $\dot{\underset{\sim}{E}} = \dot{E}_{11} \begin{pmatrix} 1 & 0 & 0 \\ 0 & -1 & 0 \\ 0 & 0 & 0 \end{pmatrix}$ with $\dot{E}_{11} = 0.5\text{s}^{-1}$. The following results report the evolution of overall stress Σ_{11} as a function of the macroscopic applied strain E_{11}, and the evolutions of mean phase stresses $\sigma_{11}^{(r=I,M)}$ (i.e. the first order moment of stresses) as functions of mean phase strains $\varepsilon_{11}^{(r=I,M)}$ (i.e. the first-order moment of strains). The same materials parameters and microstructures were studied in Lahellec and Suquet (2007) using a time incremental variational method with a "Composite Cylinder Assemblage" (CCA) microstructure and Hashin–Shtrikman (HS) estimates. Based on a time-incremental variational method, these authors developed a "SEC (EIV+HS)" estimate, which relies on an incremental effective potential with effective viscoplastic strain rates as internal variables (i.e. "EIV" for "Effective Internal Variable") using a "modified secant" approach to derive the "effective secant viscosity". The full-field FFT results reported in Lahellec and Suquet (2007) are used as reference solutions for these fiber-reinforced or weakened composites with respective fiber volume fractions of $f = 0.21$ and $f = 0.41$.

E (GPa)	ν	$\dot{\varepsilon}_0(s^{-1})$	$\sigma_0^{(M)}$ (GPa)	$\sigma_0^{(I)}$ (GPa)	m
100	0.45	1	1	5 (Hard) or 0.2 (Soft)	0.1 to 1

Table 3.1. *Material parameters for simulations of two-phase fiber-reinforced or weakened elasto-viscoplastic composites*

First of all, the numerical results for the effective behavior with "hard fibers" and the phase responses given by the "TF-AFF" and the "TF-MODIF SEC" are reported in Figure 3.1 and are compared to the FFT result (Lahellec and Suquet 2007). For $f = 0.21$, the predicted responses from the "TF-AFF" and "TF-MODIF SEC" methods are reported for $m = 1$ (linear viscosity) (Figure 3.1(a)) and for $m = 0.2$ (nonlinear viscosity) (Figure 3.1(b)). In the linear case, it is seen that the TF method (both "TF-AFF" and "TF-MODIF SEC" give same per phase and effective responses) well predicts the effective behavior of the composite. The overall responses are superimposed with the FFT result. In the nonlinear case, some differences occur in comparison with the FFT response in the transition between elastic and viscoplastic states and at large times when the overall behavior tends to the viscoplastic asymptotic state. Compared to the "TF-AFF" prediction, the estimate given by the "TF-MODIF SEC" method is seen to be the closest to the full-field FFT result. The "TF-AFF", which is based on first-order moment of stress, overestimates the effective behavior of the composite. The effective response given by the "TF-MODIF SEC" is much softer than the TF-AFF's response and is close to the FFT result, even at large times. For $f = 0.41$, the predicted responses from the "TF-AFF" and "TF-MODIF SEC" methods are reported for $m = 1$ (linear viscosity) (Figure 3.1(c)) and for $m = 0.2$ (nonlinear viscosity) (Figure 3.1(d)). In the linear case, it is seen that the MT estimate (which also provides a lower bound) becomes less efficient for this finite volume fraction of fibers due to the influence of spatial interactions between fibers. Therefore, the asymptotic state at large times is underestimated by the TF method due to the MT scheme. In the nonlinear case, the "TF-AFF" overestimates the effective behavior of the composite for the same reason as explained before and the difference with the FFT result is even increased due to a higher volume fraction f. In contrast, the "TF-MODIF SEC" estimate remains relatively close to the FFT, in spite of a compensation of errors due to the MT prediction. From the phase responses in the nonlinear case reported in Figures 3.1(b) and (d) for the matrix phase (dashed lines) and for fibers (dotted lines), we first notice that the phase stress heterogeneity is always larger in the case of the "TF-MODIF SEC" estimate compared to the "TF-AFF" estimate. Second, the matrix response is softer in the case of the "TF-MODIF SEC" estimate compared to the "TF-AFF" estimate. These differences are mainly due to the incorporation of "effective secant viscosity" based on the second-order moment of the stress field in the "TF-MODIF SEC".

Regarding the case of "soft fibers", the effective behavior and the phase responses given by the "TF-AFF" and the "TF-MODIF SEC" are reported in Figure 3.2 and compared to the FFT result (Lahellec and Suquet 2007). The results show that for $m = 1$ (linear viscosity) and $f = 0.21$, the FFT result is well reproduced (Figure 3.2(a)). However, for $f = 0.41$ (Figure 3.2(c)), the effective response given by the TF method is overestimated due to the MT scheme, which is less adapted for $f = 0.41$ and provides an upper bound in this case. For $m = 0.2$ (nonlinear viscosity), it is seen that compared to the "TF-AFF" estimate, the "TF-MODIF SEC" estimate gives better predictions for both $f = 0.21$ (Figure 3.2(b)) and $f = 0.41$ (Figure 3.2(b)). However, the overall stress is overestimated due to three additional errors: first, the MT scheme gives an upper bound (which is observable for $f = 0.41$); second, intra-phase field fluctuations are not fully estimated even with the "TF-MODIF SEC" method. From the phase responses in the nonlinear case reported in Figures 3.2(b) and 3.2(d) for the matrix phase (dashed lines) and for fibers (dotted lines), it is observed that the matrix response is still softer in the case of the "TF-MODIF SEC" estimate compared to the "TF-AFF" estimate for both volume fractions.

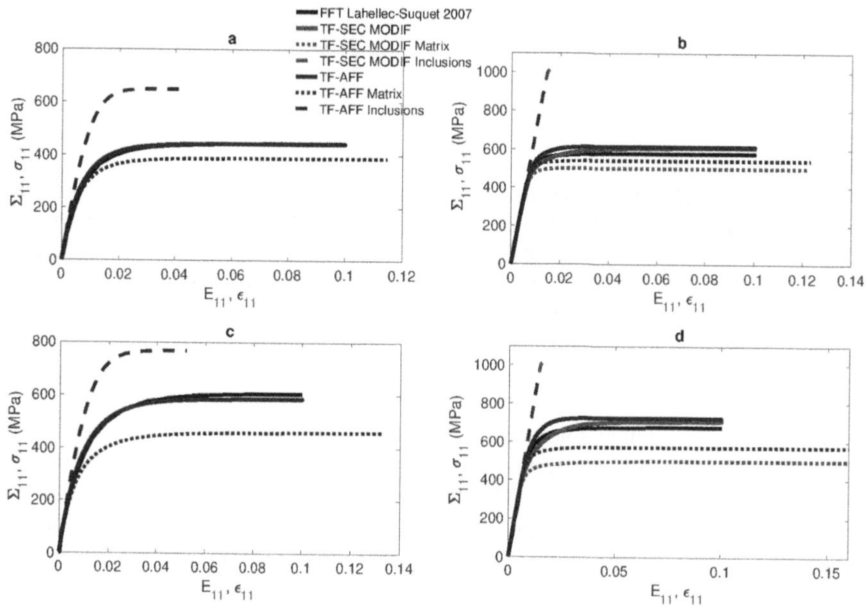

Figure 3.1. *Overall and phase average stress responses of two-phase linear and nonlinear elasto-viscoplastic fiber-reinforced composites (case referred to in the text as "hard fibers") subjected to in-plane shear loading perpendicular to the fibers for different fiber volume fractions and different viscous nonlinearities: (a)* $f = 0.21$ *and* $m = 1/n = 1$*, (b)* $f = 0.21$ *and* $m = 1/n = 0.2$*, (c)* $f = 0.41$ *and* $m = 1/n = 1$*, and (d)* $f = 0.41$ *and* $m = 1/n = 0.2$

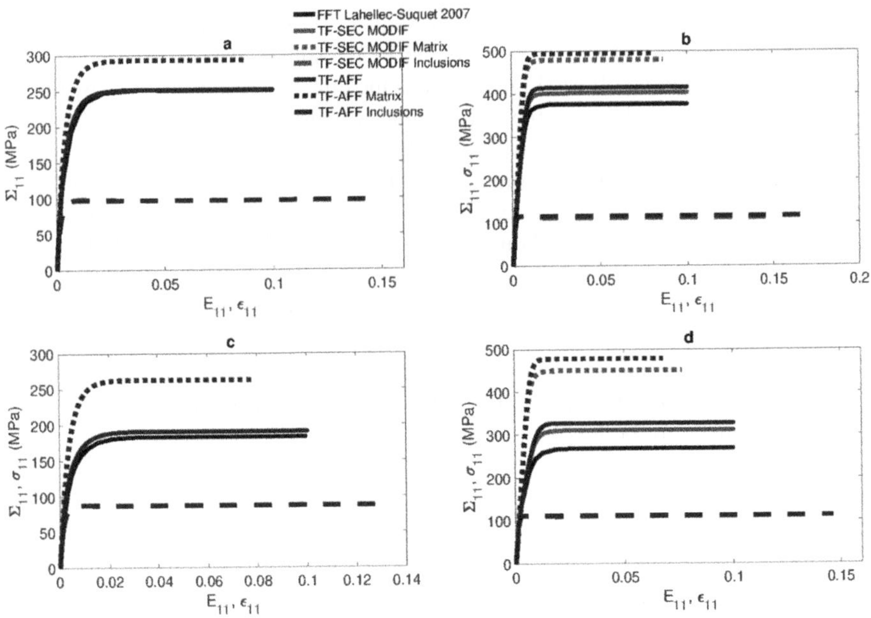

Figure 3.2. *Overall and phase average stress responses of two-phase linear and nonlinear elasto-viscoplastic fiber-weakened composites (case referred to in the text as "soft fibers") subjected to in-plane shear loading perpendicular to the fibers for different fiber volume fractions and different viscous nonlinearities: (a)* $f = 0.21$ *and* $m = 1/n = 1$*, (b)* $f = 0.21$ *and* $m = 1/n = 0.2$*, (c)* $f = 0.41$ *and* $m = 1/n = 1$*, and (d)* $f = 0.41$ *and* $m = 1/n = 0.2$

In addition to comparisons with full-field FFT, it is interesting to report the effective behavior predicted by the "TF-AFF" and "TF-MODIF SEC" estimates in comparisons with the classical "TF-SEC" estimate (Paquin et al. 1999) and the modified "SEC (EIV+HS)" estimate (Lahellec and Suquet 2007). Figures 3.3 and 3.4 report such comparisons for "hard fibers" and "soft fibers", respectively. In Figure 3.3, it is seen that the classical "TF-SEC" estimate (pink solid line) gives a too stiff response, especially at larger times. The "TF-AFF" estimate (blue solid line) is better than the "TF-SEC" one, especially at larger times where it approaches the "TF-MODIF SEC" estimate (red solid line). The best estimate is the one given by the new "TF-MODIF SEC" method, for which the asymptotic state is superimposed with the "SEC (EIV+HS)" (green solid line). In Figure 3.4, an even larger improvement is obtained with the second-order moment stress based "TF-MODIF SEC" method (red solid line) to predict the overall response of the two-phase composite with "soft fibers" compared to first-order moment stress based methods: "TF-SEC" (pink solid line) and "TF-AFF" (blue solid line).

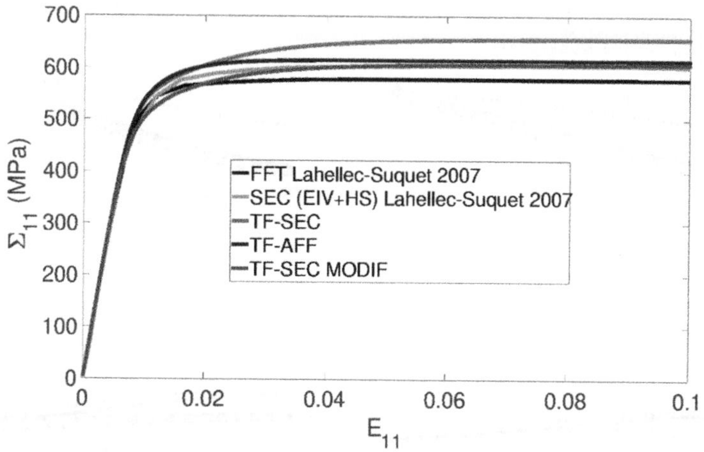

Figure 3.3. *Simulation of the effective behavior of a two-phase nonlinear elasto-viscoplastic fiber-reinforced composite ("hard fibers" with* $f = 0.21$ *and* $m = 0.2$*) with different TF estimates: "TF-SEC", "TF-AFF" and "TF-MODIF SEC" and comparisons with full-field FFT and the "SEC (EIV+HS)" variational method (Lahellec and Suquet 2007)*

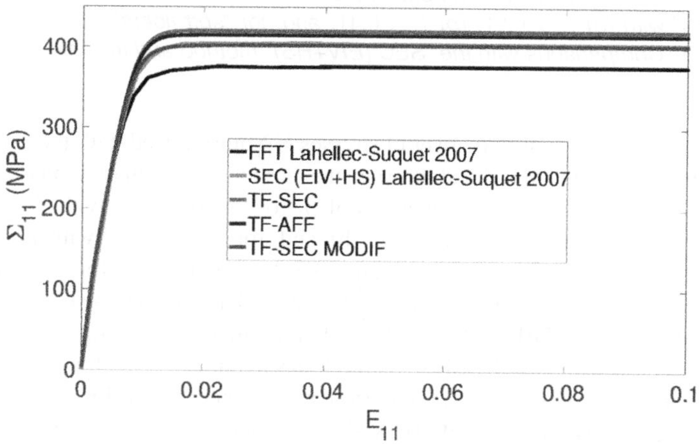

Figure 3.4. *Simulation of the effective behavior of a two-phase nonlinear elasto-viscoplastic fiber-weakened composite ("soft fibers" with* $f = 0.21$ *and* $m = 0.2$*) with different TF estimates: "TF-SEC", "TF-AFF" and "TF-MODIF SEC" and comparisons with full-field FFT and the "SEC (EIV+HS)" variational method (Lahellec and Suquet 2007)*

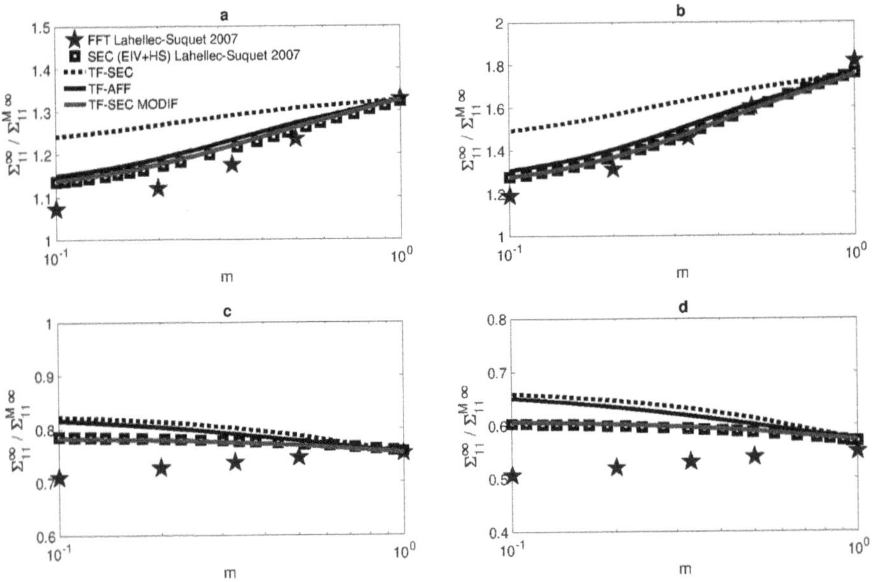

Figure 3.5. *The estimates of the overall asymptotic stresses Σ_{11}^{∞}, given by "TF-SEC", "TF-AFF" and "TF-MODIF SEC" methods, normalized with the asymptotic stresses obtained in the matrix alone (without fibers) $\Sigma_{11}^{M\infty}$ plotted as functions of m: for "hard fibers" with (a) $f = 0.21$, (b) $f = 0.41$, and, for "soft fibers" with (c) $f = 0.21$, (d) $f = 0.41$. Comparisons with the "SEC (EIV+HS)" method and full-field FFT*

In Figure 3.5, the asymptotic overall stresses Σ_{11}^{∞} normalized with the asymptotic stresses obtained in the matrix alone (i.e. the homogeneous matrix phase without fiber) and $\Sigma_{11}^{M\infty}$ are reported as functions of m for "hard fibers" with $f = 0.21$ in Figure 3.5(a), with $f = 0.41$ in Figure 3.5(b), and, for "soft fibers" with $f = 0.21$ in Figure 3.5(c), and with $f = 0.41$ in Figure 3.5(d). For fiber-reinforced composites ("hard fibers"), the estimates of $\Sigma_{11}^{\infty}/\Sigma_{11}^{M\infty}$ shows that the "TF-AFF" (black solid line) and the "TF-MODIF SEC" (red solid line) methods drastically reduce the stiffness observed with the "TF-SEC" method (dashed lines). Even for high nonlinear viscoplasticity linearity ($m = 0.1$), the error with respect to the FFT is not so high and the same results as the "SEC (EIV+HS)" (black squares) are found with the "TF-MODIF SEC" method. For fiber-weakened composites ("soft fibers"), the differences between the "TF-AFF" (black solid line) and the "TF-SEC" (dashed lines) approaches are not too high for all strain-rate sensitivities. The advantage of using the "TF-MODIF SEC" (red solid line) method over the "TF-AFF" method for these fiber-weakened composites is proved in Figures 3.5(c) and 3.5(d) where the discrepancies with the FFT results (blue diamonds) are decreased for $m = 0.1$. However, the present "TF-MODIF SEC" estimate (as the "SEC (EIV+HS)" (black squares)) still tends to overestimate the effective behavior of fiber-weakened composites with respect to exact results.

In the case of polycrystalline materials, the local behavior of each crystal, described by $\underset{\approx}{c}$, $\underset{\approx}{b}$ and $\underset{\sim}{\dot{\varepsilon}}^{v0}$, depends on the crystallographic orientation distributions because of the anisotropy of elastic and viscoplastic properties. Therefore, the overall behavior is dependent on the inter-granular stresses generated during a given loading path due to the mechanical interactions between the different grains (or crystals). Here, for the application of both "SC-TF-SEC" and "SC-TF-AFF" methods, the local viscoplastic behavior is described considering f.c.c. crystals with a power law. The plastic flow rule is established for an f.c.c. crystal with slip systems belonging to the $\{111\} < 110 >$ family, which corresponds to a maximum number of slip systems $N = 12$. The single crystal constitutive equations are the following:

$$
\begin{aligned}
&\underset{\sim}{\dot{\varepsilon}}^{vp} = \sum_{s=1}^{N} \underset{\sim}{R}^{(s)} \dot{\gamma}^{(s)} \\
&\underset{\sim}{R}^{(s)} = \sum_{s=1}^{N} \frac{1}{2} \left(\underline{m}^{(s)} \otimes \underline{n}^{(s)} + \underline{n}^{(s)} \otimes \underline{m}^{(s)} \right) \\
&\dot{\gamma}^{(s)} = \dot{\gamma_0}^{(s)} \left(\frac{\left| \tau^{(s)} \right|}{\tau_0{}^{(s)}} \right)^{n^{(s)}} \text{sign} \left(\tau^{(s)} \right) \\
&\tau^{(s)} = \underset{\sim}{\sigma} : \underset{\sim}{R}^{(s)}
\end{aligned}
\qquad [3.42]
$$

where $\underset{\sim}{R}^{(s)}$ is the Schmid tensor for each slip system (s) in each grain, and $\underline{m}^{(s)}$ and $\underline{n}^{(s)}$ are unit vectors in the slip direction and normal to the slip plane for each slip system (s). The crystalline materials parameters $\tau_0{}^{(s)}$, $\dot{\gamma_0}^{(s)}$ and $n^{(s)}$ are identical for all slip systems and equal to τ_0, $\dot{\gamma_0}$ and n. Materials crystalline parameters are reported in Table 3.2. The elastic properties are homogeneous and given in Table 3.2 (E is elastic Young's modulus and ν is the elastic Poisson ratio).

E (GPa)	ν	τ_0 (MPa)	$\dot{\gamma_0}(s^{-1})$	n
200	1/3	200	1	1 and 5

Table 3.2. *Material parameters for uniaxial tension simulations with "SC-TF-SEC" and "SC-TF-AFF" methods*

Then, using equation [3.42] together with equation [3.2] in the case of the SEC linearization gives the following local anisotropic secant viscoplastic compliance tensor $\underset{\approx}{m}$ and the back-extrapolated strain rate $\underset{\sim}{\dot{\varepsilon}}^{v0}$ used in equations [3.29]–[3.32]:

$$\underset{\approx}{m} = \sum_{s=1}^{N} \frac{\dot{\gamma}_0}{\tau_0} \left(\frac{\left|\tau^{(s)}\right|}{\tau_0} \right)^{n-1} \underset{\sim}{R}^{(s)} \otimes \underset{\sim}{R}^{(s)} \qquad [3.43]$$

$$\underset{\sim}{\dot{\varepsilon}}^{v0} = \underset{\sim}{0}$$

In the case of the AFF linearization, the local anisotropic tangent viscoplastic compliance tensor $\underset{\approx}{m}$ and the back-extrapolated strain rate $\underset{\sim}{\dot{\varepsilon}}^{v0}$, used in equations [3.29]–[3.30], are also obtained using equation [3.42] together with equation [3.2]:

$$\underset{\approx}{m} = \sum_{s=1}^{N} n \frac{\dot{\gamma}_0}{\tau_0} \left(\frac{\left|\tau^{(s)}\right|}{\tau_0} \right)^{n-1} \underset{\sim}{R}^{(s)} \otimes \underset{\sim}{R}^{(s)}$$

$$\underset{\sim}{\dot{\varepsilon}}^{v0} = \sum_{s=1}^{N} (1-n)\, \dot{\gamma}_0 \left(\frac{\left|\tau^{(s)}\right|}{\tau_0} \right)^{n} \operatorname{sgn}\left(\tau^{(s)}\right) \underset{\sim}{R}^{(s)} \qquad [3.44]$$

Both SEC and AFF linearization schemes consider the first-order moment of the stress field, that is, the average stress over each crystal $\underset{\sim}{\sigma}$ as introduced in equation [3.42]. As an application of both "SC-TF-SEC" and "SC-TF-AFF" methods, the case of an f.c.c. polycrystal subjected to uniaxial tension is considered. A constant macroscopic strain rate $\dot{E}_{11} = 1\ \mathrm{s}^{-1}$ is imposed in the x_1 direction. Numerical results are shown for both "SC-TF-SEC" and "SC-TF-AFF" methods, which are compared to the results given by the FFT-based method (Lebensohn et al. 2012) (labeled as "EVP-FFT"). In the present case, the "EVP-FFT"-based method uses the augmented Lagrangian algorithm proposed by Michel et al. (2001). In order to assess the validity of the proposed model, the results of the spectral method are taken as reference, since this full-field numerical homogenization method provides an "exact" solution to the Lippmann–Schwinger–Dyson integral equation without neglecting non-local interactions. For FFT calculations, a 3D unit cell with periodic microstructure was generated using Voronoi tessellations. The unit cell was then discretized into a 128 × 128 × 128 regular grid. The RVE used for the comparisons of the "SC-TF-SEC" and "SC-TF-AFF" methods with the "EVP-FFT" results, consists of 200 randomly oriented spherical grains. The values of the n exponent are either $n = 1$ (linear viscoplasticity) or $n = 5$ (nonlinear viscoplasticity) (see Table 3.2). These different n values allow testing of the efficiency of the "SC-TF-SEC" and "SC-TF-AFF" linearization procedures as a function of nonlinearity.

The macroscopic stress–strain curves obtained with the different estimates ("SC-TF-SEC" and "SC-TF-AFF") are plotted in Figure 3.6 for $n = 1$ and in Figure 3.7 for $n = 5$. In the case of linear viscoelasticity (i.e. $n = 1$), no difference is observed between "SC-TF-SEC" and "SC-TF-AFF" estimates, since they both yield the same exact definition of the viscoplastic compliances tensor and $\dot{\underset{\sim}{\varepsilon}}^{v0} = \underset{\sim}{0}$. The results obtained from the SC-TF models are very similar to the reference solution calculated with the "EVP-FFT" formulation in both the transient and asymptotic regimes. The good description of the transient regime indicates that the SC approximation of the integral equation provides a realistic representation of the interactions associated with the elasto-viscoplastic space-time couplings. When nonlinearity is present ($n = 1/m = 5$), both "SC-TF-SEC" and "SC-TF-AFF" models tend to overestimate the overall stiffness of the polycrystal. However, the prediction of the "SC-TF-AFF" method is always closer to the reference solution (see Figure 3.7).

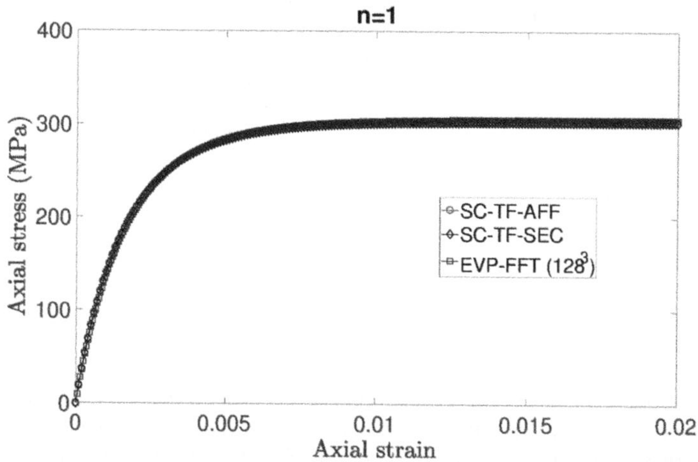

Figure 3.6. *Case of an f.c.c. polycrystal made up of 200 isotropic grain orientations with $n = 1$ subjected to uniaxial tension. Comparisons of the overall tensile stress/strain responses given by the "SC-TF-SEC" and "SC-TF-AFF" models with the "EVP-FFT" reference solution*

For the specific case of $n = 5$, the macroscopic responses of the "SC-TF-SEC" and "SC-TF-AFF" models are compared in Figure 3.8 to those obtained from the SC hereditary affine formulation and reported in Masson and Zaoui (1999) (labeled as "Masson-Zaoui") from the Taylor–Lin approximation (Taylor 1938; Lin 1957) (labeled as "Taylor") and from the Kröner–Weng model (Kröner 1961; Weng 1981) (labeled as "Kröner-Weng"). For numerical simulations, 2016 randomly oriented spherical grains were used as in Masson and Zaoui (1999). It is seen that

"SC-TF-AFF" estimate is very close to the one given by the "Masson-Zaoui" estimate, whereas the "SC-TF-SEC" leads to a too stiff response, which tends close to the "Taylor" estimate (upper bound at large times). It is noteworthy that the "Masson-Zaoui" estimate, which is based on the correspondence principle and an affine linearization procedure of the viscoplastic flow rule, requires a complex numerical implementation. Indeed, the solution is obtained from Laplace–Carson transforms whose inversion to the real time space is generally not trivial. Therefore, the proposed "SC-TF-AFF" model is an interesting alternative solution, since it provides very similar results to the hereditary approach of Masson and Zaoui (1999), while the implementation and the application to complex loading paths are much easier.

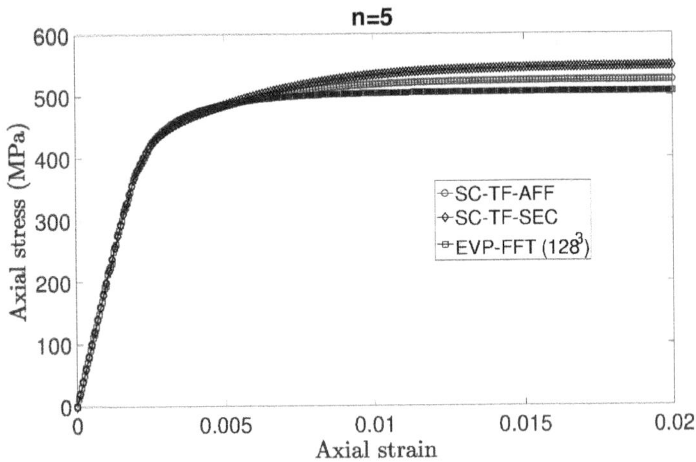

Figure 3.7. *Case of an f.c.c. polycrystal made up of* 200 *isotropic grain orientations with* $n = 5$ *subjected to uniaxial tension. Comparisons of the overall tensile stress/strain responses given by the "SC-TF-SEC" and "SC-TF-AFF" models with the "EVP-FFT" reference solution*

To understand the differences between the models, Figure 3.9 ($n = 1$) and Figure 3.10 ($n = 5$) report the axial stress/axial strain states given by the "SC-TF-SEC" and "SC-TF-AFF" models for all of the 200 grains, which are compared with the corresponding averaged fields per grain obtained with the "EVP-FFT" simulations at the same macroscopic uniaxial strain of $E_{11} = 0.02$. It is observed that the dispersion of the axial stress and strain fields is underestimated by both "SC-TF-SEC" and "SC-TF-AFF" models. For $n = 5$, the axial strains are almost equal to the prescribed macroscopic strain with the "SC-TF-SEC" model, which confirms the previous results (see Figure 3.8). Thus, the formulation based on the classical secant linearization provides results close to the upper bound estimate, that is, $\varepsilon_{11} = E_{11}$, which is obtained from Taylor–Lin's model (Taylor 1938; Lin 1957). On the contrary, the affine formulation yields a more realistic distribution of the axial stress and strain fields and a larger distribution of axial strains. However, important discrepancies with respect to the reference FFT solution persist. The reason for these differences lies in the lack of description of intra-granular field fluctuations in the "SC-TF-AFF" model, which remains based on first order moment of stresses, while the spectral method explicitly accounts such effects (Lebensohn 2001; Lebensohn et al. 2012). The calculations of second-order moments of strain/stress fields for elastic-viscoplastic polycrystals in the case of the SC model remains difficult and is still an open question (see, for example, the recent contribution of Zecevic and Lebensohn (2020)), where second moments of stress and strain rate within grains were estimated. Thus, the extension of the SC scheme based on the present TF incorporating both inter- and intra-granular field fluctuations for polycrystals is left for a future work.

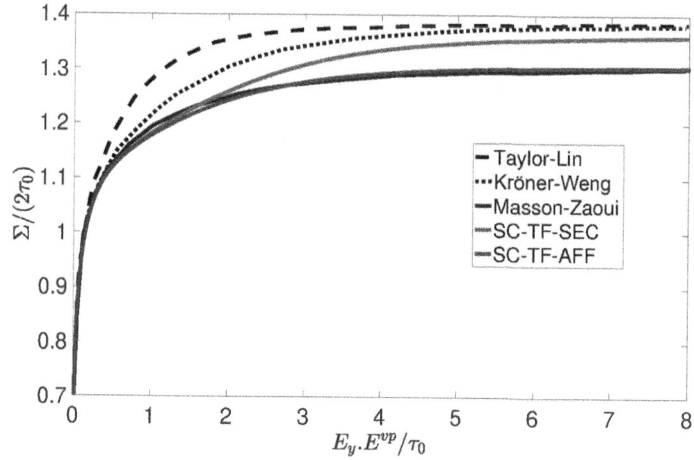

Figure 3.8. *Case of an f.c.c. polycrystal made of* 2016 *isotropic grain orientations with* $n = 5$ *subjected to uniaxial tension. Comparisons of the overall tensile stress/strain responses given by the "SC-TF-SEC" and "SC-TF-AFF" models with different formulations: "Taylor–Lin" (Taylor 1938; Lin 1957), "Kröner–Weng" (Kröner 1961; Weng 1981) and "Masson–Zaoui" (Masson and Zaoui 1999)*

3.4.2. *Applications of β-models to elastoviscoplastic polycrystals at finite deformations*

Homogenization methods for polycrystals have been extensively used for more than 60 years to predict the evolution of crystallographic texture during forming processes. Viscoplastic approaches, notably the viscoplastic SC schemes, were very successful in this endeavor (Molinari et al. 1987). The proposed Eulerian approach considers the strain rate and spin tensors. The spin tensor contains a plastic contribution due to lattice rotation rate during the deformation process. The elasto-viscoplastic deformation of single crystals is best described by Mandel's celebrated model, which is based on the multiplicative decomposition of the deformation gradient (see Mandel (1973), where a triad of directors is attached to the crystal lattice). Mandel until the end of his life sought to determine such a triad of directors for the polycrystal (see Mandel 1983). These thoughts gave rise to a sophisticated SC scheme at finite elastic-plastic deformations based on the multiplicative decomposition for each grain by Lipinski et al. (1990), which represents one the most accurate and thermodynamically consistent finite deformation homogenization scheme for polycrystals.

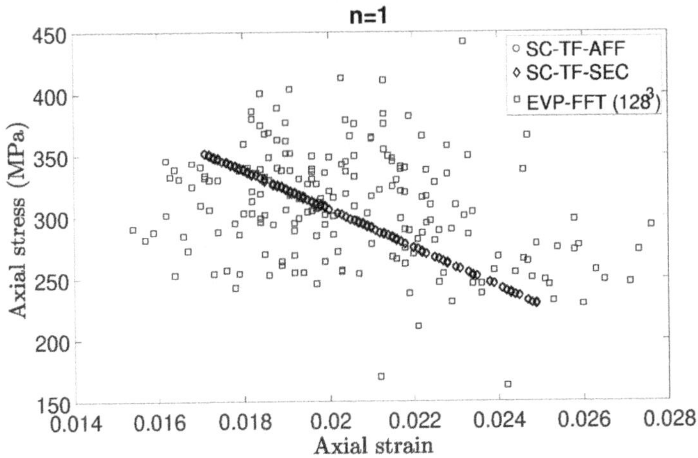

Figure 3.9. *Case of an f.c.c. polycrystal made of* 200 *isotropic grain orientations with* $n = 1$ *subjected to uniaxial tension. Comparisons of axial stress/axial strain states given by the "SC-TF-SEC" and "SC-TF-AFF" models for all grains with the corresponding averaged fields per grain with "EVP-FFT" at a macroscopic uniaxial strain of* $E_{11} = 0.02$

A finite deformation extension of the β-model approach was proposed by Forest and Pilvin (1999) by resorting to the concept of local objective rotating frames recommended in Sidoroff and Dogui (2001). A relative rotation of the crystal lattice with respect to the corotational frame is introduced in this model. Strain and spin rates are additively decomposed in the corotational frame circumventing the use of the multiplicative decomposition. It was shown that Mandel's original model and the proposed formulation deliver essentially the same results as long as elastic strains remain small, as it is the case in general in metals and alloys. An illustration of the predictions of this single crystal model is shown in Figure 3.11, where the stress components induced by simple glide are drawn for a BCC single crystal, including 12 or 24 slip systems. The typical periodic response with respect to the shear strain is apparent and was predicted by Boukadia and Sidoroff (1988), Bertram and Kraska

(1995) and Forest and Pilvin (1999) for FCC crystals using Mandel's single crystal plasticity model. Remarkably, for the very special orientation considered in Figure 3.11, and in the absence of hardening, an endless rotation of the lattice is observed at the corotational rate. The potential slip systems are alternatively activated and deactivated to produce the periodic stress response. The smoothing effect of the pencil glide mechanism is highlighted.

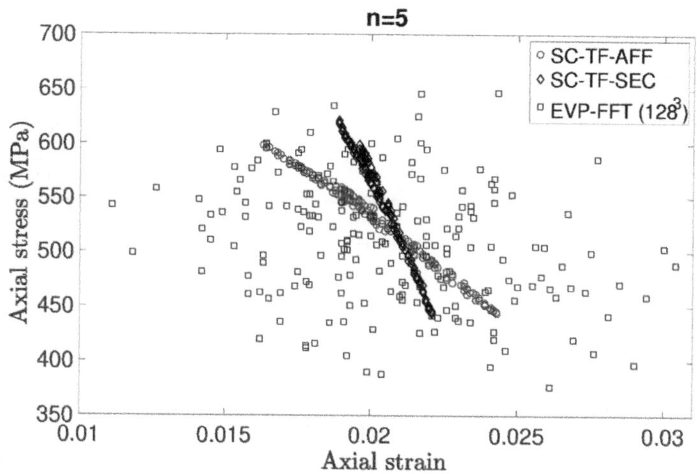

Figure 3.10. *Case of an f.c.c. polycrystal made of* 200 *isotropic grain orientations with* $n = 5$ *subjected to uniaxial tension. Comparisons of axial stress/axial strain states given by the "SC-TF-SEC" and "SC-TF-AFF" models for all grains with the corresponding averaged fields per grain with "EVP-FFT" at a macroscopic uniaxial strain of* $E_{11} = 0.02$

The polycrystal model adopts a β-rule for the stress tensor and is limited to the Taylor assumption regarding material rotation.

This model was revisited recently by Le et al. (2020) to incorporate general anisotropic elasticity and the pencil glide mechanism active in BCC crystals. The Eulerian velocity gradient is decomposed as usual into its symmetric and skew–symmetric parts:

$$\underset{\sim}{L} = \underset{\sim}{D} + \underset{\sim}{W} \qquad [3.45]$$

which involves the strain rate and material spin tensor. The time-dependent rotation $\underset{\sim}{Q}^c(t)$ linking the corotational space frame to the current one is defined as

$$\dot{\underset{\sim}{Q}}^c \underset{\sim}{Q}^{cT} = \underset{\sim}{W} \qquad [3.46]$$

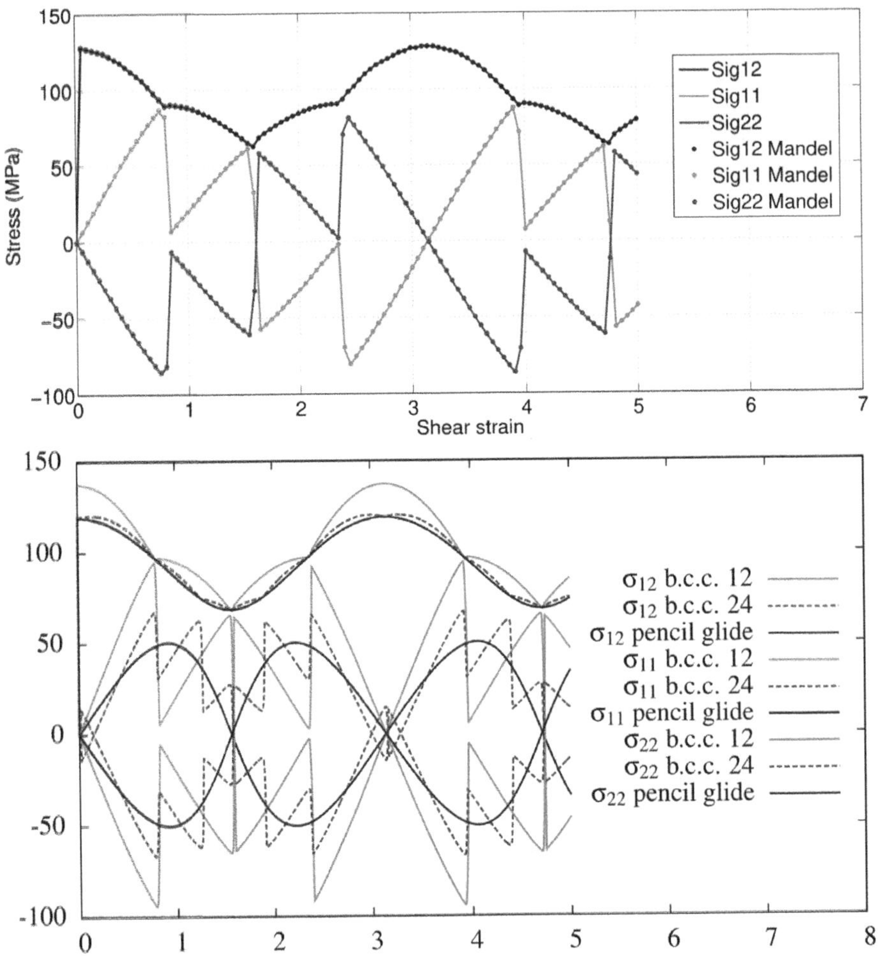

Figure 3.11. *Cauchy stress components for a BCC single crystal under prescribed simple glide F_{12}. The loading axes are initially aligned with the $< 100 >$ crystal directions. The periodic responses in the case of perfect plasticity (no hardening) differ depending on the considered slip system families: 12 slip systems (top) and 24 slip systems or pencil glide mechanism (bottom)*

The strain rate tensor is transported into the corotational frame and splits into elastic and plastic strain rate contributions:

$$\dot{\underset{\sim}{e}} = \underset{\sim}{Q}^{cT} \underset{\sim}{D} \underset{\sim}{Q}^{c} = \dot{\underset{\sim}{e}}^{e} + \dot{\underset{\sim}{e}}^{p} \qquad [3.47]$$

The plastic strain rate observed in the corotational frame for each grain is the result of the contributions of all slip systems in the form:

$$\dot{\underset{\sim}{e}}^{pg} = \sum_{s} \dot{\gamma}^{s} \underline{m}^{cgs} \overset{sym}{\otimes} \underline{n}^{cgs} \qquad [3.48]$$

where $\underline{m}^{cgs}, \underline{n}^{cgs}$, respectively, denote the slip direction and the normal to the slip plane for the slip system s in grain g with respect to the corotational frame. The symmetric part of the orientation tensor, also called Schmid tensor, $\underline{m}^{csg} \otimes \underline{n}^{csg}$ is used.

It is assumed that the spin tensor $\underset{\sim}{W}$ is the same for all the grains g (Taylor assumption for the spin). The definition of the corotational frame is such that the spin tensor transported into the corotational frame vanishes:

$$\underset{\sim}{W}^{c} = \dot{\underset{\sim}{Q}}^{cT} \underset{\sim}{Q}^{c} + \underset{\sim}{Q}^{cT} \underset{\sim}{W} \underset{\sim}{Q}^{c} = 0 = \underset{\sim}{W}^{cge} + \underset{\sim}{W}^{cgp} \qquad [3.49]$$

with the following expression of the corotational plastic spin tensor for each grain g:

$$\underset{\sim}{W}^{cgp} = \sum_{s} \dot{\gamma}^{s} \underline{m}^{cgs} \overset{skew}{\otimes} \underline{n}^{cgs}, \qquad [3.50]$$

that is, involving the skew–symmetric part of the orientation tensor. The link between the lattice space frame of each grain and the corotational frame is the rotation $\underset{\sim}{Q}^{\sharp g}$ with the following evolution equation:

$$\dot{\underset{\sim}{Q}}^{\sharp g} \underset{\sim}{Q}^{\sharp gT} = \underset{\sim}{W}^{cge} = -\underset{\sim}{W}^{cgp} = -\sum_{s} \dot{\gamma}^{s} \underline{m}^{cgs} \overset{skew}{\otimes} \underline{n}^{cgs} \qquad [3.51]$$

The initial value $\underset{\sim}{Q}^{\sharp g}(t_0)$ depends on the initial orientation of the grain g. Crystallographic directions are known in the lattice frame: $\underline{m}^{cgs} = \underset{\sim}{Q}^{\sharp g} \underline{m}^{gs}$ and $\underline{n}^{cgs} = \underset{\sim}{Q}^{\sharp g} \underline{n}^{gs}$.

The corotational stress tensor is defined as follows:

$$\underset{\sim}{\sigma}^c = (\det \underset{\sim}{F}) \underset{\sim}{Q}^{cT} \underset{\sim}{\overline{\sigma}} \underset{\sim}{Q}^c \qquad [3.52]$$

from the macroscopic Cauchy stress tensor. It is computed from the elastic strain using the effective moduli:

$$\underset{\sim}{\sigma}^c = \underset{\approx}{C}^{effc} : \underset{\sim}{e}^e \qquad [3.53]$$

The stress tensor in each grain is computed by means of a modified scale transition rule [3.38] as follows:

$$\underset{\sim}{\sigma}^{(cg)} = \underset{\approx}{B}^{(r)} : \left(\underset{\sim}{\sigma}^c + \underset{\approx}{C}^{eff} : (\underset{\approx}{I} - \underset{\approx}{S}^{Esh}) : (\underset{\sim}{\beta} - \underset{\sim}{\beta}^{(g)}) \right) \qquad [3.54]$$

where the involved tensors are defined as in section 3.3. The β-variables are subjected to the evolution equation [3.35]. This evolution equation includes a fourth-order tensor of parameters that must be calibrated from some specific homogenization model or from full–field simulation of polycrystalline aggregates. The macroscopic β-variable and the macroscopic plastic strain variable are still given by the relations [3.40] and [3.41], respectively.

The applications of the polycrystal model at finite deformations presented by Le et al. (2020) deal with the evolution of crystallographic texture for initially isotropy iron at room temperature under various loading conditions: tension, compression, rolling and shear, with comparison to experimental results from the literature. An example of final texture after 100% logarithmic rolling strain is shown in Figure 3.12. The results are compared in the case of 24 slip systems, often considered for BCC materials, and in the case of the pencil glide mechanism.

Other applications of the β-model can be mentioned. β-Transition rules can be nested to represent multiscale microstructures, like martensitic variants in a polycrystal (see Martin et al. 2011). Twinning and displacive phase transformation can be incorporated in the presented framework, as demonstrated in the case of magnesium polycrystals under complex cyclic loading conditions by Guillemer et al. (2011). Phase transformation can be addressed by this approach, together with transformation-induced plasticity accompanying this process (Fischlschweiger et al. 2012). Phase transformation is an essential ingredient in the modelling of new processing techniques like additive manufacturing. A β-model was proposed in Longuet et al. (2009) to predict residual stresses resulting from complex multiscale thermomechanical processes during additive manufacturing.

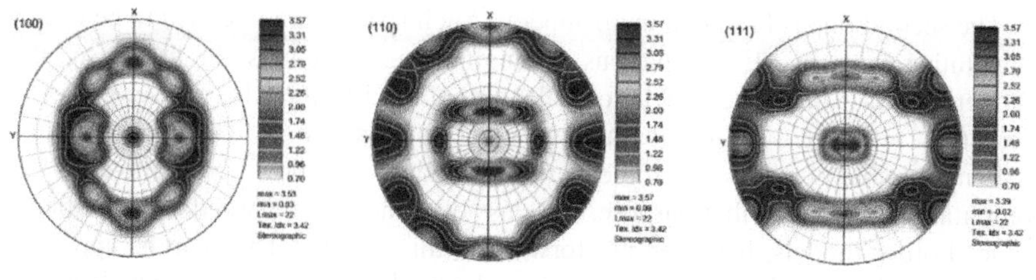

(a) {110} + {112} slip systems.

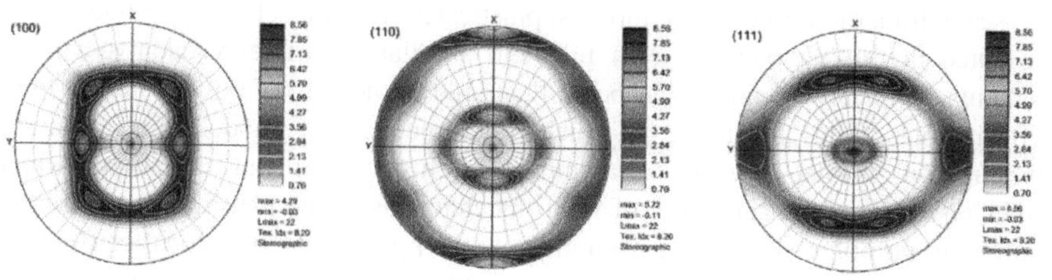

(b) Pencil glide.

Figure 3.12. *Rolling texture after 100% logarithmic strain: in the case of (a) 24 slip systems and (b) the pencil glide mechanism (after Le et al. 2020)*

3.5. Concluding remarks

Micromechanical approaches are essentially used to develop local criteria for plasticity, damage or fracture making use of the estimated stress–strain values and internal variables for individual phases. In this chapter, we have presented two different complementary analytical mean field micromechanical models for elasto-viscoplastic heterogeneous materials.

The first one is based on the TF, for which different classes of analytical homogenization estimates (MT or Hashin–Shtrickmann estimates, SC estimate) can be easily applied to composite materials with ellipsoidal inhomogeneities or polycrystals. The method was seen to be efficient compared to full-field numerical results (FFT) and can give improved estimates considering the second moment of stresses in the local constitutive behavior of the phases through modified linearization procedures. As a major advantage of the method, there is no need to use Laplace–Carson transforms and numerically treat their inversions in order to derive the evolutions of both the per phase (or grain) mean fields and the overall responses. The extension of the "SC-TF-AFF" model to finite strains is on-going, following the formulations reported in Nemat-Nasser and Obata (1986) and Wang et al. (2010). The full implementation of the "SC-TF-AFF" model as an elasto-viscoplastic polycrystalline constitutive law for FE structural calculations is under development.

The second one is based on the β-models. Such models are intended to be used as constitutive laws in FE calculations of engineering structures. In that case, they compete with usual phenomenological viscoplasticity models whose power of prediction is limited to the parameter identification range. This is however a challenge, since many micromechanical models requires prohibitively long computation times. For that purpose, the class of β-models, regarded as simplified or reduced order models, has been put forward in this chapter. They can be applied to arbitrary nonlinear behavior of the phases and complex loading conditions. However, they require a preliminary (off-line) calibration (training) of specific parameters of the scale transition rule from more sophisticated micromechanical models, also presented in this chapter, or from full-field simulations of representative volume elements. This class of model has been illustrated in the case of polycrystals at finite deformations. The advantage of such micromechanical models for structural computations is that they do not assume explicit forms of yield functions, which rather result from the activation of multi-mechanisms at the phase scale. This is in contrast to complex anisotropic phenomenological potentials, which suffer from the limitations of the large number of parameters to be calibrated from experiments and from the absence of yield surface distorsion during straining.

Analytical micromechanical approaches should be increasingly used in engineering applications, in combination with recent types of reduced order models and machine learning methods.

3.6. References

Agoras, M., Avazmohammadi, R., Ponte Castañeda, P. (2016). Incremental variational procedure for elasto-viscoplastic composites and application to polymer and metal matrix composites reinforced by spheroidal elastic particles. *International Journal of Solids and Structures*, 98, 668–686.

Amouzou, K.E.K., Richeton, T., Roth, A., Lebyodkin, M., Lebedkina, T. (2016). Micromechanical modeling of hardening mechanisms in commercially pure alpha-titanium in tensile condition. *International Journal of Plasticity*, 80, 222–240.

Badulescu, C., Lahellec, N., Suquet, P. (2015). Field statistics in linear viscoelastic composites and polycrystals. *Eur. J. Mech. A Solids*, 49, 329–344.

Benveniste, Y. (1987). A new approach to the application of Mori-Tanaka's theory in composite materials. *Mech. Mater.*, 6(2), 147–157.

Berbenni, S. (2021). A time-incremental homogenization method for elastoviscoplastic particulate composites based on a modified secant formulation. *International Journal of Solids and Structures*, 229, 111136.

Berbenni, S. and Capolungo, L. (2015). A Mori-Tanaka homogenization scheme for non-linear elastoviscoplastic heterogeneous materials based on translated fields: An affine extension. *Comptes Rendus de Mécanique*, 343, 95–106.

Berbenni, S., Favier, V., Lemoine, X., Berveiller, M. (2004). Micromechanical modeling of the elastic viscoplastic behavior of polycrystalline steels having different microstructures. *Materials Science Engineering A*, 372, 128–136.

Berbenni, S., Favier, V., Berveiller, M. (2007). Impact of the grain size distribution on the yield stress of heterogeneous materials. *Int. J. Plast.*, 23(1), 114–142.

Berbenni, S., Dinzart, F., Sabar, H. (2015). A new internal variables homogenization scheme for linear viscoelastic materials based on an exact eshelby interaction law. *Mech. Mater.*, 81, 110–124.

Bertram, A. and Kraska, M. (1995). Determination of finite plastic deformations in single crystals. *Archives of Mechanics*, 47, 203–222.

Berveiller, M. and Zaoui, A. (1979). An extension of the self-consistent scheme to plastically flowing polycrystals. *Journal of the Mechanics and Physics of Solids*, 26, 325–344.

Besson, J., Cailletaud, G., Chaboche, J.-L., Forest, S., Blétry, M. (2009). *Non-Linear Mechanics of Materials, Solid Mechanics and Its Applications*. Springer, Berlin, Heidelberg.

Boukadia, J. and Sidoroff, F. (1988). Simple shear and torsion of a perfectly plastic single crystal in finite transformations. *Archives of Mechanics*, 40, 497–513.

Brassard, L., Stainier, L., Doghri, I., Delannay, L. (2012). Homogenization of elasto (visco)plastic composites based on an incremental variational principle. *International Journal of Plasticity*, 36, 86–112.

Cailletaud, G. (1992). A micromechanical approach to inelastic behaviour of metals. *Int. J. Plast.*, 8(1), 55–73.

Cailletaud, G. (2009). An overview of the anatomy of crystal plasticity models. *Adv. Eng. Mater.*, 11(9), 710–716.

Cailletaud, G. and Coudon, F. (2016). Scale transition rules applied to crystal plasticity. In *Materials with Internal Structure*, Trovalusci, P. (ed.). Springer International Publishing, Cham.

Cailletaud, G. and Pilvin, P. (1994). Utilisation de modèles polycristallins pour le calcul par éléments finis. *Rev. Eur. Élém. Finis*, 3(4), 515–541.

Cailletaud, G. and Sai, K. (2008). A polycrystalline model for the description of ratchetting: Effect of intergranular and intragranular hardening. *Mater. Sci. Eng. A Struct. Mater.*, 480(1–2), 24–39.

Cotelo, J., Das, S., Ponte Castañeda, P. (2020). A differential homogenization method for estimating the macroscopic response and field statistics of particulate viscoelastic composites. *Int. J. Solids Struct.*, 204-205, 199–219.

Coudon, F., Cailletaud, G., Cormier, J., Marcin, L. (2019). A multiscale model for nickel-based directionally solidified materials. *Int. J. Plast.*, 115, 1–17.

Czarnota, C., Kowalczyk-Gajewska, K., Salahouelhadj, A., Martiny, M., Mercier, S. (2015). Modeling of the cyclic behavior of elastic–viscoplastic composites by the additive tangent Mori–Tanaka approach and validation by finite element calculations. *Int. J. Solids Struct.*, 56–57, 96–117.

Das, S. and Ponte Castañeda, P. (2021). Differential variational estimates for the macroscopic response and field statistics of elasto-viscoplastic polycrystals. *J. Mech. Phys. Solids*, 147, 104202.

Doghri, I., Adam, L., Bilger, N. (2010). Mean-field homogenization of elasto-viscoplastic composites based on a general incrementally affine linearization method. *Int. J. Plast.*, 26(2), 219–238.

Eshelby, J.D. and Peierls, R.E. (1957). The determination of the elastic field of an ellipsoidal inclusion, and related problems. *Proceedings of the Royal Society of London. Series A. Mathematical and Physical Sciences*, 241(1226), 376–396 [Online]. Available at: https://royalsocietypublishing.org/doi/abs/10.1098/rspa.1957.0133.

Fischlschweiger, M., Cailletaud, G., Antretter, T. (2012). A mean-field model for transformation induced plasticity including backstress effects for non-proportional loadings. *Int. J. Plast.*, 37, 53–71.

Forest, S. and Pilvin, P. (1996). Modelling the cyclic behaviour of two-phase single crystal nickel-base superalloys. In *IUTAM Symposium on Micromechanics of Plasticity and Damage of Multiphase Materials*. Springer, Dordrecht.

Forest, S. and Pilvin, P. (1999). Modelling finite deformation of polycrystals using local objective frames. *Z. Angew. Math. Mech.*, 79(S1), 199–202.

Gilormini, P. (1995). Insuffisance de l'extension classique du modèle autocohérent au comportement non linéaire. *C. R. Acad. Sci. Paris*, 320, 115–122.

Gu, T., Medy, J.-R., Klosek, V., Castelnau, O., Forest, S., Hervé-Luanco, E., Lecouturier-Dupouy, F., Proudhon, H., Renault, P.-O., Thilly, L. et al. (2019). Multiscale modeling of the elasto-plastic behavior of architectured and nanostructured Cu-Nb composite wires and comparison with neutron diffraction experiments. *Int. J. Plast.*, 122, 1–30.

Guillemer, C., Clavel, M., Cailletaud, G. (2011). Cyclic behavior of extruded magnesium: Experimental, microstructural and numerical approach. *Int. J. Plast.*, 27(12), 2068–2084.

Hashin, Z. (1969). The inelastic inclusion problem. *Int. J. Eng. Sci.*, 7(1), 11–36.

Hill, R. (1965). A self-consistent mechanics of composite materials. *J. Mech. Phys. Solids*, 13(4), 213–222.

Hill, R. (1967). The essential structure of constitutive laws for metal composites and polycrystals. *J. Mech. Phys. Solids*, 15(2), 79–95.

Hlilou, A., Naceur, I. B., Saï, K., Gérard, C., Forest, S., Cailletaud, G. (2009). Generalization of the polycrystalline β-model: Finite element assessment and application to softening material behavior. *Comput. Mater. Sci.*, 45(4), 1104–1112.

Hu, G. (1996). A method of plasticity for general aligned spheroidal void or fiber-reinforced composites. *Int. J. Plast.*, 12(4), 439–449.

Kowalczyk-Gajewska, K. and Petryk, H. (2011). Sequential linearization method for viscous/elastic heterogeneous materials. *Eur. J. Mech. A Solids*, 30(5), 650–664.

Kröner, E. (1961). Zur plastischen verformung des vielkristalls. *Acta Metall.*, 9(2), 155–161.

Kröner, E. (1990). Modified green functions in the theory of heterogeneous and/or anisotropic linearly elastic media. In *Micromechanics and Inhomogeneity*, Weng, G.J., Taya, M., Abé, H. (eds). Springer, New York.

Lahellec, N. and Suquet, P. (2007). On the effective behavior of nonlinear inelastic composites: I. Incremental variational principles. *J. Mech. Phys. Solids*, 55(9), 1932–1963.

Lahellec, N. and Suquet, P. (2013). Effective response and field statistics in elasto-plastic and elasto-viscoplastic composites under radial and non-radial loadings. *Int. J. Plast.*, 42, 1–30.

Laws, N. and McLaughlin, R. (1978). Self-consistent estimates for the viscoelastic creep compliances of composite materials. *Proc. R. Soc. Lond.*, 359(1697), 251–273.

Le, L.T., Ammar, K., Forest, S. (2020). Efficient simulation of single and polycrystal plasticity based on the pencil glide mechanism. *Comptes Rendus Mécanique*, 202, 532–551.

Lebensohn, R.A. (2001). N-site modeling of a 3D viscoplastic polycrystal using fast fourier transform. *Acta Materialia*, 49, 2723–2737.

Lebensohn, R.A., Kanjarla, A.K., Eisenlohr, P. (2012). An elasto-viscoplastic formulation based on fast fourier transforms for the prediction of micromechanical fields in polycrystalline materials. *Int. J. Plast.*, 32–33, 59–69.

Lhadi, S., Berbenni, S., Gey, N., Richeton, T., Germain, L. (2018). Micromechanical modeling of the effect of elastic and plastic anisotropies on the mechanical behavior of beta-ti alloys. *International Journal of Plasticity*, 109, 88–107.

Li, J. and Weng, G. (1997). A secant-viscosity approach to the time-dependent creep of an elastic-viscoplastic composite. *Journal of the Mechanics and Physics of Solids*, 45, 1069–1083.

Lin, T.H. (1957). Analysis of elastic and plastic strains of a face-centered cubic crystal. *Journal of the Mechanics and Physics of Solids*, 5, 143–149.

Lipinski, P., Krier, J., Berveiller, M. (1990). Elastoplasticité des métaux en grandes déformations : comportement global et évolution de la structure interne. *Revue Phys. Appl.*, 25(4), 361–388.

Longuet, A., Robert, Y., Aeby-Gautier, E., Appolaire, B., Mariage, J.F., Colin, C., Cailletaud, G. (2009). A multiphase mechanical model for Ti–6Al–4V: Application to the modeling of laser assisted processing. *Comput. Mater. Sci.*, 46(3), 761–766.

Luo, M. and Rousselier, G. (2014). Modeling of large strain multi-axial deformation of anisotropic metal sheets with strength-differential effect using a reduced texture methodology. *Int. J. Plast.*, 53, 66–89.

Mandel, J. (1966). *Cours de mécanique des milieux continus*. Gauthiers-Villars, Paris.

Mandel, J. (1973). Equations constitutives et directeurs dans les milieux plastiques et viscoplastiques. *Int. J. Solids Struct.*, 9(6), 725–740.

Mandel, J. (1983). Sur la definition de la vitesse de deformation elastique en grande transformation elastoplastique. *Int. J. Solids Struct.*, 19(7), 573–578.

Mareau, C. and Berbenni, S. (2015). An affine formulation for the self-consistent modeling of elasto-viscoplastic heterogeneous materials based on the translated field method. *Int. J. Plast.*, 64, 134–150.

Mareau, C., Favier, V., Berveiller, M. (2009). Micromechanical modeling coupling time-independent and time-dependent behaviors for heterogeneous materials. *Int. J. Solids Struct.*, 46(2), 223–237.

Martin, G., Andnazé, L., Cailletaud, G. (2011). Numerical multi-scale simulations of the mechanical behavior of β-metastable titanium alloys ti5553 and ti17. *11th International Conference on the Mechanical Behavior of Materials (ICM11)*, 10, 1803–1808.

Martin, G., Ochoa, N., Saï, K., Hervé-Luanco, E., Cailletaud, G. (2014). A multi-scale model for the elastoviscoplastic behavior of directionally solidified alloys: Application to FE structural computations. *International Journal of Solids and Structures*, 51, 1175–1187.

Masson, R. and Zaoui, A. (1999). Self-consistent estimates for the rate-dependent elastoplastic behaviour of polycrystalline materials. *Journal of the Mechanics and Physics of Solids*, 47, 1543–1568.

Masson, R., Brenner, R., Castelnau, O. (2012). Incremental homogenization approach for ageing viscoelastic polycrystals. *Comptes Rendus Mécanique*, 340(4-5), 378–386.

Masson, R., Seck, M.E.B., Fauque, J., Gărăjeu, M. (2020). A modified secant formulation to predict the overall behavior of elasto-viscoplastic particulate composites. *J. Mech. Phys. Solids*, 137(103874), 103874.

Mercier, S. and Molinari, A. (2009). Homogenization of elastic–viscoplastic heterogeneous materials: Self-consistent and Mori-Tanaka schemes. *Int. J. Plast.*, 25(6), 1024–1048.

Mercier, S., Molinari, A., Berbenni, S., Berveiller, M. (2012). Comparison of different homogenization approaches for elastic–viscoplastic materials. *Model. Simul. Mat. Sci. Eng.*, 20(2), 024004.

Michel, J.C., Moulinec, H., Suquet, P. (2001). A computational scheme for linear and non-linear composites with arbitrary phase contrast. *Int. J. Numer. Methods Eng.*, 52(12), 139–160.

Molinari, A. (2002). Averaging models for heterogeneous viscoplastic and elastic-viscoplastic materials. *Journal of Engineering Materials and Technology, Trans. ASME*, 124, 62–70.

Molinari, A., Canova, G.R., Ahzi, S. (1987). A self consistent approach of the large deformation polycrystal viscoplasticity. *Acta Metall.*, 35(12), 2983–2994.

Molinari, A., Ahzi, S., Kouddane, R. (1997). On the self-consistent modeling of elastic-plastic behavior of polycrystals. *Mech. Mater.*, 26(1), 43–62.

Mori, T. and Tanaka, K. (1973). Average stress in matrix and average elastic energy of materials with misfitting inclusions. *Acta Metall.*, 21(5), 571–574.

Nemat-Nasser, S. and Obata, M. (1986). Rate-dependent, finite elasto-plastic deformation of polycrystals. *Proc. Math. Phys. Eng. Sci.*, 407(1833), 343–375.

Nicaise, N., Berbenni, S., Wagner, F., Berveiller, M., Lemoine, X. (2011). Coupled effects of grain size distributions and crystallographic textures on the plastic behaviour of IF steels. *Int. J. Plast.*, 27(2), 232–249.

Paquin, A., Sabar, H., Berveiller, M. (1999). Integral formulation and self-consistent modelling of elastoviscoplastic behavior of heterogeneous materials. *Arch. Appl. Mech.*, 69(1), 14–35.

Paquin, A., Berbenni, S., Favier, V., Lemoine, X., Berveiller, M. (2001). Micromechanical modeling of the elastic–viscoplastic behavior of polycrystalline steels. *Int. J. Plast.*, 17(9), 1267–1302.

Parenteau, T., Bertevas, E., Ausias, G., Stocek, R., Grohens, Y., Pilvin, P. (2014). Characterisation and micromechanical modelling of the elasto-viscoplastic behavior of thermoplastic elastomers. *Mech. Mater.*, 71, 114–125.

Pierard, O. and Doghri, I. (2006). An enhanced affine formulation and the corresponding numerical algorithms for the mean-field homogenization of elasto-viscoplastic composites. *International Journal of Plasticity*, 69, 131–157.

Pilvin, P. (1996). The contribution of micromechanical approaches to the modelling of inelastic behaviour of polycrystals. In *Fourth Int. Conf. on Biaxial/Multiaxial Fatigue and Design*. ESIS, Mechanical Engineering Publications, London, 21, 3–19.

Pilvin, P., Onimus, F., Brenner, R., Pascal, S., Feaugas, X., Saï, K. (2017). Finite element assessment of an affine self-consistent model for hexagonal polycrystals. *Eur. J. Mech. A Solids*, 61, 345–356.

Ponte Castañeda, P. (1991). The effective mechanical properties of nonlinear isotropic composites. *J. Mech. Phys. Solids*, 39(1), 45–71.

Ponte Castañeda, P. and Suquet, P. (2002). Nonlinear composites and microstructure evolution. In *Mechanics for a New Mellennium*, Aref, H. and Phillips, J.W. (eds). Springer, Dordrecht.

Qiu, Y.P. and Weng, G.J. (1992). A theory of plasticity for porous materials and particle-reinforced composites. *J. Appl. Mech.*, 59(2), 261–268.

Ricaud, J.-M. and Masson, R. (2009). Effective properties of linear viscoelastic heterogeneous media: Internal variables formulation and extension to ageing behaviours. *Int. J. Solids Struct.*, 46(7-8), 1599–1606.

Robert, C. and Mareau, C. (2015). A comparison between different numerical methods for the modeling of polycrystalline materials with an elastic–viscoplastic behavior. *Comput. Mater. Sci.*, 103, 134–144.

Rougier, Y., Stolz, C., Zaoui, A. (1994). Self-consistent modelling of elastic–viscoplastic polycrystals. *Comptes Rendus Académie des Sciences Paris*, 318, 145–151.

Rousselier, G. and Leclercq, S. (2006). A simplified "polycrystalline" model for viscoplastic and damage finite element analyses. *Int. J. Plast.*, 22(4), 685–712.

Sabar, H., Berveiller, M., Favier, V., Berbenni, S. (2002). A new class of micro–macro models for elastic-viscoplastic heterogeneous materials. *International Journal of Solids and Structures*, 39, 3257–3276.

Sai, K., Cailletaud, G., Forest, S. (2006). Micromechanical modeling of the inelastic behavior of directionally solidified materials. *Mechanics of Materials*, 38, 203–217.

Segurado, J., Lebensohn, R.A., LLorca, J., Tomé, C.N. (2012). Multiscale modeling of plasticity based on embedding the viscoplastic self-consistent formulation in implicit finite elements. *Int. J. Plast.*, 28(1), 124–140.

Sidoroff, F. and Dogui, A. (2001). Some issues about anisotropic elastic–plastic models at finite strain. *Int. J. Solids Struct.*, 38(52), 9569–9578.

Soho, K., Abed Meraim, F., Lemoine, X., Zahrouni, H. (2014). Multiscale finite element simulation of forming processes based on crystal plasticity. *Key Eng. Mater.*, 611–612, 545–552.

Suquet, P. (1987). Elements of homogenization for inelastic solid mechanics. In *Homogenization Techniques for Composite Media*, Sanchez-Palencia, E. and Zaoui, A. (eds). Springer, Berlin.

Suquet, P. (1995). Overall properties of nonlinear composites: A modified secant moduli theory and its link with ponte castañeda nonlinear variational procedure. *C. R. Acad. Sci. Paris Serie IIb*, 320, 563–571.

Taylor, G.I. (1938). Plastic strain in metals. *Journal Institute of Metals*, 61, 307–324.

Terentyev, D., Xiao, X., Dubinko, A., Bakaeva, A., Duan, H. (2015). Dislocation-mediated strain hardening in tungsten: Thermo-mechanical plasticity theory and experimental validation. *J. Mech. Phys. Solids*, 85, 1–15.

Wang, H., Wu, P.D., Tomé, C.N., Huang, Y. (2010). A finite strain elastic–viscoplastic self-consistent model for polycrystalline materials. *J. Mech. Phys. Solids*, 58(4), 594–612.

Weng, G.J. (1981). Self-consistent determination of time-dependent behavior of metals. *J. Appl. Mech.*, 48(1), 41–46.

Weng, G.J. (1984). Some elastic properties of reinforced solids, with special reference to isotropic ones containing spherical inclusions. *Int. J. Eng. Sci.*, 22(7), 845–856.

Zecevic, M. and Lebensohn, R.A. (2020). New robust self-consistent schemes for elastic-viscoplastic polycrystals. *International Journal of Solids and Structures*, 202, 434–453.

4

Vertex and Front-Tracking Methods for the Modeling of Microstructure Evolution at the Solid State

Marc BERNACKI
CNRS UMR 7635, Centre for Material Forming (CEMEF), MINES Paris, PSL Research University, Sophia Antipolis, France

4.1. Introduction

When dealing with the modeling of microstructural evolutions at the mesoscopic scale, there are typically two main families of methods to describe complex moving or interfaces network: front-capturing (FC) and front-tracking (FT). This terminology is ultimately common to all themes where the monitoring of a front or a network of interfaces is central (de Sousa et al. 2004). As illustrated in following chapters, FC models, such as phase-field/multiphase field (see Chapter 5) or level-set frameworks (see Chapter 6), implicitly (indirectly) define interfaces by capturing the change of state of a field variable. In contrast, FT models explicitly (directly) define interfaces using a set of interconnected segments (in 2D) or surfaces (in 3D).

By this definition, Vertex models (Fullman 1952) are FT models, as the vertices are connected between them to form the segments that define the boundaries between domains. It will be illustrated that the boundary between the definition of Vertex and FT methods, and even the definition of FT methods themselves, has evolved over time compared with the first articles on the subject, and that can sometimes be rather vague.

Digital Materials,
coordinated by Marc BERNACKI and Samuel FOREST.

Typically, approaches such as Monte Carlo (MC) or cellular automaton (CA) (see Chapter 7) can be considered from the point of view of recent authors as FT-type approaches since, in these frameworks, the grain boundaries can be explicitly described by a collection of edges of pixels/voxels used to discretize the computational domain. Historically, it was rather the fact that the physics of the mechanism was implemented solely through the migration of the interface that distinguished the FT approaches from the preexisting approaches (MC and CA) (Frost and Thompson 1996).

The notion of an FT approach is thus today often associated with a Lagrangian movement of interfaces to describe the movement of modeled interfaces networks and for which the kinetics are derived from geometrical properties, which are only dependent of the interfaces. This can therefore be done mainly by using techniques based on FE meshing/remeshing techniques or by approaches limiting the interfaces discretization and thus the resolution cost (as Vertex approaches). A common interest to these approaches is to gain a degree of spatial resolution compared to other existing approaches by considering surface geometries in 3D and line ones in 2D. The flip side of the coin is that these methods are generally not suitable for modeling intragranular mechanisms. Thus, the most often modeled mechanism concerns grain growth problems without heterogeneous plastic-stored energy field in the core of the grains and where only the surface energy of the interfaces network needs to be minimized. The difficulty of managing topological operations at multiple junctions is also a complex algorithmic subject, especially in 3D. Again, this explains the limited number of studies dedicated to the use of these approaches in 3D for a large number of interfaces. However, recent work illustrates more than ever the interest of these methods in terms of computational time optimization, precise discussion of the mobility and energy properties at multiple junctions, and their possible extension for the consideration of intragranular phenomena. These different elements will be presented in this chapter with regard to the state of the art.

4.2. Vertex frameworks

In the early 1950s, John von Neumann introduced a kinetic law for 2D soap froths, establishing a connection between the growth rate of individual bubbles' areas and their number of neighbors (von Neumann 1952). This concept was later expanded by Mullins (1956) to encompass two-dimensional grain growth structures characterized by isotropic boundary energies. Known today as the von Neumann–Mullins law, this equation was further extended to three dimensions 50 years later by MacPherson and Srolovitz (2007). Interestingly, the appearance of the first Vertex framework in the context of 2D grain growth has followed the same logical way as the model presented by Soares et al. (1985) is based on the model of Weaire and Kermode (1983) that was applied to the evolution of soap froths. These models consider that the evolution of an interface network can be simulated from the movement of its triple junctions (multiple junctions with three connections), a proposition first suggested in Fullman (1952) (but with arbitrary hypothesis concerning the velocity definition). This choice allows us to avoid the computation of curvature by the use of straight lines as grain boundaries. The motion equation of the triple junctions used by Soares et al. was derived from a thermodynamics study (unlike the one proposed in Fullman (1952)) and can be summarized as:

$$\vec{v} = \mu \sum \vec{\varepsilon}_i, \qquad [4.1]$$

where $\vec{v}$ and μ are, respectively, the velocity and mobility of a Vertex and $\vec{\varepsilon}_i$ denotes each of the line tensions acting at a Vertex. Figure 7.1 illustrates a 2D grain growth simulation performed by Soares et al. for few grains thanks to this method. In this model, the mobility was assumed to be homogeneous and vertices at all times only have three possible connections, avoiding higher order multiple junctions configurations by means of unitary topological changes (only the transformations **T1** and **T2** in Figure 4.2 were considered). Moreover, μ and $|\varepsilon_i| = \varepsilon$ were assumed to be homogeneous in space and constant in time; therefore, the model was tested only under isotropic conditions. In these early works, the absence of actual curvature calculations, and thus the manipulation of exclusively straight interfaces, was crucial in proposing an efficient method compared to the computational means available at that time. Interestingly, Soares et al. exhibited in this article a very good agreement with the Burke and Turnbull law (Burke and Turnbull 1952) seemingly without being aware of it.

A few years later, by keeping the idea of grain interfaces described by straight lines between evolving multiple junctions (the vertices), that is, by the fact that the grain boundary network is defined by the positions $\vec{r}$ of the vertices and their velocities $\vec{v}$, a more general model was proposed in Kawasaki et al. (1989). Here, the authors proposed to associate the polycrystalline structure with a potential term $\nu(\vec{r})$ that describes the total surface energy and a dissipative term $R(\vec{r}, \vec{v})$ that describes the dissipation due to the movement of the GB network. These two terms can be summarized as:

$$\nu(\vec{r}) = \int_{\Gamma} \gamma \, d\Gamma, \qquad [4.2]$$

$$R(\vec{r}, \vec{v}) = \frac{1}{2} \int_{\Gamma} \frac{\|\vec{v}\|^2}{\mu} \, d\Gamma, \qquad [4.3]$$

where γ corresponds to the grain boundary energy and $\Gamma = \partial G$ depicts here the union of the GBs, that is, the GB network. In this scenario, the equations of motion are obtained by deriving them from the Lagrange function that includes a dissipative term but does not include a term for kinetic energy:

$$\frac{\partial R}{\partial \vec{v}} = -\frac{\partial \nu}{\partial \vec{r}}. \qquad [4.4]$$

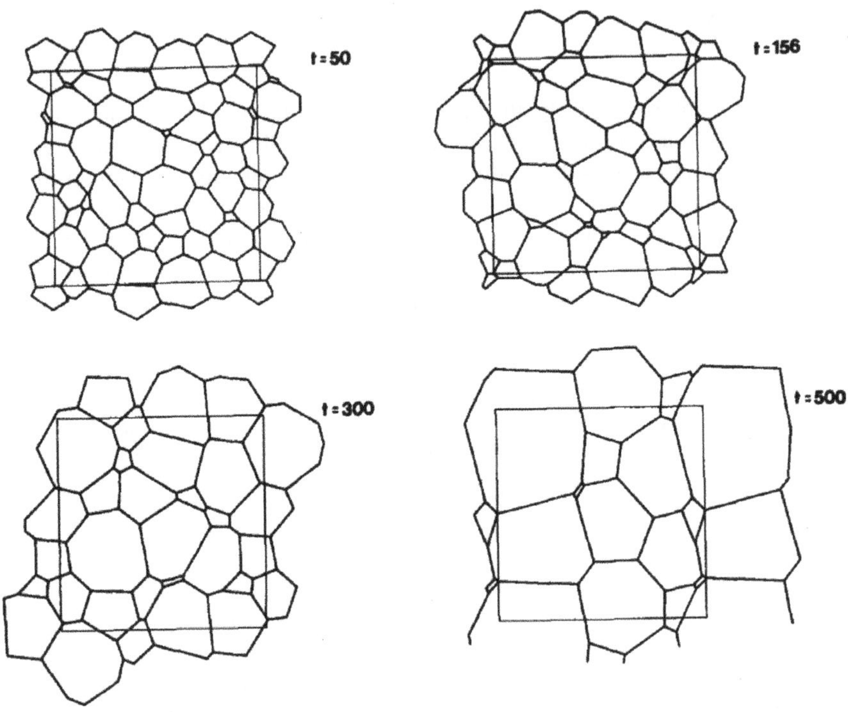

Figure 4.1. *Successive stages in the growth of a 2D grain structure with the number of steps (proportional to the physical time). Simulation from Soares et al. (1985)*

The Vertex method finally consists of transform equations [4.2]–[4.4] in a discretized way by considering the vertices and the edges between vertices (i.e. the GB) as the geometric support of the discretization. First of all, by considering N_v as the total number of vertices, N_i as the number of neighboring vertices to the vertice i $\forall i \in [\![1, N_v]\!]$, γ_{ij} as the GB energy in the ij segment and $\vec{r}_{ij} = \vec{r}_i - \vec{r}_j$:

$$\nu(\vec{r}) = \frac{1}{2}\sum_{i=1}^{N_v}\sum_{j=1}^{N_i} \gamma_{ij} \|\vec{r}_{ij}\|. \qquad [4.5]$$

Now, considering the fact that the velocity at a point k of the segment ij whose position is defined by:

$$\vec{r} = \underbrace{\frac{\|\vec{r}_{kj}\|}{\|\vec{r}_{ij}\|}}_{\xi}\vec{r}_i + \underbrace{\frac{\|\vec{r}_{ik}\|}{\|\vec{r}_{ij}\|}}_{1-\xi}\vec{r}_j,$$

is $\vec{v}(k) = ((\xi\vec{v}_i + (1-\xi)\vec{v}_j)\cdot\vec{n}_{ij})\,\vec{n}_{ij}$ with $\vec{n}_{ij}$ a unitary orthogonal vector to $\vec{r}_{ij}$; we can operate the calculation of $R(\vec{r},\vec{v})$:

$$R(\vec{r},\vec{v}) = \frac{1}{4}\sum_{i=1}^{N_v}\sum_{j=1}^{N_i}\frac{1}{\mu_{ij}}\int_0^1 \left(\xi\vec{v}_i\cdot\vec{n}_{ij} + (1-\xi)\,\vec{v}_j\cdot\vec{n}_{ij}\right)^2 \|\vec{r}_{ij}\|\,d\xi, \qquad [4.6]$$

$$R(\vec{r},\vec{v}) = \sum_{i=1}^{N_v}\sum_{j=1}^{N_i}\frac{\|\vec{r}_{ij}\|}{12\mu_{ij}}\left((\vec{v}_i\cdot\vec{n}_{ij})^2 + (\vec{v}_j\cdot\vec{n}_{ij})^2 + (\vec{v}_i\cdot\vec{n}_{ij})\,(\vec{v}_j\cdot\vec{n}_{ij})\right). \qquad [4.7]$$

Thus, the N_v equations of movement are obtained by verifying equation [4.4] in each vertices and by using equations [4.5] and [4.7]:

$$\frac{\partial R}{\partial \vec{v}_i} = -\frac{\partial \nu}{\partial \vec{r}_i}, \qquad [4.8]$$

$$\frac{\partial \nu}{\partial \vec{r}_i} = \frac{1}{2}\sum_{j=1}^{N_i}\gamma_{ij}\frac{\vec{r}_{ij}}{\|\vec{r}_{ij}\|}, \qquad [4.9]$$

$$\frac{\partial R}{\partial \vec{v}_i} = \sum_{j=1}^{N_i}\frac{\|\vec{r}_{ij}\|}{12\mu_{ij}}\left(2\,(\vec{v}_i\cdot\vec{n}_{ij})\,\vec{n}_{ij} + (\vec{v}_j\cdot\vec{n}_{ij})\,\vec{n}_{ij}\right), \qquad [4.10]$$

By using the fact that $\vec{n}\otimes\vec{n}\vec{u} = (\vec{u}\cdot\vec{n})\,\vec{u}$, the last equation can be rewritten as:

$$\frac{\partial R}{\partial \vec{v}_i} = \sum_{j=1}^{N_i}\frac{\|\vec{r}_{ij}\|}{6\mu_{ij}}\vec{n}_{ij}\otimes\vec{n}_{ij}\left(\vec{v}_i + \frac{1}{2}\vec{v}_j\right). \qquad [4.11]$$

Finally, by combining equations [4.8], [4.9] and [4.11], we obtain the well-known equation from Kawazaki et al.:

$$\frac{1}{3}\sum_{j=1}^{N_i}\frac{\|\vec{r}_{ij}\|}{\mu_{ij}}\vec{n}_{ij}\otimes\vec{n}_{ij}\left(\vec{v}_i + \frac{1}{2}\vec{v}_j\right) = -\sum_{j=1}^{N_i}\gamma_{ij}\frac{\vec{r}_{ij}}{\|\vec{r}_{ij}\|}, \qquad [4.12]$$

that is,

$$\mathcal{D}_i\vec{v}_i = \vec{f}_i - \frac{1}{2}\sum_{j=1}^{N_i}\mathcal{D}_{ij}\vec{v}_j, \text{ with,} \quad [4.13]$$

$$\mathcal{D}_{ij} = \frac{\|\vec{r}_{ij}\|}{3\mu_{ij}}\vec{n}_{ij}\otimes\vec{n}_{ij},\ \mathcal{D}_i = \sum_{j=1}^{N_i}\mathcal{D}_{ij}, \text{ and } \vec{f}_i = -\sum_{j=1}^{N_i}\gamma_{ij}\frac{\vec{r}_{ij}}{\|\vec{r}_{ij}\|}.$$

As highlighted by Kawazaki et al. (1989) 35 years ago, this equation is not very comfortable to handle since the determination of $\vec{v}_i$ cannot be calculated directly from the positions of $\vec{r}_i$ and the neighboring vertices. Thus, they proposed to deal with simplified formulations known as Model-I and Model-II. In Model-II, the anisotropy is neglected and an average is done over all directions of $\vec{n}_{ij}$ and $\vec{v}_j$ leading to simplified dissipative term, D_i, and kinetic equations:

$$R^{II}(\vec{r},\vec{v}) = \sum_{i=1}^{N_v}\sum_{j=1}^{N_i}\frac{\|\vec{r}_{ij}\|}{12\mu}\|\vec{v}_i\|^2,\ \mathcal{D}_i = \frac{1}{6\mu}\sum_{j=1}^{N_i}\|\vec{r}_{ij}\|, \quad [4.14]$$

$$\mathcal{D}_i\vec{v}_i = \vec{f}_i, \text{ that is, } \vec{v}_i = \frac{6\mu}{\sum_{j=1}^{N_i}\|\vec{r}_{ij}\|}\vec{f}_i. \quad [4.15]$$

Moreover, in Model-I, the term $\sum_{j=1}^{N_i}\|\vec{r}_{ij}\|$ is approximated as $3r(t)$ with $r(t)$ the mean value of the edge length, leading to:

$$\vec{v}_i = \frac{2\mu}{r(t)}\vec{f}_i. \quad [4.16]$$

The Model-I and Model-II were tested by Kawasaki et al. (1989) by considering the T1 and T2 transformations introduced earlier in Weaire and Kermode (1983) (see the top of Figure 4.2).

4.3. From Vertex to front-tracking then to enriched Vertex frameworks

It is interesting to note that almost simultaneously with the first Vertex articles for grain growth modeling (Soares et al. 1985; Weaire and Kermode 1983), another point of view, based on an enrichment of the description of interfaces, was proposed. The viewpoint being ultimately radically different: modeling the physical equation describing the kinetics of grain boundaries by discretizing the GB, used time-explicit Euler scheme to update the node positions at each time step, while imposing at each time the position of the triple junctions to respect a 120° equilibrium. This idea was first introduced by Frost et al. (1988) as a more precise alternative to the Vertex or MC approaches. Following the von Neumann–Mullins kinetic law for grain growth, Frost et al. (1988) considered $\vec{v} = -\tilde{\mu}\kappa\vec{n}$ to evaluate the GB migration rate with κ the local curvature and $\tilde{\mu}$ a proportionality constant that would today be called reduced mobility, that is to say the product of GB mobility and energy. This methodology is illustrated in Figure 4.3.

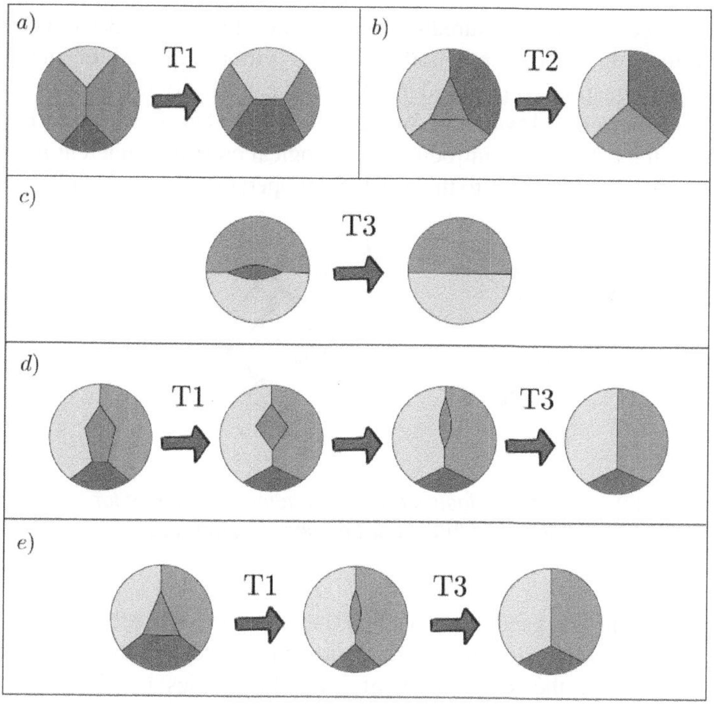

Figure 4.2. *2D Topological transformations: (a) recombination (T1), (b) annihilation of small three-sided grain (T2), (c) elimination of a grain with two triple points (T3), (d) an example of the occurrence of a T3 transformation and (e) T2 transformation shown as equivalent to a sequence of T1 and T3 transformations. Source: From Florez (2020) as inspired from Weygand et al. (1998a). For a color version of this figure, see www.iste.co.uk/bernacki/digital.zip*

Following these works, we can estimate that some confusion has arisen in the state-of-the-art between what differentiates a Vertex approach from a FT type approach in general: the fact of subdividing grain boundaries? The explicit consideration of curvature in simulations? The question of verifying a kinetic equation or considering an approach with dissipative and potential terms? In the seminal work of Frost et al. (1988), the first point is highlighted as essential in order to propose a more realistic description of GB, that is, curved. The rest of the story, however, is less clear-cut. Indeed, very quickly, new approaches based on the Kawasaki et al. principle emerged while trying to better take into account curvature (Marsh and Pande 1995). Finally, the works of Weygand et al. (1998a) extended the concept of Kawasaki et al. by implementing a Vertex model capable of modeling curved grain boundaries. The model used *real* and *virtual* vertices, where virtual vertices were used to discretize the grain boundaries between triple junctions (real vertices). Weygand et al. also directly used equation [4.13] in order to compute the velocity values of all vertices (both real and virtual) instead of the Model-II approximation equation [4.14] used by Kawasaki et al. (1989) in their study. An illustration of the seminal works of Weygand et al. (1998a) is shown in Figure 4.4. Furthermore, the authors also introduce a topological operation inherent in the use of virtual vertices, corresponding to the topological operation **T3** given in Figure 4.2.

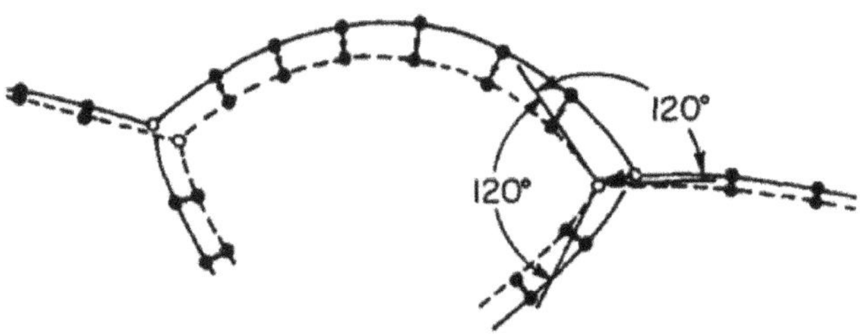

Figure 4.3. *Schematic illustration of simulation techniques for modeling boundary migration and triple point motion. Source: Frost et al. (1988)*

Additionally, in Weygand et al. (1998a) the notion of microstepping was applied for the first time. Microstepping is necessary given the unstable nature of some vertices that disrupt the smoothness of grain boundaries (vertices that are too off-positioned regarding the tendency of their contiguous vertices) and which may appear after a topological change or another numerical phenomenon. Microstepping was then implemented using the following reasoning: no segment should shrink nor extend more than a fraction $f = dl/l$ of its length l (where the reported value for f was 0.5), leading to a maximum time step allowed per Vertex. Finally, the movement of Vertex i was performed using a time-explicit Euler scheme both for the hole steps and for the micro steps. Using this model, Weygand et al. (1998a) found that multiple junctions evolve to fulfill their equilibrium angles.

Today, ambiguity sometimes persists, as some articles refer to the use of the Kawasaki principle to discuss the Vertex approach. There are numerous Vertex papers discussing kinetic approaches based on curvature to describe interfaces, themselves discretized. From my perspective, these approaches align more closely with front-tracking methodologies, as originally conceptualized by Frost et al. (1988), rather than the original Vertex methods.

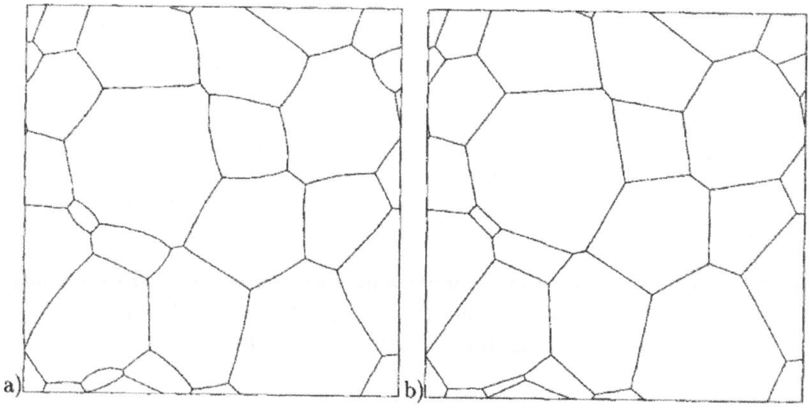

Figure 4.4. *Grain structure obtained by Weygand et al. (1998a) by considering virtual vertices along GBs (a) and without virtual vertices (b)*

4.4. Enrichment of the Vertex approach

The works of Weygand et al. (1998b, 1999b) then lead to a series of publications in which the influence of the reduced mobility at multiple junctions and grain boundaries on grain growth was studied. However, a point quickly discussed in the literature was the ability, compared to large-scale Potts model simulations (Anderson et al. 1984), to extend the existing 2D Vertex formalism to 3D while retaining the appeal of the approach, that is, its accuracy and reasonable computational cost. The first substantial work on the subject can be attributed to Nagai et al. (1990), who, starting from the Kawasaki model applied to 2D surfacic triangulation, focused on extending the possible topological operations to multiple junctions as illustrated in Figure 4.5 (left side), but producing quite limited 3D simulations as illustrated in Figure 4.5 (right side). Based on the work of Nagai et al. (1990), first representative simulations were proposed in Kazuhiro Fuchizaki and Kawasaki (1995) and Weygand et al. (1999a). The size, representativeness and complexity of the calculations then increased over the years, as illustrated by the work of Barrales Mora et al. (2008) on the subject. These works of increasing complexity are illustrated in Figure 4.6.

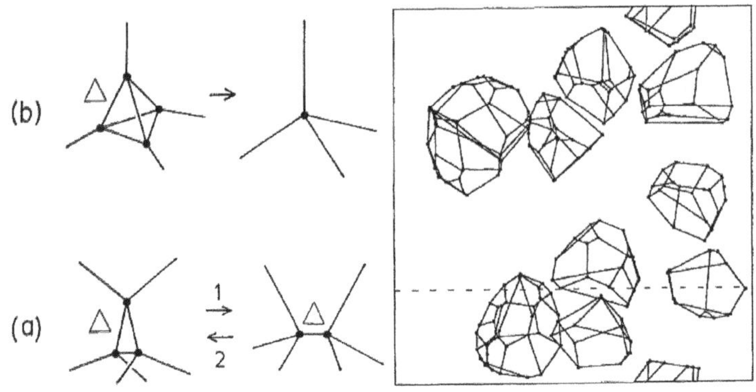

Figure 4.5. *Left side: Three-dimensional elementary processes: (a) recombination process; (b) tetrahedron annihilation. Right side: First 3D Vertex simulations. Source: Nagai et al. (1990)*

Vertex models continue to be developed given their very attractive way of handling phenomenon such as grain boundary anisotropy. Indeed, one of the strengths of this approach lies in its ability to precisely describe the behavior of multiple junctions and to incorporate more or less elaborate models to describe the heterogeneity of mobility and GB energy. Typically, as illustrated by equation 4.12, even in the earlier model of Kawasaki et al. (1989), μ and γ through their local GB values μ_{ij} and γ_{ij}, respectively, could be defined as dependent on the misorientation tensor. Moreover, the addition of virtual vertices also allows for consideration of the evolution of inclination along grain boundaries as an additional parameter to describe interface energy. Thus, the usual 5 degrees of freedom for describing interface energy can be immediately taken into account in these approaches. A step forward consists of being able to take into account torque terms in the grain growth driving pressure when the reduced mobility is assumed to be inclination dependent. In the context of the Vertex method, this development was first proposed by Barrales Mora as an improvement of the Vertex framework in Weygand et al. (Barrales Mora 2010). Figure 4.7 illustrates the predictions of the Barrales Mora (2010) Vertex model for the evolution of a triple junction with different levels of heterogeneity regarding reduced mobility.

Smith–Zener pinning mechanism by static second phase particles (SPP) was initially introduced by Smith (1948) and further elaborated by Zener (1949) a year later. Smith and Zener have highlighted how precipitates serve as barriers to grain boundaries movement, potentially stopping grain growth which presents significant industrial interest in the metal forming processes. In specific scenarios, SPP may anchor the microstructure, ultimately resulting in a characteristic limiting mean grain size (MGS) as a result of what is known as Smith–Zener pinning. This subject has thus been studied with all of the existing full-field methods, and the Vertex approach is no exception to the rule.

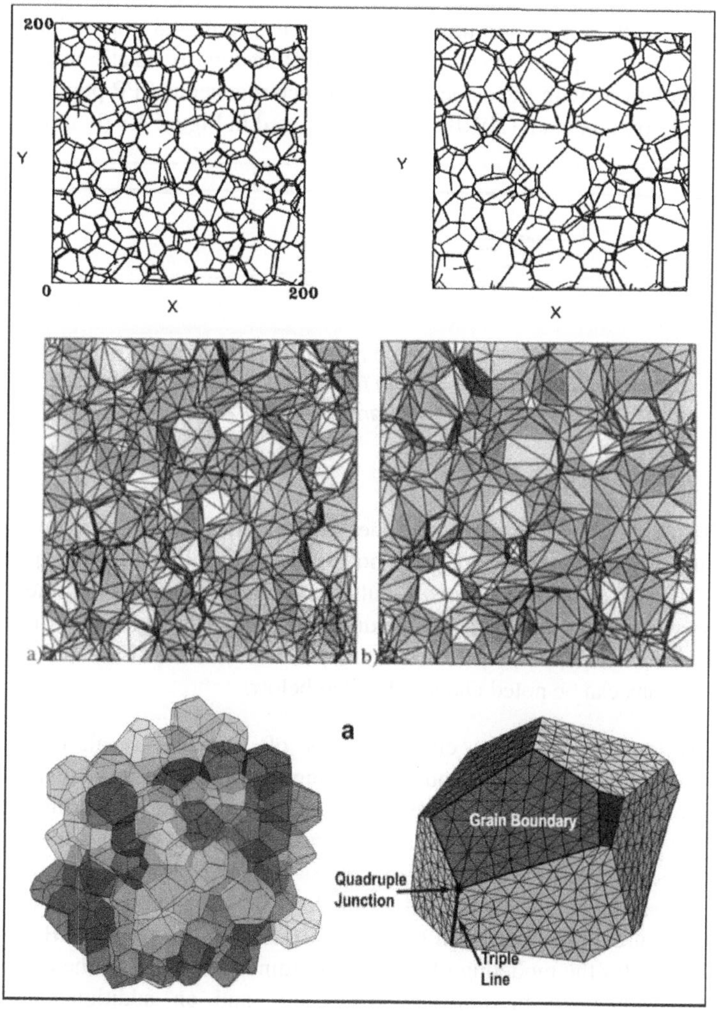

Figure 4.6. *First 3D representative simulations realized with the Vertex method: (top) Kazuhiro Fuchizaki and Kawasaki (1995); (middle) Weygand et al. (1999a); (bottom) Barrales Mora et al. (2008). For a color version of this figure, see www.iste.co.uk/bernacki/digital.zip*

Based on its own Vertex model (Weygand et al. 1998a), the first Vertex attempt was proposed by Weygand et al. (1999c) by considering the SPP as static vertices and a pinning position for the evolving GB (Weygand et al. 1999c). An unpinning force was proposed in agreement with the Smith–Zener model in order to model unpinning events. Figure 4.8 (top left side) illustrates the elementary process for unpinning GBs from ponctual SPP proposed in this model, whereas Figure 4.8 (top middle and right side) illustrates a 2D grain growth case with SPP leading to a static microstructure. Another simulations with the same model and heterogeneous SPP populations were proposed in Lépinoux et al. (2010) and are illustrated in Figure 4.8 (bottom).

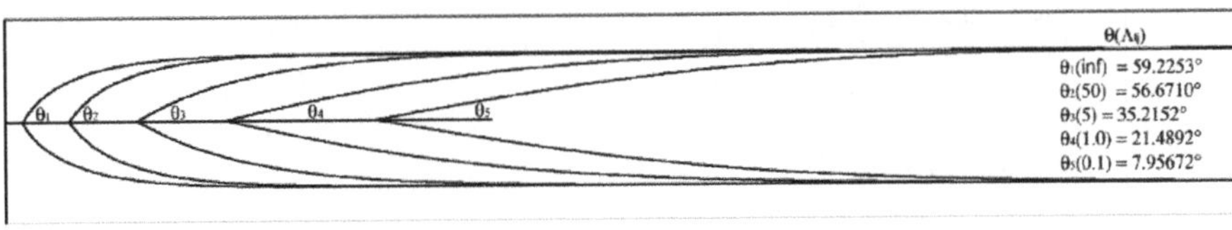

Figure 4.7. *The equilibrium shape of the grain boundaries as a function of the triple junction mobility in the Barrales Mora Vertex model. Adapted from Barrales Mora (2010)*

The basic concept of Vertex approaches is to gain a spatial resolution dimension by focusing only on the surface of grain boundaries. Thus, this method is particularly suited to the problem of grain growth, but is much less suited to other metallurgical mechanisms involving intragranular features. As such, this method has rarely been extended to other mechanisms, such as recrystallization or solid-state phase changes. Some exceptions can be noted and are detailed below.

As first tentative, we can cite the work of Weygand et al. (2000) where post-dynamic recrystallization through a bulging mechanism is investigated with the introduction of a heterogeneous value of reduced mobility, but where, finally, driving pressure due to stored plastic energy is not discussed nor considered. A more physical approach was proposed by Piękos et al. (2008a and 2008b) a few years later in the context of post-dynamic recrystallization modeling. This approach is in fact more a mixture of a pure Vertex approach (without virtual vertices) for the GB description and an MC one for the modeling of vertices coordinates evolution. The plastic stored energy is assumed to be constant per grain, and an additional topological transformation (TA) must be taken into account due to the additional driving pressure potentially leading to the evolution of a grain boundary in the direction opposite to its curvature center. This transformation is illustrated in Figure 4.9 (top) and a recrystallization case is shown in Figure 4.9 (bottom). We cite, as advanced work, the model proposed in Mellbin et al. (2015), where crystal plasticity model is coupled with a graph-based Vertex algorithm. The crystal plasticity model is utilized within a classical finite element framework, enabling to evaluate the plastic stored energy within the polycrystal microstructure while simultaneously allowing the crystal lattices within the grains to reorient. This process impacts the advancement of recrystallization because nucleation takes place at locations where enough plastic stored energy is present, considering that the mobility and energy of grain boundaries can change on the basis of the crystallographic misorientation along the boundaries. The introduced graph-based Vertex model enables to model the TAs that occur in the grain microstructure and maintains awareness of the connectivity between grains.

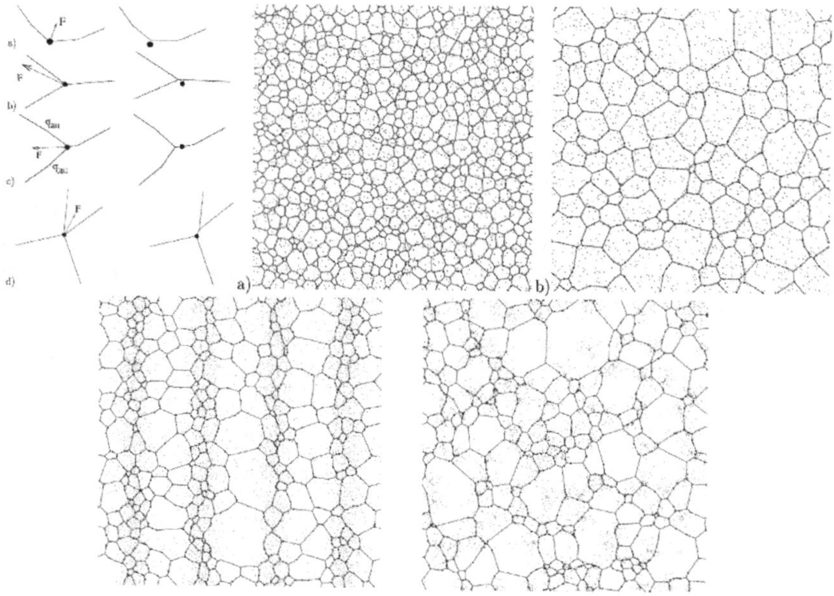

Figure 4.8. *Top: elementary process for unpinning process as proposed by Weygand et al. (1999c) and a first Vertex simulation taken into account SPP with the initial and the final stable microstructure. Bottom: Vertex GG simulation with an heterogeneous population of SPP proposed by Lépinoux et al. (2010) and based on the model in Weygand et al. (1999c)*

Finally, we can also mention the versatile and collaborative Elle Numerical Simulation Platform initiated by Jessell et al. (2001), primarily used today in geology to model deformation mechanisms, recrystallization and grain growth in minerals. It is interesting to note that the original formalism proposed in Elle to describe the migration of grain boundaries is very close to a Vertex approach in the sense of Frost et al. (1988). To the author knowledge, post-dynamic recrystallization has never been investigated in 3D by the Vertex approach, nor has continuous or discontinuous dynamic recrystallization.

Further comments are necessary upon reviewing the state of the art. First, surprisingly, it seems that Vertex approaches have not been or have rarely been confronted with the modeling of surface mechanisms involving derivatives of the curvature such as in surface diffusion mechanism (e.g. in sintering or globularization). Second, the aspects of parallelization and parallel computing seem to have been little studied in the Vertex context. This can be understood in 2D given the method's efficiency for polycrystals of reasonable size, but it seems more surprising in the quest for massive 2D or 3D simulations. Third, if, as detailed previously, the issues of Smith–Zener pinning have been addressed in a Vertex context, it remains that the proposed approaches suffer from numerous limitations in terms of representativeness. Indeed, from the way to introduce SPP by static vertices, the SPP/GB interactions are poorly described, and the assumptions made seem realistic only for particle sizes that are negligible compared to the grain size, which is not representative of numerous metallic alloys. Finally, it seems that the approach has also not been adapted for the description of more realistic experimental microstructures considering the morphology of regular grain boundaries or special grain boundaries like coherent twins. All of these comments have probably contributed to the development of FT methods, sometimes incorporating some ideas from Vertex in managing multiple junctions but diverging in the description of grain boundaries and the modeling of their kinetics. These methods are introduced in the following section.

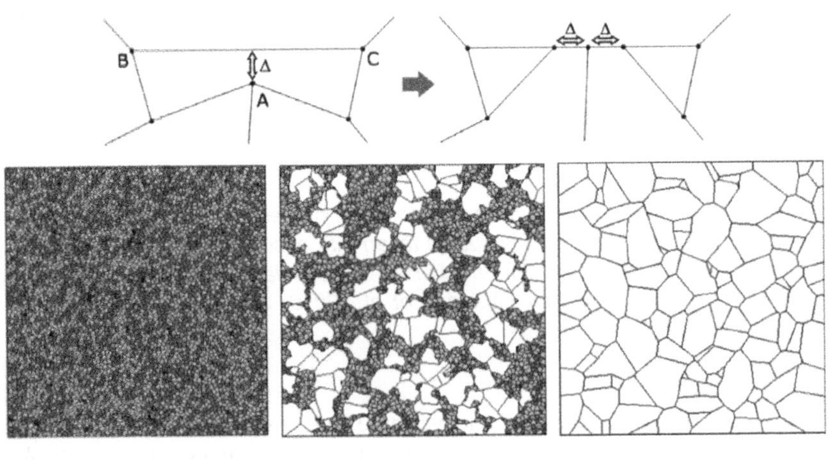

Figure 4.9. *Top: new TA topological transformation proposed by Piękos et al. (2008b) Bottom from left to right: evolution of the microstructure predicted by the Piękos et al. (2008b) model. (a) Initial microstructure made of 4000 grains, (b) after 50% of recystallization and (c) fully recrystallized. The color code depicts the stored energy. Adapted from Piękoś et al. (2008b)*

4.5. Other front-tracking frameworks for the modeling of microstructure evolution

FT approaches different from the Vertex model can be found in the literature. In Brakke (1992), the authors introduce, in the context of 2D GG, a FT methodology where instead of using a dissipative equation of motion, a variational approximation of the curvature of the interface is preferred to compute the velocity at every Vertex except for those being triple junctions (see section 6.4 of Brakke (1992) for more information). In fact, triple junctions are placed to respect the equilibrium in a similar manner to the works of Frost et al. (1988). Moreover, the study presented in Brakke (1992) is dedicated not only to the modeling of 2D GG but also to the more general context of surface energy minimization, for which the author gives many examples in 2D and 3D through a dedicated code called "SurfaceEvolver".

Kuprat proposed in 1998 a new code "GRAIN3D" to simulate grain growth on a volumetric mesh through the gradient weighted moving finite element (GWFE) method (Kuprat 2000). However, the method's approach to topological transitions may not align with physical realities, potentially leading to significant deviations in microstructural development. On a different note, Shya and Weygand (2009) developed a technique called "generalized Vertex dynamics model" for modeling grain growth on a surface mesh while managing topological changes by breaking them down into simple steps.

Another example of FT models is given in Couturier et al. (2003), where a finite element model, based on a variational formulation for grain boundary motion by viscous drag, is used to solve the equations governing grain boundary motion of an arbitrary-shaped surface and its interaction with SPP in 3D. The model was later extended in Couturier et al. (2004, 2005, 2003) to take into account motion by curvature flow but in the context of a single grain boundary and led to impressive simulations of Smith–Zener pinning mechanisms comparatively to the state of the art, as illustrated in Figure 4.10(top). However, real polycrystal structures were not tested with this approach. A similar discussion was recently proposed in Mohles (2020) with comparable simulations, but a different approach to describe the interaction between SPP and grain boundary allowing to consider a richer description of SPP in terms of shape and energy relationship with the matrix as illustrated in Figure 4.10(bottom).

Another FT model was proposed in Becker et al. (2008) where the movement of grain boundaries is defined through the minimization of the total energy of the system by moving each node along the energy gradient toward a lower total energy state. In this model, the energy term can be defined via stored energy, grain boundary energy and nonconservative reaction terms. The evaluation of the energy gradient field is determined for every node via a local finite-difference method centered at the node in question, and the movement is made in the direction of the maximal energy reduction. The authors tested this model in the context of isotropic and anisotropic grain boundary migration, where melting was simulated through a second phase, attributing an "energy-based" melting-fraction function, where the more the melting fraction deviates from the target value, the more the melting/crystallization energy dominates. However, no information regarding the grain boundary anisotropy function was given, even though some results show anisotropic behavior between the solid–solid and solid–liquid boundaries.

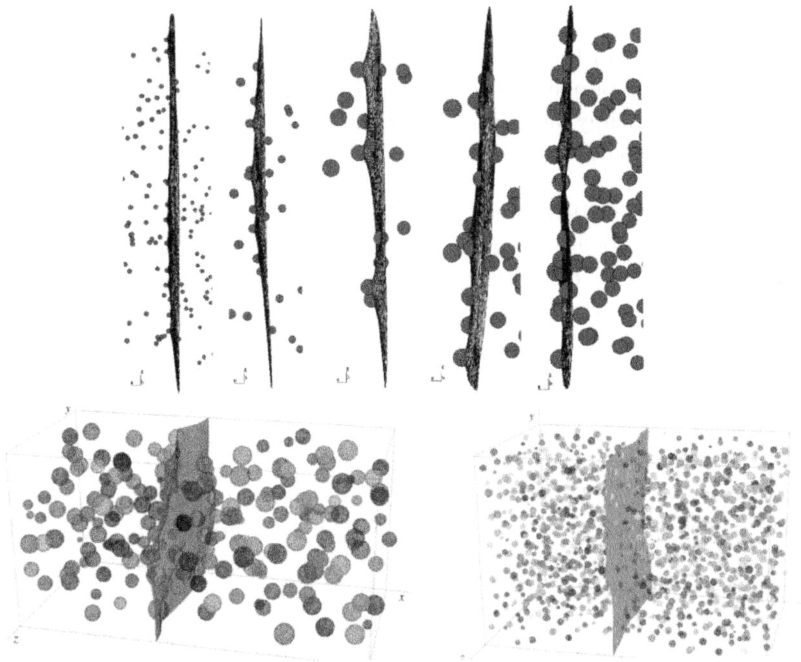

Figure 4.10. *(Top) Time evolution in 3D of one grain boundary evolving in a cloud of spherical static particles (from Couturier et al. (2005)), and (bottom) similar simulation from Mohles (2020). For a color version of this figure, see www.iste.co.uk/bernacki/digital.zip*

In 2011, Lazar et al. (2011) introduced a method based on a discretized version of the MacPherson Srolovitz relationship (MacPherson and Srolovitz 2007) to simulate ideal grain growth on a surface mesh and with uniform grain boundary characteristics, necessitating only a limited number of topological changes. Figure 4.11 illustrates a 3D ideal grain growth simulation performed with this method.

Recently, Eren and Mason (2021) and Eren et al. (2022) proposed a new 3D approach based on 3D finite element mesh. This approach presents the interest to promote optimal fitted mesh in terms of numerical cost to model grain growth (Eren et al. 2022). Moreover, a large investigation of 3D topological transitions was performed in the context of the use of unstructured finite element mesh (Eren and Mason 2021). This choice also enables to use efficient and parallelized finite-element libraries for FE resolution and remeshing operations. Today, it is the most advanced work in the context of 3D mesh-based FT approaches for grain growth.

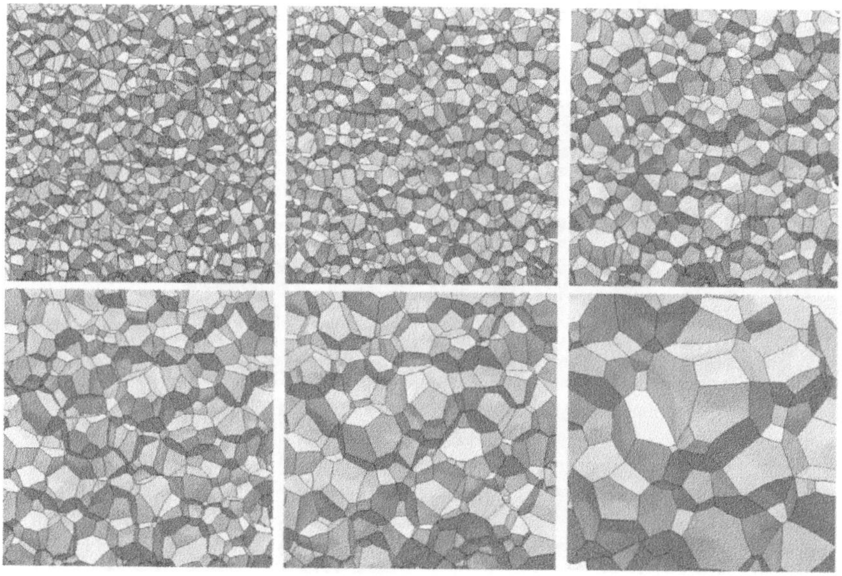

Figure 4.11. *From top to bottom and left to right: time evolution of a 3D pure grain growth case realized with a MacPherson Srolovitz based front-tracking method. Adapted from Lazar et al. (2011)*

In the context of 2D generic FT models for numerous mechanisms, we can also cite the recent developments of Florez et al. concerning the ToRealMotion (TRM) code for "topological remeshing in Lagrangian framework for large interface motion" (Florez et al. 2020). The TRM approach introduces the concept of using unstructured FE meshes for the detailed representation of grain interiors. This strategy is adopted for several key reasons: first, to incorporate intragranular data such as stored energy to model recrystallization; second, to accurately simulate significant domain deformations and discontinuous dynamic recrystallization mechanisms (Florez et al. 2021a); and third, to enhance the parallel processing capabilities of the method, surpassing those of conventional FT models (Florez et al. 2021b). This method was also improved, considering the treatment of multiple junctions proposed by Barrales Mora (2010), to deal with anisotropic reduced mobility (Florez et al. 2021c) and torque terms (Florez et al. 2022).

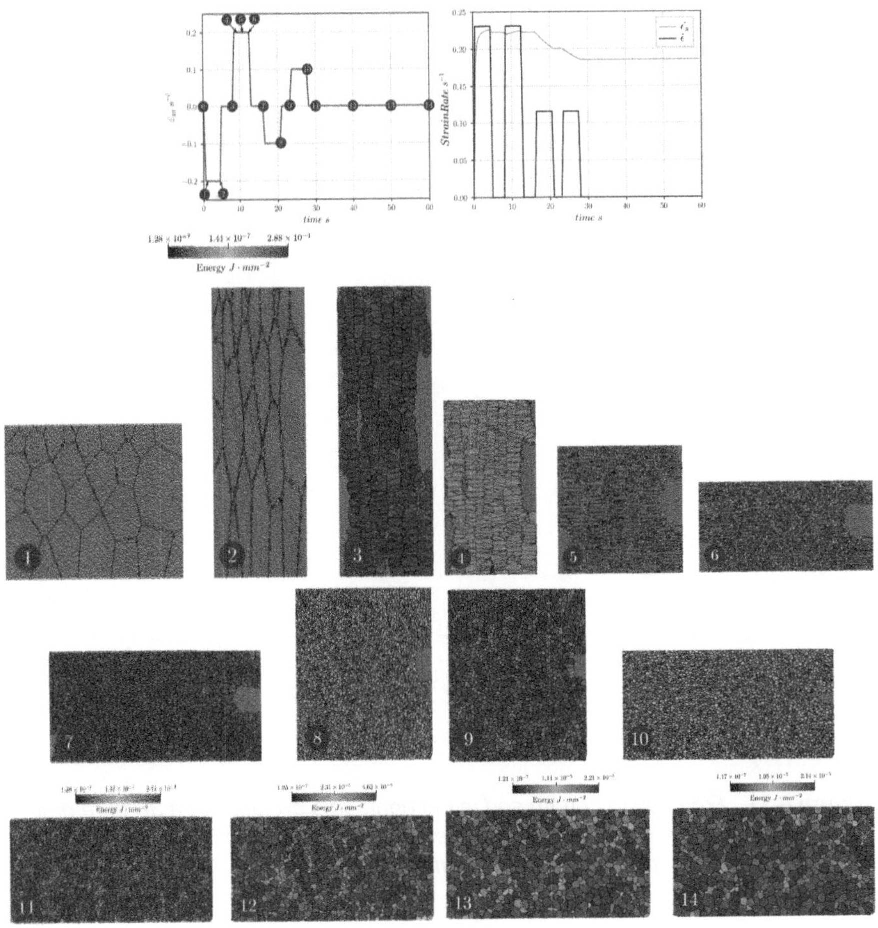

Figure 4.12. *This figure illustrates the use of TRM code in the context of the modeling of a complex industrial thermomechanical path applied to a 304L stainless steel and described at the top (strain and strain rate, the temperature being constant and equal to 1080°C). The predicted microstructures at each numerated position in the path are described by exhibiting the stored energy field and the grain boundaries in black. This simulation involves discontinuous dynamic recrystallization, meta-dynamic recrystallization and grain growth mechanisms. Adapted from Florez et al. (2021a). For a color version of this figure, see www.iste.co.uk/bernacki/digital.zip*

Figure 4.12 illustrates the use of TRM in the context of discontinuous dynamic recrystallization modeling with a complex industrial thermomechanical treatment that includes large deformation and involves discontinuous dynamic recrystallization, meta-dynamic recrystallization and grain growth mechanisms.

4.6. Conclusion

In conclusion, Vertex approaches represent a critical methodology in the field of computational materials science. These methods, which take advantage of the precise discretization of grain boundaries, offer a sophisticated framework for understanding and predicting the evolution of microstructures under various conditions. Vertex models stand out for their ability to accurately model complex grain boundary movements, including curvature-driven growth with anisotropic grain boundary properties. Furthermore, their computational efficiency makes them particularly suitable for study of large polycrystals compared to FC methods such as multi-phase field and level-set ones. Due to some challenges such as the need for more generic models to capture all relevant phenomena and the difficulty to develop 3D models, Vertex approaches continue to evolve mainly through new FT approaches where some ingredients of the original Vertex approaches are used such as the treatment of the multiple junctions. With ongoing developments in computational power and numerical techniques, these methods are set to provide even deeper insights into grain boundary modeling, supporting advancements in materials design and engineering.

4.7. References

Anderson, M., Srolovitz, D., Grest, G., Sahni, P. (1984). Computer simulation of grain growth kinetics. *Acta Metallurgica*, 32(5), 783–791.

Barrales Mora, L.A. (2010). 2D vertex modeling for the simulation of grain growth and related phenomena. *Mathematics and Computers in Simulation*, 80(7), 1411–1427. doi: 10.1016/j.matcom.2009.08.005.

Barrales Mora, L.A., Gottstein, G., Shvindlerman, L. (2008). Three-dimensional grain growth: Analytical approaches and computer simulations. *Acta Materialia*, 56(20), 5915–5926.

Becker, J.K., Bons, P.D., Jessell, M.W. (2008). A new front-tracking method to model anisotropic grain and phase boundary motion in rocks. *Computers and Geosciences*, 34(3), 201–212.

Brakke, K.A. (1992). The surface evolver. *Experimental Mathematics*, 1(2), 141–165.

Burke, J. and Turnbull, D. (1952). Recrystallization and grain growth. *Progress in Metal Physics*, 3, 220–292 [Online]. Available at: https://www.sciencedirect.com/science/article/pii/0502820552900099.

Couturier, G., Maurice, C., Fortunier, R. (2003). Three-dimensional finite-element simulation of Zener pinning dynamics. *Philosophical Magazine*, 83(30), 3387–3405.

Couturier, G., Maurice, C., Fortunier, R., Doherty, R., Driver, J.H. (2004). Finite element simulations of 3D Zener pinning. *Materials Science Forum*, 467, 1009–1018.

Couturier, G., Doherty, R., Maurice, C., Fortunier, R. (2005). 3D finite element simulation of the inhibition of normal grain growth by particles. *Acta Materialia*, 53(4), 977–989.

Eren, E. and Mason, J.K. (2021). Topological transitions during grain growth on a finite element mesh. *Phys. Rev. Mater.*, 5, 103802.

Eren, E., Runnels, B., Mason, J. (2022). Comparison of evolving interfaces, triple points, and quadruple points for discrete and diffuse interface methods. *Computational Materials Science*, 213, 111632.

Florez, S. (2020). Towards highly efficient massive-multidomain simulations in the context of microstructural evolutions. PhD Thesis, MINES Paris PSL, Paris.

Florez, S., Alvarado, K., Muñoz, D.P., Bernacki, M. (2020). A novel highly efficient lagrangian model for massively multidomain simulation applied to microstructural evolutions. *Computer Methods in Applied Mechanics and Engineering*, 367, 113107.

Florez, S., Alvarado, K., Bernacki, M. (2021a). A new front-tracking lagrangian model for the modeling of dynamic and post-dynamic recrystallization. *Modelling and Simulation in Materials Science and Engineering*, 29(3), 035004.

Florez, S., Fausty, J., Alvarado, K., Murgas, B., Bernacki, M. (2021b). A 2D front-tracking Lagrangian model for the modeling of anisotropic grain growth. *Materials*, 14(15), 4219.

Florez, S., Fausty, J., Alvarado, K., Murgas, B., Bernacki, M. (2021c). Parallelization of an efficient 2D-Lagrangian model for massive multi-domain simulations. *Modelling and Simulation in Materials Science and Engineering*, 29(6), 065005.

Florez, S., Alvarado, K., Murgas, B., Bozzolo, N., Chatain, D., Krill, C.E., Wang, M., Rohrer, G.S., Bernacki, M. (2022). Statistical behaviour of interfaces subjected to curvature flow and torque effects applied to microstructural evolutions. *Acta Materialia*, 222, 117459.

Frost, H.J. and Thompson, C.V. (1996). Computer simulation of grain growth. *Current Opinion in Solid State and Materials Science*, 1(3), 361–368 [Online]. Available at: https://www.sciencedirect.com/science/article/pii/S135902869680026X.

Frost, H.J., Thompson, C.V., Howe, C.L., Whang, J. (1988). A two-dimensional computer simulation of capillarity-driven grain growth: Preliminary results. *Scripta Metallurgica*, 22(1), 65–70.

Fullman, R. (1952). Boundary migration during grain growth. *Metal Interfaces*, 179–207.

Jessell, M., Bons, P., Evans, L., Barr, T., Stüwe, K. (2001). Elle: The numerical simulation of metamorphic and deformation microstructures. *Computers & Geosciences*, 27(1), 17–30.

Kawasaki, K., Nagai, T., Nakashima, K. (1989). Vertex models for two-dimensional grain growth. *Philosophical Magazine B*, 60(3), 399–421. doi: 10.1080/13642818908205916.

Kazuhiro Fuchizaki, T.K. and Kawasaki, K. (1995). Computer modelling of three-dimensional cellular pattern growth. *Philosophical Magazine B*, 71(3), 333–357.

Kuprat, A. (2000). Modeling microstructure evolution using gradient-weighted moving finite elements. *SIAM Journal on Scientific Computing*, 22(2), 535–560.

Lazar, E.A., Mason, J.K., MacPherson, R.D., Srolovitz, D.J. (2011). A more accurate three-dimensional grain growth algorithm. *Acta Materialia*, 59(17), 6837–6847.

Lépinoux, J., Weygand, D., Verdier, M. (2010). Modeling grain growth and related phenomena with vertex dynamics. *Comptes Rendus Physique*, 11(3–4), 265–273 [Online]. Available at: https://linkinghub.elsevier.com/retrieve/pii/S1631070510000800

MacPherson, R.D. and Srolovitz, D.J. (2007). The von Neumann relation generalized to coarsening of three-dimensional microstructures. *Nature*, 446(7139), 1053–1055.

Marsh, S.P. Masumaru, R.A., Pande, C.S. (1995). A curvature-driven vertex model for two-dimensional grain growth. *Philosophical Magazine Letters*, 72(6), 429–434.

Mellbin, Y., Hallberg, H., Ristinmaa, M. (2015). A combined crystal plasticity and graph-based vertex model of dynamic recrystallization at large deformations. *Modelling and Simulation in Materials Science and Engineering*, 23(4), 045011.

Mohles, V. (2020). 3-D front tracking model for interfaces with anisotropic energy. *Computational Materials Science*, 176, 109534.

Mullins, W.W. (1956). Two-dimensional motion of idealized grain boundaries. *Journal of Applied Physics*, 27(8), 900–904. doi: 10.1063/1.1722511.

Nagai, T., Ohta, S., Kawasaki, K., Okuzono, T. (1990). Computer simulation of cellular pattern growth in two and three dimensions. *Phase Transitions*, 28(1–4), 177–211.

von Neumann, J. (1952). *Metal Interfaces*. American Society for Metals, Cleveland.

Piękoś, K., Tarasiuk, J., Wierzbanowski, K., Bacroix, B. (2008a). Generalized vertex model of recrystallization – Application to polycrystalline copper. *Computational Materials Science*, 42(4), 584–594.

Piękoś, K., Tarasiuk, J., Wierzbanowski, K., Bacroix, B. (2008b). Stochastic vertex model of recrystallization. *Computational Materials Science*, 42(1), 36–42.

Smith, C.S. (1948). Introduction to grains, phases, and interfaces – An interpretation of microstructure. *Transactions of the American Institute of Mining and Metallurgical Engineers*, 175(2), 15–51.

Soares, A., Ferro, A., Fortes, M. (1985). Computer simulation of grain growth in a bidimensional polycrystal. *Scripta Metallurgica*, 19(12), 1491–1496 [Online]. Available at: https://www.sciencedirect.com/science/article/pii/0036974885901577.

de Sousa, F.S., Mangiavacchi, N., Nonato, L.G., Castelo, A., Tomé, M.F., Ferreira, V.G., Cuminato, J.A., McKee, S. (2004). A front-tracking/front-capturing method for the simulation of 3D multi-fluid flows with free surfaces. *Journal of Computational Physics*, 198(2), 469–499.

Syha, M. and Weygand, D. (2009). A generalized vertex dynamics model for grain growth in three dimensions. *Modelling and Simulation in Materials Science and Engineering*, 18(1), 015010.

Weaire, D. and Kermode, J.P. (1983). Computer simulation of a two-dimensional soap froth. *Philosophical Magazine B*, 48(3), 245–259.

Weygand, D., Bréchet, Y., Lépinoux, J. (1998a). A vertex dynamics simulation of grain growth in two dimensions. *Philosophical Magazine B*, 78(4), 329–352. doi: 10.1080/13642819808206731.

Weygand, D., Bréchet, Y., Lépinoux, J. (1998b). Influence of a reduced mobility of triple points on grain growth in two dimensions. *Acta Materialia*, 46(18), 6559–6564.

Weygand, D., Bréchet, Y., Lépinoux, J. (1999a). Reduced mobility of triple nodes and lines on grain growth in two and three dimensions. *Interface Science*, 7(3), 285–295.

Weygand, D., Bréchet, Y., Lépinoux, J. (1999b). Zener pinning and grain growth: A two-dimensional vertex computer simulation. *Acta Materialia*, 47(3), 961–970 [Online]. Available at: https://linkinghub.elsevier.com/retrieve/pii/S1359645498003838.

Weygand, D., Bréchet, Y., Lépinoux, J., Gust, W. (1999c). Three-dimensional grain growth: A vertex dynamics simulation. *Philosophical Magazine B: Physics of Condensed Matter; Statistical Mechanics, Electronic, Optical and Magnetic Properties*, 79(5), 703–716.

Weygand, D., Bréchet, Y., Lépinoux, J. (2000). On the nucleation of recrystallization by a bulging mechanism: A two-dimensional vertex simulation. *Philosophical Magazine B: Physics of Condensed Matter; Statistical Mechanics, Electronic, Optical and Magnetic Properties*, 80(11), 1987–1996.

Zener, C. (1949). Theory of growth of spherical precipitates from solid solution. *Journal of Applied Physics*, 20(10), 950–953.

5

Phase Field: Theory, Numerical Implementation and Applications

Ingo STEINBACH and Oleg SHCHYGLO
Interdisciplinary Centre for Advanced Materials Simulation (ICAMS), Ruhr University Bochum, Germany

5.1. Introduction

Phase field – we will use this phrase for the whole discipline of phase-field theory – modeling and simulations, as well as applications, stems from basically two different routes: (a) thermodynamics and interfaces; (b) wave mechanics. The first route is based on the Ginzburg–Landau theory of phase transformations, where an order parameter ϕ is introduced characterizing the ordered state $\phi = 1$ in contrast to the disordered state $\phi = 0$. A well-known example is magnetism, where the ordered state is called "ferromagnetic" with preferentially parallel elementary spins in contrast to the disordered "paramagnetic" state. Also, we may consider solidification, where liquid is disordered while the solid will show crystalline order. In phase field, we treat this order parameter as a field in space $\vec{x}$ and time t, $\phi \rightarrow \phi(\vec{x}, t)$: each material point can have an ordering state different from its neighboring point, and the state can change in time. This approach to phase field will be detailed in section 5.3.

The second route to phase field bases on wave mechanics, the so-called soliton solution (Scott et al. 1973) of the Korteveg–De Vries equation, is a nonlinear wave equation. The soliton describes a wave front which is self-similar in an

Digital Materials,
coordinated by Marc BERNACKI and Samuel FOREST.

eigen-coordinate $\vec{\xi} = \vec{x} - \vec{v}t$ moving with velocity v, also called traveling wave solution. This will be detailed in section 5.2. The reason for using this solution is that we want to describe the evolution of domains of different order in a material in time $t_1 \to t_2$: $\phi(\vec{x}, t_1) \to \phi(\vec{x}, t_2)$.

What have the two approaches in common? Nothing. Ginzburg–Landau tells us that materials like to form domains of the same order (due to minimization of the total free energy of the system) and that there must be an interface between the domains with an energy penalty. The system evolves to minimize its total free energy, that is, domain walls (interfaces) must evolve, must move. The soliton solution tells us how such an interface between domains can move without changing its characteristics. This may seem trivial to the reader, but it is the main challenge in a numerical simulation of evolving domain or microstructures in materials. The different methods presented in this book, namely, Vertex (see Chapter 4), level-set (see Chapter 6) and cellular automata (see Chapter 7), and phase field represent different approaches. We seek a stable description of the interface between domains: stable against numerical errors in a simulation. This is not a trivial task, since the respective transport equation,

$$\dot{\phi}(\vec{x}, t) = \vec{v} \cdot \vec{\nabla}\phi(\vec{x}, t), \qquad [5.1]$$

is of hyperbolic type with nasty numerical properties up to shock wave solutions. Of course, there are good numerical procedures to cope with this problem, in particular in fluid mechanics. Phase-field offers an elegant alternative procedure by transferring the equation to parabolic type: equation [5.1] will be transformed into a "diffusion" type equation. This is hard to believe, but we will see.

What have the two methods, Ginzburg–Landau and Korteveg–De Vries, in common? The solution! Phase field!

This solution will also provide a numerically cheap method for the evaluation of the local curvature of the front, that is, if the wave front represents a non-planar interface, in 2D or 3D. Curvature of the front is related to capillarity in materials science, which is a dominating mechanism, in particular in nano-sized microstructures.

Here, we shall stress the similarity of phase field and level set with respect to the propagation of a self-similar wave representing an interface. The methods, however, are quite distinct in the numerical approach or evaluation of curvature.

The application of phase-field mostly addresses microstructural evolution in metallic materials, from dendritic and eutectic solidification, solid state transformation and precipitation to grain growth and recrystallization. We will highlight examples from the authors' own research in section 5.7. Special phase-field models address viscous fingering in fluid mechanics (Nguyen et al. 2010), fracture (Hakim and Karma 2009; Miehe et al. 2010) and shape optimization (Jantos et al. 2016). Also, applications in biology (Lázaro et al. 2015) are known. The references above are very elusive without any effort to give a complete and balanced (with respect to different branches of the theory) literature review, but are put here to demonstrate the spectrum of applications beyond the focus of the present review. Also, a quantization of the theory has also been attempted by the first author, which is recommended for interested readers (Steinbach 2017; Kundin and Steinbach 2020).

5.2. The soliton solution of a propagating wave front

We start from the transport equation [5.1] in 1D (Steinbach and Salama 2023). The velocity v will be defined later by a so-called driving pressure Δg with a mobility parameter M^ϕ: $v = M^\phi \Delta g$. Both quantities, M^ϕ and Δg, will be defined as complicated functions of the, for example, crystalline state, orientation, temperature, solute, pressure and magnetic forces. We need sophisticated constitutive material models for this in order to relate our simulations to real materials. For the numerical problem, which we treat within this chapter, we assume they are given. For now, to start easy, we consider them as constants. As said before, this hyperbolic type of equation [5.1] is difficult to solve numerically by finite differences or finite element techniques. How can we remedy this problem? The original derivation of the phase-field equation has, most probably, been different from what we will do in the following to illustrate how phase field works: we combine the transport equation [5.1] with the minimum solution of a functional $F = F(\phi, \nabla\phi)$. F is called the Ginzburg–Landau functional, from a physical perspective. From a mathematical perspective, it simply penalizes "interfaces". What is an interface? It is a material state which is not bulk phase α and not bulk phase β. The bulk phases shall be the thermodynamic minimum states of the material, but if the material state (in temperature pressure and composition of an alloy) is within a two-phase region, it must separate into two phases, and these will be separated by an interface. Also, the miss-orientation between two different grains of a multigrain structure will necessitate the formation of an interface. The traditional Ginzburg–Landau functional we start with is (in 1D):

$$\begin{aligned} F^{intf} &= \int_{-\infty}^{+\infty} dx\, f^{intf} \\ f^{intf} &= \frac{\epsilon}{2}\left(\frac{\partial\phi}{\partial x}\right)^2 + \frac{\gamma}{4}\phi^2(1-\phi)^2. \end{aligned} \qquad [5.2]$$

The phase-field ϕ is normalized between 0 and 1 where $\phi = 1$ may correspond to phase (or orientation) α and $\phi = 0$ to phase (or orientation) β. Correspondingly, the potential operator $f^{DW} = \frac{\gamma}{4}\phi^2(1-\phi)^2$ has two minima at 0 and 1. The potential function, also termed Landau functional, has the form of a double well with the height of the obstacle between the minima of $\frac{\gamma}{16}$ at $\phi = \frac{1}{2}$.

In the phase-field and related literature, there is little debate about the "gradient" operator $(\frac{\partial \phi}{\partial x})^2$. It penalizes variations in the ordering state ϕ with $\frac{\epsilon}{2}$.

The potential operator is somehow arbitrary, depending on the application. As said, traditionally it has the form of a double well $\phi^2(1-\phi)^2$. Later in section 5.4, where the multi-phase fields is introduced, we will switch to a different potential, the so-called double obstacle potential $\frac{\gamma^{DO}}{2}|\phi(1-\phi)|$ for reasons given there. We may also use the so-called top-hat potential: $f^{TH} = \gamma^{TH}[\Theta(\phi) - \Theta(\phi - 1)]$. Θ is the step-function $\Theta(\phi) = 0$ for $\phi < 0$; $\Theta(\phi) = 1$ for $\phi \geq 0$. For completeness, an obstacle to cut off non-physical values of $\phi < 0$ and $\phi > 1$ has to be added, which is omitted here for simplicity. For details, see the appendix of Steinbach (2009). This potential has the advantage that the phase-field profile as the minimum solution of the functional with this potential (in 1D) is linear in space, that is, it can be directly compared to a level-set method (see Chapter 4). To the best knowledge of the authors, it has, however, never been used in the phase-field context. For now, we simply state that details of the special choice of the potential function shall not influence the physical interpretation of the model besides the contour of the phase-field within the interface. The first term in [5.2] is the famous gradient contribution of a Ginzburg–Landau functional, penalizing gradients of the field. The minimum solution is found:

$$0 = -\frac{\delta F^{intf}}{\delta \phi} = \epsilon \frac{\partial^2 \phi}{\partial x^2} - \gamma \phi(1-\phi)(\frac{1}{2} - \phi); \qquad [5.3]$$

$$\phi(x) = \frac{1}{2}\left\{1 - \tanh(\sqrt{\frac{\gamma}{8\epsilon}}\ x)\right\}. \qquad [5.4]$$

We easily prove the solution by inserting [5.4] into [5.3]; for more details, see the appendix of Steinbach (2009). We have adopted, without loss of generality, that the phase-field is $\phi = 1$ at the left boundary and $\phi = 0$ at the right boundary (see Figure 5.1, red contour for $\phi(x)$). The width of the transition region of the hyperbolic tangent, the interface width η of the phase-field contour between $\phi = 5\%$ and $\phi = 95\%$, is given by the relation between the parameter ϵ and γ, $\eta = \sqrt{\frac{72\epsilon}{\gamma}}$. We readily derive:

$$\frac{\partial \phi}{\partial x} = -\frac{6}{\eta}\phi(1-\phi), \qquad [5.5]$$

$$\frac{\partial^2 \phi}{\partial x^2} = \frac{72}{\eta^2}\phi(1-\phi)(\frac{1}{2} - \phi). \qquad [5.6]$$

In Figure 5.1, three distinct contributions to the phase-field functional are depicted. The first two contributions are related to the interface energy f^{intf}. Clearly, they vanish in the bulk, outside of the transition region of width η. Therefore, both parameters ϵ and γ have to be proportional to the interface energy σ. In fact, for the minimum solution (equation [5.4]), both contributions $\frac{\epsilon}{2}(\frac{\partial \phi}{\partial x})^2$ and $\frac{\gamma}{4}\phi^2(1-\phi)^2$ are identical (see equation [5.6]). The integral in normal direction through the interface $\int_{-\infty}^{+\infty} dx f^{intf} = \sigma$ and both contributions amount to one half of the interface energy.

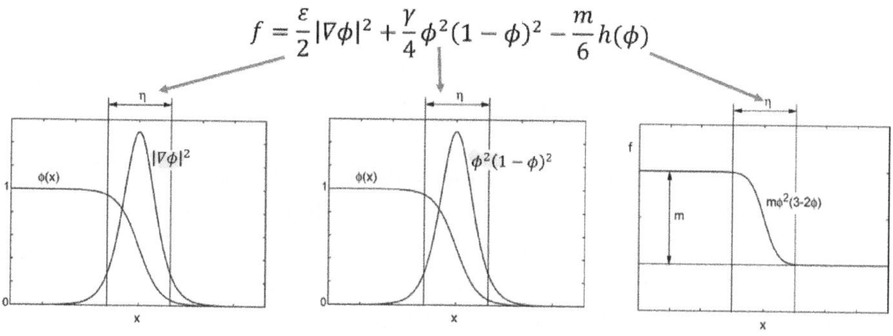

Figure 5.1. *Three contributions to the phase-field functional. The phase-field contour (equation [5.4]) is depicted in red with the width of the interface region η. For a color version of this figure, see www.iste.co.uk/bernacki/digital.zip*

The last term in Figure 5.1, $\frac{m}{6}h(\phi)$, is discontinuous between the phases. It will be related to the bulk free energy density between the phases below in section 5.3. Note that this term, in principle, contains the full bulk thermodynamics of the local material point. In most applications known in the literature, it is approximated by a simple regular solution model or by a linearized phase diagram. However, if you have a good thermodynamic model for your alloy, use it! It does not at all restrict the application of the phase-field functionality to propagate the interface.

For now, let us return to the transport equation [5.1]. As said before, this is nasty to solve numerically and we want to transform it into a parabolic type of equation. We do this in two steps:

– insert an "intelligent 0", equation [5.3];

– replace the gradient term in the transport equation [5.1] by its local counterpart for the traveling wave solution equation [5.5].

$$\begin{aligned} \dot{\phi} &= v \frac{\partial \phi}{\partial x} + 0 \\ &= v \frac{\partial \phi}{\partial x} + M^{\phi}\left[\epsilon \frac{\partial^2 \phi}{\partial x^2} - \gamma\phi(1-\phi)(\frac{1}{2}-\phi)\right] \end{aligned} \qquad [5.7]$$

M^ϕ is a positive material constant known as the interface mobility.

The derivation up to here was done in one dimension. In three dimensions, the space derivative is replaced by the gradient $\frac{\partial \phi}{\partial x} \to \vec{\nabla}\phi$ and the second derivative by the Laplacian $\frac{\partial^2 \phi}{\partial x^2} \to \nabla^2\phi$. The Laplacian of any field is sensitive to the curvature of the field, that is, the intelligent 0, equation [5.3], is recovered only for a planar wave front, grain boundary or interface in our context. We explicitly calculate the local curvature of the front κ (Beckermann et al. 1999):

$$\kappa = \vec{\nabla} \cdot \vec{n} = \vec{\nabla} \cdot \frac{\vec{\phi}}{|\vec{\phi}|} = \frac{1}{|\vec{\nabla}\phi|}\left[\nabla^2\phi - \frac{72}{\eta^2}\phi(1-\phi)(\frac{1}{2}-\phi)\right]. \qquad [5.8]$$

We see our intelligent 0 now has an additional property. Besides stabilizing the contour of the interface in normal direction $\vec{n}$ through the interface, $\vec{n} = \frac{\vec{\nabla}\phi}{|\vec{\nabla}\phi|}$, it additionally evaluates the local curvature of the front. The local curvature of the interface then is related to capillarity, that is, the local equilibrium between phases must be corrected by the curvature undercooling $\Delta g^{cap} = \sigma^*\kappa$, where $\sigma^* = \sigma + \sigma''$ is the so-called interface stiffness and σ'' is the Herring torque (Herring 1952). We find the analogy to the so-called Gibbs–Thomson equation for the motion of a transformation front:

$$\vec{v} = \vec{n}\ M^\phi \left[\sigma^*\kappa + \Delta g^{bulk}\right]. \qquad [5.9]$$

This can be related to the phase-field equation in three dimensions:

$$\vec{v} = \frac{d}{dt}\phi\frac{\vec{n}}{|\vec{\nabla}\phi|}.$$

$$\frac{d}{dt}\phi = \vec{v}\ \vec{\nabla}\phi$$

$$= M^\phi\left\{\sigma^*\left[\nabla^2\phi - \frac{72}{\eta^2}\phi(1-\phi)(\frac{1}{2}-\phi)\right] - \frac{6}{\eta}\phi(1-\phi)\Delta g^{bulk}\right\} \qquad [5.10]$$

We have used equation [5.8] to express the curvature by the phase-field contour. Equation [5.10] is the final version of the phase-field equation for the double well potential in three dimensions. We replaced the model parameters ϵ and γ by the materials parameter interface energy and interface width, $\sigma = \frac{\epsilon}{\eta} = \frac{\gamma\eta}{72}$ and also added the Herring torque $\sigma \to \sigma^*$. The driving force Δg is taken linearly to the velocity. Equation [5.10] is the famous phase-field equation. It is parabolic type, a sort of diffusion equation. The hyperbolic term or first spacial derivative $\frac{\partial \phi}{\partial x}$ is replaced by the local (in terms of numerical implementation) contribution $\frac{6}{\eta}\phi(1-\phi)$. Of course, this works only if the contour is right. However, in a numerical calculation, if the contour is not right, the terms in the square bracket do not cancel to 0, but push the contour back to what it should. See also section 5.7 below where numerical implications are discussed. The indexSharp Interface Comparison shows the equivalence to the phenomenological Gibbs–Thomson equation [5.9] of the phase-field equation [5.10]. Now, the Gibbs–Thomson equation has no entity for the interface width η. It is a "sharp interface" equation. This means we can choose the interface width freely, within special bounds (see section 5.7). It is not bound to the atomistic scale of an interface between two crystalline phases, or a solid and its melt, which is of the order of 1 nm. This is of course good for a numerical simulation of microstructural evolution at the micrometer scale. We call this a mesoscopic phase-field model, agnostic of the atomistic scale of the interface width. In contrast, so-called microscopic phase-field models are also popular, where the interface width is treated as a physical entity at the nanoscale.

5.3. Thermodynamically consistent derivation of the phase-field equations

This section will be very short, since most of the relevant concepts have already been introduced in section 5.2. We start out from the so-called Clausius–Duhem relation for the evolution of a state variable ϕ in agreement with entropy production, where the entropy, $S = \int_\Omega d^3x \; s$, is defined within the domain Ω by the entropy density $s = s(\phi, t)$ as function of time and the state variable ϕ[1]:

$$
\begin{aligned}
&\frac{dS}{dt} = \frac{\partial \phi}{\partial t}\frac{\delta S}{\delta \phi} > 0 \\
&\frac{\partial \phi}{\partial t} \propto \frac{\delta S}{\delta \phi} = -\frac{\delta F}{\delta \phi}.
\end{aligned}
\qquad [5.11]
$$

In the following, we will use instead of the entropy the free energy F for isothermal problems, and we will accept that $\frac{\delta S}{\delta \phi} = -\frac{\delta F}{\delta \phi}$. To build a model for the free energy F, we simply combine the interface functional F^{intf} with a bulk free energy functional $F^{bulk} = \int_\Omega d^3x \; f^{bulk}$, where the α-phase is characterized by $\phi = 1$ and the β-phase by $\phi = 0$:

$$
f^{bulk} = h(\phi) f_\alpha + (1 - h(\phi)) f_\beta. \qquad [5.12]
$$

1 The reader may not be puzzled too much by the different symbols, d for the total derivative, ∂ for the partial derivative and δ for the variational derivative of a function, $\phi(\vec{x}, t)$ in the present case.

Now we have everything we need. The minimum solution $\frac{\delta F}{\delta \phi} = 0$ is our 0 in [5.3] for equal bulk energies in both phases $\frac{\delta F^{bulk}}{\delta \phi} = 0$. If these energies are not equal, the material will transform from the higher energetic state to the lower and we have a finite driving force for transformation Δg, that is, movement of the interface, with the convention $\phi = \phi_\alpha = 1 - \phi_\beta$ and $h'(\phi) = \frac{\partial h(\phi)}{\partial \phi}$:

$$\Delta g = \frac{\delta F^{bulk}}{\delta \phi} = \frac{\partial f^{bulk}}{\partial \phi} = h'(\phi)(f_\alpha - f_\beta), \qquad [5.13]$$

For low driving pressures Δg, we assume the velocity is linear with the driving pressure with the interface mobility M^ϕ as a proportionality constant.

$$\tau \frac{\partial \phi}{\partial t} = -\frac{\delta F}{\delta \phi} \qquad [5.14]$$

$$\frac{\partial \phi}{\partial t} = -M^\phi \left[\frac{\partial^2 \phi}{\partial x^2} - \frac{72}{\eta^2}\phi(1-\phi)(\frac{1}{2} - \phi) - 6\eta\phi(1-\phi)\Delta g \right] \qquad [5.15]$$

Thus, we have reconstructed the soliton equation [5.7] from the Ginzburg–Landau free energy of a real system with a relaxation constant τ inversely proportional to the interface mobility M^ϕ, $\tau = \frac{\eta}{M^\phi}$ and with the special choice of the coupling function $h(\phi) = \frac{1}{\eta}(2\phi^3 - 3\phi^2)$. Real systems comprise many phases, many grains, many interfaces, triple lines formed by three interfaces, quadruple junctions formed by four triple lines, etc. Therefore, we need a multi-phase fields, as detailed in section 5.4.

5.4. The multi-phase fields approach

Technical materials are usually multi-grain and multi-phase structures. This means, we need thousands of phase fields to characterize their microstructure, and we want to study the evolution of these microstructures, from solidification, thermomechanical processing until degradation under service conditions, may be sometimes repair and rejuvenation. Figure 5.2 displays a multi-grain structure. Facets between grains are not straight, but curved. Each grain "communicates" with all neighboring grains. The main impact on the evolution of a grain structure, however, is junctions, where more than two phase-fields overlap (blue lines in Figure 5.2, right). Three facets form a triple line, and four triple lines form a quadruple junction. In reality, more than four triple lines may join in a junction of higher order than four. This poses a challenge to the method to solve the evolution of these structures with their complicated network of facets, lines and junctions.

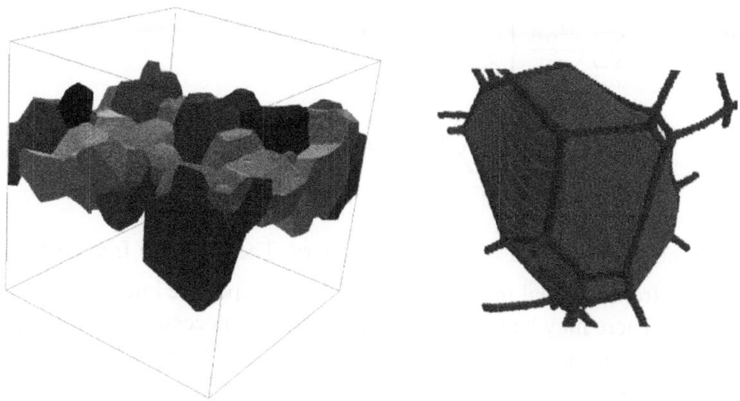

Figure 5.2. *Cutout of a multi-grain structure (left) and a single grain with triple lines forming a cage around the grain (right). Phase-field simulations using OpenPhase. For a color version of this figure, see www.iste.co.uk/bernacki/digital.zip*

A first statement is that the relation between two phase fields, as discussed in section 5.1, $\phi_\beta = 1 - \phi_\alpha$, is replaced by the sum constraint:

$$\sum_{\alpha=1}^{\tilde{N}} \phi_\alpha = 1 \qquad [5.16]$$

$\tilde{N}$ is the number of phase fields $\phi_\alpha(\vec{x}, t)$ at space $\vec{x}$ and time t with non-zero values: $\tilde{N} = 1$ for material points in the bulk, $\tilde{N} = 2$ in interfaces, $\tilde{N} = 3$ in triple lines, $\tilde{N} = 4$ in quadruple junctions, etc. In practical simulations, $\tilde{N} > 10$ are not seldom. This has to be handled efficiently on a technical level. From a theoretical point of view, there are three important consequences (which are somehow related):

– The field ϕ_α are not independent: $\frac{\partial \phi_\alpha}{\partial \phi_\beta} \neq 0$.

– What is the mobility of a junction formed by three or more fields?

– What is the interface energy of a junction formed by three or more fields?

The first item relates to the question: if one grain in contact to two other grains is growing, from which of the other grains does it take its gain? Or how much from each of these? The other two items relate to physical questions: How do we define materials properties of junctions which are almost impossible to measure? The multi-phase fields method (Steinbach et al. 1996; Steinbach and Pezzola 1999), as developed by the first author of this chapter and several other co-workers, offers an elegant (we think a beautiful!) solution to these problems. Not a rigorous or "final" solution. It is a "model": We decompose all multiple transformations $\tilde{N} > 2$ into pairs of transformations:

$$\frac{d\phi_\alpha}{dt} = -\sum_{\alpha=1}^{\tilde{N}} \frac{M^\phi_{\alpha\beta}}{\tilde{N}} \left\{ \frac{\delta}{\delta\phi_\alpha} - \frac{\delta}{\delta\phi_\beta} \right\} F \qquad [5.17]$$

$$F = \int_\Omega d^3x \left[\sum_{\beta=1}^{\tilde{N}} \left(\sum_{\gamma=\beta+1}^{\tilde{N}} f^{intf}_{\beta\gamma} + \phi_\beta f^{bulk}_\beta \right) \right] \qquad [5.18]$$

$M^\phi_{\alpha\beta}$ is the mobility of a α-β grain boundary. The interface free energy density $f^{intf}_{\beta\gamma}$ is taken from the dual interface [5.2] with the respective interface energy of the β-γ boundary. There may be higher order contributions necessary from a physical, or a numerical point of view, as discussed by Miyoshi and Takaki (2016). In any case, we will have for each phase field α to run a loop over all fields β and γ which are non-zero at the respective materials point. So, junctions are costly with the power $\tilde{N}^3$! The third loop (over γ) can be avoided if all interface energies are equal and the equilibrium angles between all interfaces are equal (see Guo et al. 2011). However, in general they are not (Salama et al. 2020).

Now we come back to the question of the potential function used in the interface free energy density $f^{intf}_{\beta\gamma}$. This potential penalizes interfaces and is responsible for phase separation.

An important feature of the multi-phase fields theory (Steinbach and Pezzola 1999) is the sum constraint in junctions [5.16]. For the double-well potential, we write the potential (without prefactors) f^{DW}_0:

$$f^{\text{DW}} = \sum_{\alpha=1}^{\tilde{N}} \sum_{\beta=\alpha+1}^{\tilde{N}} \phi^2_\alpha \phi^2_\beta. \qquad [5.19]$$

The maximum of the potential in the center of the junction, where $\phi^\alpha = \phi^\beta = \ldots = \frac{1}{\tilde{N}}$ for all $\alpha, \beta, \ldots$, is then evaluated:

$$f^{\text{DW}}_0 = \binom{\tilde{N}}{2} \left(\frac{1}{\tilde{N}}\right)^4 \propto \frac{1}{\tilde{N}^2} \quad \text{for } \tilde{N} \gg 1. \qquad [5.20]$$

This means that for $\tilde{N} > 3$, the energy of the junction decreases with the order $\tilde{N}$ and approaches 0 for large $\tilde{N}$. This must be termed "nonphysical", since junctions between objects lose their penalty, and the system would return to the disordered state. The double-obstacle potential is introduced to remedy this problem:

$$f^{\text{DO}} = \sum_{\alpha=1}^{\tilde{N}} \sum_{\beta=\alpha+1}^{\tilde{N}} |\phi_\alpha \phi_\beta|. \qquad [5.21]$$

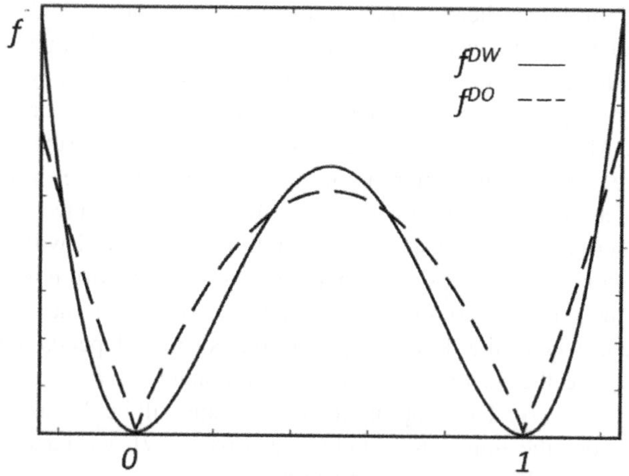

Figure 5.3. *Double-well versus double obstacle potential in dimensionless units. They have minima at* 0 *and* 1 *and show the same area within these minima. This area is related to the interface energy*

This has the same topology as the double-well potential (see Figure 5.3) but a maximum power of two. We calculate the maximum potential of the junction $f_m^{\rm DO}$:

$$f_0^{\rm DO} = \binom{\tilde{N}}{2}(\frac{1}{\tilde{N}})^2 \propto 1 \;\; \text{for} \;\; \tilde{N} \gg 1. \qquad [5.22]$$

This means that the energy of the junction increases with the order $\tilde{N}$ and approaches a constant for large $\tilde{N}$, as it should. The main drawback of this potential is the non-analytical form with the absolute signs. However, this is technical. The use of a top hat potential, which is, as discussed, related to the level set approach, would even give a stronger penalty to junctions. We should give it a try.

5.5. Multi-component alloy transformation

Up to now, we were only discussing interface related-problems, where the microstructure evolves by the reduction of interface energy. This is important for coarsening of a multi-grain structure. For recrystallization, we may define a driving force Δg based on stored dislocation energy after deformation. In alloy transformation, however, the driving force $\Delta g = \Delta g(T, \vec{c}, P, \vec{\vec{\sigma}}, \vec{M}, ...)$ is a function of temperature T, composition $\vec{c}$, pressure P, local stress $\vec{\vec{\sigma}}$, maybe magnetization $\vec{M}$ and other fields. This is particularly important, since all these fields are not constant, but related to the evolution of the phase fields. If a phase transformation is going on, we have release of latent heat, which influences temperature, redistribution of solutes, which in turn influences the local composition, the change of materials density, which changes pressure and causes micro-stress, and the magnetic state of the material may vary from ferro- to paramagnetic, piezoelectric; or our material might evaporate. All these phenomena are, more or less, well described in materials science literature. They cause morphological unstable interface motion, as in dendritic solidification or viscous fingering in oil drilling. We have to couple the phenomena at interfaces with transport in the bulk phases. Here, I will only discuss the role of multi-component diffusion on phase transformation within the phase-field framework and the coupling to stress and strain. To set the frame, we discuss a materials problem where the free energy of the system depends on the local composition $\vec{c}$, where the vector runs over n_c components $c_i, i = 1...n_c$, and the phase state ϕ_α, $\alpha = 1...N$. The coupling to stress–strain will be discussed in section 5.6. For now, besides the multi-phase fields equation [5.17], we have to solve the diffusion equation using Einstein sum convention[2]:

$$\frac{dc_i}{dt} = \nabla M_{ij}^c \nabla \frac{\partial f}{\partial c_j} = \nabla M_{ij}^c \frac{\partial^2 f}{\partial c_j c_k} \nabla c_k = \nabla D_{ik} \nabla c_k, \quad [5.23]$$

where $\frac{\partial^2 f}{\partial c_j c_k}$ is the thermodynamic factor and $D_{ik} = M_{ij}^c \frac{\partial^2 f}{\partial c_j c_k}$ the diffusion matrix.

This is a well-posed problem from a theoretical point of view. It is an ill-posed problem from a numerical point of view. The reason for this is that $f^{bulk} = f^{bulk}(\vec{c})$, according to the definition [5.12] for an α-β interface, is non-convex in $\vec{c}$. For illustration, we plot the bulk free energy f^{bulk} of a binary alloy as function of the composition in Figure 5.4, and as a function of the normal coordinate n through the interface.

The non-convex bulk free energy contributes with a chemical interface energy, as defined by the integral over the mixture bulk free energy in normal direction through the interface. This chemical interface energy is well known from the Cahn–Hilliard model of spinodal composition (Cahn and Hilliard 1958), though formulated for a second-order phase transformation there. It leads to the obvious consequence that the thermodynamic factor or the diffusion coefficient becomes negative, that is, any implicit solution of this equation becomes impossible. There is a whole community in continuum mechanics who deal with such problems. Depending on their interest, the reader may dig into this problem or continue with us. The solution, as proposed by one of the authors and co-workers in 1998 (Tiaden et al. 1998) (see also Eiken et al. 2006), is simply to convexify the chemical bulk free energy. Originally, the chemical interface

2 The treatment here is a bit sloppy with respect to conservation of the total concentration and the Gibbs–Duhem relation. For a thorough treatment, see Kundin et al. (2021).

energy simply had been omitted. It was assumed to be small compared to the structural interface energy in a first-order phase transformation. But, of course, there is no reason not to consider it in a properly scaled version as a contribution to the total interface energy σ within the interface part of the phase-field functional (see Steinbach 2013). Why "properly scaled?" It is because the chemical-free energy in a Cahn–Hilliard type phase-field model scales with the interface width (see Figure 5.4). If we want to scale the interface width to its maximum allowed by the proper resolution of the microstructure, and not fix it to the atomistic scale of $\approx 1nm$, we have to consider this scaling. There is no other choice.

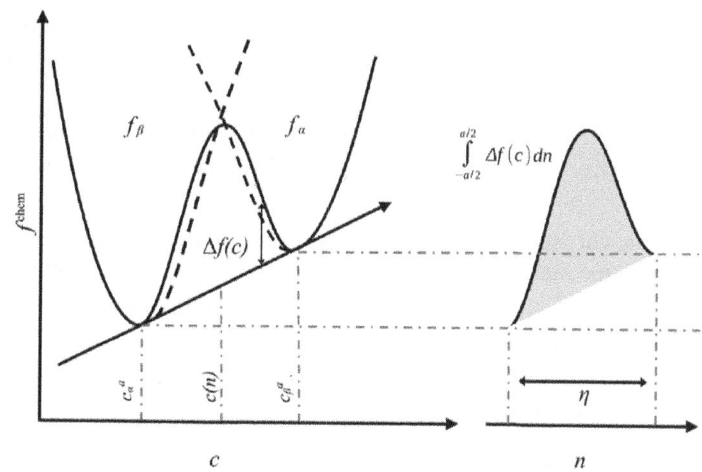

Figure 5.4. *The chemical bulk free energy f^{bulk} as a mixture of two convex (in composition c) free energy functions of the individual phases α and β. Since the composition varies monotonously in normal direction n through the interface, we can as well plot the "chemical" interface energy as a function of n (right side of the figure)*

Returning to the problem of convexification: as proposed in Tiaden et al. (1998), we split the concentration c^i into the phase concentrations c^i_α, $c^i = \sum_\alpha \phi_\alpha c^i_\alpha$. We then define the bulk-free energy f^{bulk} as a function of the phase compositions (where we take $h(\phi) = \phi$):

$$f^{bulk} = \sum_\alpha \phi_\alpha f^{bulk}(\{c^i_\alpha\}). \qquad [5.24]$$

By construction, this is a convex function in c^i_α. There is, however, no free lunch: now we have many more variables (one more in a dual interface $c \to c_\alpha, c_\beta$, two more in a triple junctions, etc). The original approach (Tiaden et al. 1998), called the "quasi-equilibrium approach", assumed the partitioning of solute between the phases according to the equilibrium phase diagram. For a comprehensive treatment in a multicomponent alloy, see Eiken et al. (2006). An alternative treatment is the assumption of local chemical equilibrium within the interface, as proposed by Kim et al. (1999). Although this assumption is only valid at high temperatures, not valid for most situations in materials processing, this treatment became very popular and the approach to split the concentration into the phase concentration for convexification of the bulk free energy is today known as the "Kim–Kim–Suzuki (KKS)" model, from the authors of Kim et al. (1999). Today, however, it is outperformed by the so-called "grand potential approach" (Plapp 2011), which takes the chemical potential, set equal between the phases, as the principle variable. Hereby, no splitting of the variables is needed any more and we solve a type of temperature diffusion equation with a source term considering phase transformation and redistribution of solute. To conclude this discussion, we make the following statements:

– On a physical basis, the Cahn–Hilliard type of approach to multicomponent diffusion [5.23] has everything we need, also for non-equilibrium phase transformations.

– For an efficient numerical treatment of microstructure evolution and phase transformations at the mesoscopic scale, we need to convexify the problem.

– All multi-phase approaches, Tiaden (Tiaden et al. 1998), KKS (Kim et al. 1999) and Grand Potential (Plapp 2011), are equivalent if we approximate the Gibbs free energies of the bulk phases $f_\alpha(\{c^i_\alpha\})$ by parabolic functions, that is, that the chemical potential is taken linear in composition.

The latest development with regard to the treatment of phase transformations coupled to multicomponent diffusion is the so-called "finite interface dissipation" model (Steinbach et al. 2012; Zhang and Steinbach 2012). It refrains from the assumption of a quasi-equilibrium partition coefficient as well as from the assumption of equal chemical potential between the phases within the interface. Instead, it proposes a coupled set of equations for each phase composition variable c^i_α (written here only for a binary alloy with a single concentration variable and a dual α-β interface). The generalization to multiple phases is "straightforward", but technical. The equations read:

$$\phi_\alpha \dot{c}_\alpha = \vec{\nabla} \cdot (\phi_\alpha D_\alpha \vec{\nabla} c_\alpha) + P\phi_\alpha\phi_\beta(\mu_\beta - \mu_\alpha) + \phi_\alpha\dot{\phi}_\alpha(c_\beta - c_\alpha), \qquad [5.25]$$

$$\phi_\beta \dot{c}_\beta = \vec{\nabla} \cdot (\phi_\beta D_\beta \vec{\nabla} c_\beta) + P\phi_\alpha\phi_\beta(\mu_\alpha - \mu_\beta) + \phi_\beta\dot{\phi}_\beta(c_\alpha - c_\beta). \qquad [5.26]$$

By construction, both equations overlap in the interface, where two mechanisms of solute redistribution are considered: redistribution due to a chemical–potential difference $\mu_\alpha - \mu_\beta \neq 0$ with a kinetic parameter, called permeability P, and renormalization if the fraction of phases changes $\dot{\phi}_\alpha = -\dot{\phi}_\beta \neq 0$. μ_α and μ_β are the chemical potentials of the individual phases. They are monotonous functions in composition, that is, there is no "inflection". Also, the diffusion coefficients D_α and D_α are strictly positive. Expanding the chemical potential in a first-order Taylor series in the composition defines a linear set of equations for the phase-compositions c_α, which is easily solved by standard FE of CV schemes.

Limiting cases are the following: for $P \rightarrow \infty$, we recover the KKS model. For $P \rightarrow 0$, we stop redistribution of solute and the model gives complete solute trapping, partitionless transformation, or, as called in solid state, massive/martensitic transformation.

The challenge, and this is true for all descriptions in this book, is to relate the parameter functions in our model to "real materials".

5.6. Multi-phase field with elasticity

Considering the phase transformations in multi-component multi-phase systems brings up another challenge: the necessity to consider mechanical response of the system. Even in solidification, for example, eutectic solidification of many alloys where more than one solid phase is involved, the mechanical response of the solids should be considered. This is due to the different crystal symmetries and lattice parameters of various solid phases; each time two different solid phases develop an interface, and elastic stresses emerge due to the lattice misfit between the phases. We may argue that it is always possible to establish a stress-free interface between the solid phases, but in reality it will be stress free only at the specific conditions when the interface is established. If the temperature or chemical composition change with time, the stresses will develop due to the different expansion/contraction rates of different phases. The stresses will develop even between different regions of the same phase, which differ in chemical composition or temperature.

Most transformation problems in nature are controlled by the heat and chemical diffusion. These are so-called "mixed-mode transformations", as bainitic transformation are controlled by both chemical and mechanical interactions. Only the martensitic transformations are dominantly controlled by the mechanical response but even then the heat diffusion plays a key role in triggering and controlling the transformation. This allows significant simplification in considering the mechanical response of the system by assuming that the mechanical equilibrium establishes much faster than the chemical or thermal equilibrium. Thus, instead of solving the full wave equation, a simpler mechanical equilibrium equation can be considered. Only, in the case of martensitic transformation in shape memory alloys and some ceramic systems, the full wave equation solution is required to correctly describe the transformation kinetics (Steinbach and Shchyglo 2011).

Historically, the first elasticity consideration in the phase-field modeling has been proposed by Khachaturyan for elastically homogeneous systems with small strain where he used the Fourier method to solve the mechanical equilibrium equation in reciprocal space (Khachaturyan 1983). The Fourier method offers an elegant way of obtaining elastic strains developing in the system with the help of the fast Fourier transformation (FFT) routines. The method has been later transformed into an iterative scheme for elastically inhomogeneous systems (Wang et al. 2002). This family of methods is applicable to the problems characterized by relatively small transformation induced deformations. When it comes to problems involving large elastic deformations, as in shape memory polymers, or elasto-plastic, or in martensite in steel, full consideration of the finite strain mechanics is required. Here, there are two possible ways of solving the mechanical problem. The first is to use a Lagrangian description based on the appropriate hyperelastic model with suitable stress and strain formulation to describe the finite deformations. Such a model is suitable for transformations involving deformations up to around 20–30% (Shchyglo et al. 2024). Beyond this limit, other methods should be used due to severe deformation of the underlying discretization grid that affects the quality of the solution of other relevant phenomena, for example, chemical diffusion, phase-field evolution and heat transfer. To overcome this problem, a so-called arbitrary Lagrangian–Eulerian scheme (ALE) can be used. It starts by decomposing the large deformations into small, linear elastic solutions using Khachaturyan's algorithm (Lagrangian step). The full solution is obtained incrementally by integrating the elastic stresses and explicitly applying the obtained deformation increments to all relevant fields (Eulerian step). This can be achieved by a series of advection and remeshing steps allowing us to retain the undistorted discretization grid, even if the system deforms significantly (Borukhovich et al. 2015).

5.7. Numerical treatment and applications

The numerical treatment of phase field, mostly done on a fixed structured grid, is relatively easy in comparison to finite elements on an unstructured grid. Since all phase-field models are formulated in terms of partial differential equations, all appropriate numerical schemes are possible to apply. I refer here to the treatment in the OpenPhase (OpenPhase.rub.de) code as developed in the institute of the authors, now also available with commercial extensions to large strain crystal plasticity and multi-component diffusion (OpenPhase-solutions.com). The core is the multi-phase fields model for interface motion, as described in section 5.4. The multi-phase fields equation [5.17] is solved by finite differences with explicit time stepping. An implicit scheme is warranted by the strong nonlinearity of the equation, but various attempts overcoming this by operator splitting and other schemes can be found in the literature. The equation is basically a diffusion equation for the phase-fields ϕ_α with a source term considering (i) the potential operator in the phase-field functional and (ii) driving forces due to coupling to other fields, such as concentration and elasticity treated in this chapter. The challenge lies in correct book-keeping in junctions where more than two fields overlap and the consideration of the sum constraint [5.16]. Though, from a mathematical point of view, the sum is always conserved by the multi-phase fields equation [5.17], in numerics this is not always the case. Also, the individual phase-fields may overshoot $\phi_\alpha < 0$ or $\phi_\alpha > 1$. Therefore, after each solution step, the phase-field has to be renormalized, which is similar to a re-initialization step in level set. Additionally, the diffusion equations for solute [5.25] and [5.26] are currently solved explicitly with finite differences. Here, an implicit scheme might be of advantage in situations when the source terms considering redistribution at interfaces are small, that is, if mainly bulk diffusion is of interest. Elasticity (section 5.6) is solved by fast Fourier transform in a periodic setting. Fluid-flow, not considered here but implemented in OpenPhase, is solved by a Lattice–Boltzmann scheme. All different equations are coupled in a staggered scheme, if necessary with intermediate iteration. A monolithic scheme seems inappropriate to the authors because of the high nonlinearity of the coupled problem.

A last important comment, hardly discussed in the literature: the driving force for phase transformation $\Delta g = \Delta g(T, \vec{c}, P, \vec{\vec{\sigma}}, \vec{M}, ...)$ as a function of all considered fields is not constant in normal direction through the interface in general. This generates an artificial interface energy, scaling with the interface width, as discussed in section 5.5. The remedy for solutal coupling has been described from a theoretical point of view, the convexification of the chemical bulk free energy. The remedy for coupling to elasticity is even more involved (see the previous section 5.6). Here, in the section about numerics, we have to deal with another issue: in a diffuse interface region of, say, about $5\,\mu$, many properties of the material, such as local composition, local stress and strain, local temperature and all thermo-physical quantities dependent on these properties, will vary; in case of strong gradients at the interface, they may vary strongly. Therefore, the driving forces, even when considering a consistent convexification of the Gibbs energies within the interface, vary through the interface. This leads to spreading of the interface, if the driving forces in the front part of the interface are larger than in the rear part, or contraction in the opposite case. To overcome this problem, we apply in OpenPhase an averaging procedure of driving forces in normal direction through the interface. Of course, we have to keep the tangential variation of the driving forces not to hamper the morphological evolution.

Let us state here clearly that in phase field, we have a clear hand on the problem. That this problem is hardly discussed in other methods (CA, level-set) is not an indication that this problem does not exist within these methods. We have to solve it to make quantitative predictions for relevant materials problems independent of numerical artifacts.

5.8. Examples

5.8.1. *Martensitic transformation*

One of the most challenging transformations for the phase-field simulations is the martensite formation in steel. The difficulty lies in the complexity of crystallographic relation between the involved phases and in the high transformation induced deformations of up to 20%, which trigger plastic relaxation during the transformation. Despite the relatively simple Bain relation between the austenite and martensite with only three tetragonal variants of martensite, depending on the actual transformation scenario and the steel grade, we have to take into account 12 (Nishiyama–Wassermann orientation relationships) or even 24 (Kurdjumov–Sachs orientation relationships) variants of martensite. This is due to the peculiar details of the crystal lattice distortion during the martensitic transformation leading to a slight rotation of Bain variants by about 9° off their ideal orientation (Koumatos and Muehlemann 2017). Such rotation greatly multiplies the number of martensite variants and leads to the formation of a very complex hierarchical microstructure.

Interestingly, martensite formation simulation (Wang and Khachaturyan 1997) was one of the first applications of the Khachaturyan's microelasticity theory mentioned earlier in this chapter. Of course, the first attempts were limited to only three tetragonal Bain variants in small strain framework but they showed the way and led to a series of more advanced simulations in the literature. It took the community nearly two decades to advance the methodology past small strain framework. The first attempts at advancing the martensite formation simulations to the finite strain consideration were from the group of Levitas (Levitas 1998, 2013). Recently, a set of significantly more advanced simulations of the martensite formation in low alloyed steel has been presented by the group of authors, including the authors of this chapter. Figure 5.5 shows the microstructures obtained by the phase-field simulations of martensite formation in three low alloyed CxMnCr8-4 steel grades. The simulations considered 24 Kurdjumov–Sachs variants of martensite in finite strain framework using the phenomenological crystal plasticity model. The obtained results show close comparison to the experimental observations not only in the microstructure appearance but also in the transformation kinetics details. For example, the martensite start temperature has been closely predicted for all three steel grades based on the real material's data with the minimum model assumptions.

5.8.2. *Recrystallization and coarsening*

Recrystallization and coarsening of multi-grain structures is a classical application of all kinds of techniques dealing with "growth problems". The easiest, from the point of implementation, are so-called "Potts models" or cellular automata (see Chapter 7). Here, some rules are adopted under which conditions the grain index of one voxel in the microstructure changes its state. This can be the creation of a new grain in recrystallization, or the switching to the index of a neighbor grain. The latter then will grow, the former will shrink: grain growth of coarsening. In phase field or level set, the nucleation state has to be introduced in a similar manner: when a nucleation condition for a new grain is fulfilled, it is somehow planted manually into the microstructure to form a new grain. The condition may depend on the local state of

Figure 5.5. *Simulated martensite microstructure in low alloyed steel for different carbon contents. For a color version of this figure, see www.iste.co.uk/bernacki/digital.zip*

stress and strain, misorientation of neighboring grains at grain boundaries, density and gradients of stored dislocation, or simply statistics. Such a condition does not depend on the numerical technique. In a phase-field based on a Cahn–Hilliard model, some authors argue nucleation may come out automatically if the barrier in the potential is overcome by some statistical noise. We rate this as part of a statistical method. The growth problem then is mostly treated along the Gibbs–Thomson equation considering capillarity. Coupling to the alloy thermodynamics like during a phase transformation can be neglected, which makes the problem relatively "easy". There are mainly three issues to be considered:

– interface energy anisotropy in five-dimensional orientation space;

– solute and particle drag at interfaces;

– the competition between capillarity and a driving force for recrystallization, that is, reduction of stored dislocation energy (Darvishi Kamachali et al. 2015).

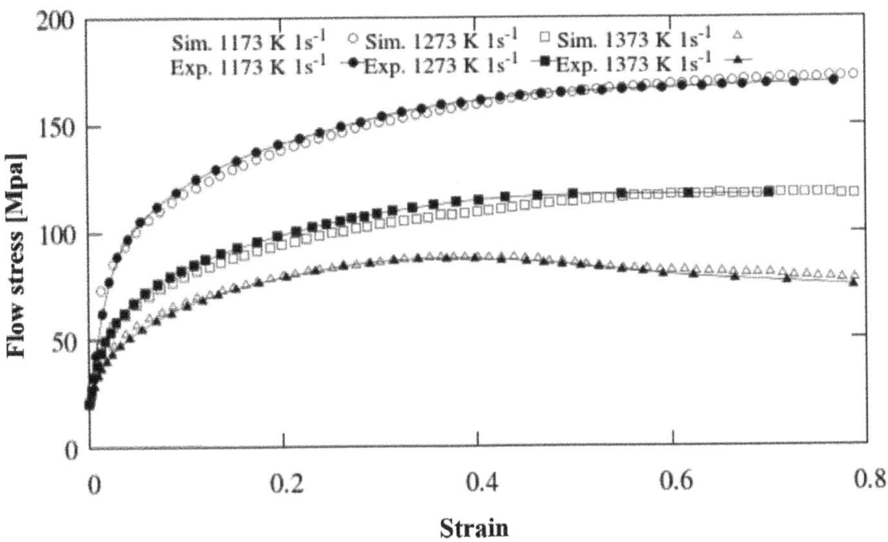

Figure 5.6. *Comparison of flow stress evolution between experiment and simulation during hot deformation and dynamic recrystallization of MS-W 120 steel at* $T = 1,173\text{K}, 1,273\text{K}$ *and* $1,373\text{K}$, *strain rate* 1s^{-1}

Figures 5.6 and 5.7 describe the application of phase field to dynamic recrystallization during hot rolling of a steel slab (Hiebeler 2017; Hiebeler et al. 2017). The phase-field solver is coupled to a phenomenological plasticity model in finite strain formulation allowing several hundred % of strain by adaptive grid refinement "on the fly". The nucleation condition for new grains is based on an energy minimum condition between reduction of local elasto-plastic energy and build

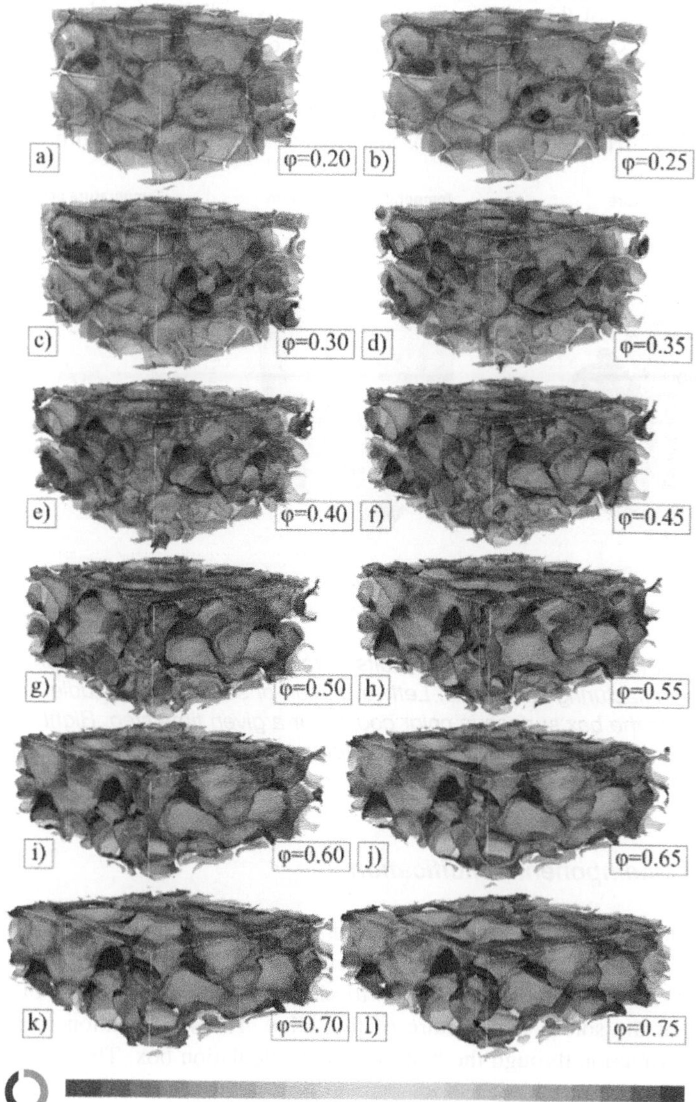

Figure 5.7. *Hot compression test of a multi-grain structure in MS-W* 120 *steel at* $T = 1,373\text{K}$ *with strain rate of* 1s^{-1}*. The maximum compression in vertical direction is* 0.75 *and the sample correspondingly expands in horizontal direction. The recrystallized grains are marked in green, while the original grains are colored with the flow stress, as given in the color bar. For a color version of this figure, see www.iste.co.uk/bernacki/digital.zip*

up of interface energy. At each material point, the flow curve is treated individually where the history of the process is stored in the local dislocation density, which, in return, governs strain hardening. Softening of the multi-grain structure then results from the birth of new dislocation free grains. In total, a realistic deformation and transformation behavior in comparison to experimental observation is reached (Hiebeler 2017; Hiebeler et al. 2017). Simulations were performed with OpenPhase.

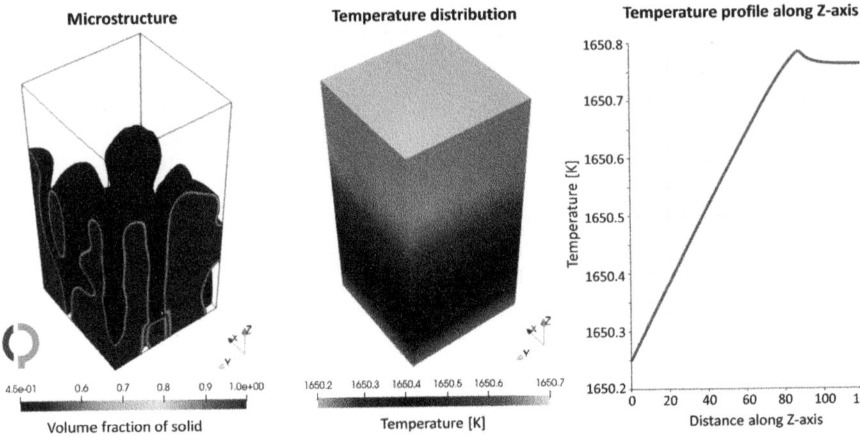

Figure 5.8. *Phase-field simulation results obtained for a solidification process under additive manufacturing conditions. Left: evolution of solid phase. Middle: temperature distribution at the box surface in color coding for a given time step. Right: line scan of temperature along the z axis with a peak temperature at the dendrite tip. For a color version of this figure, see www.iste.co.uk/bernacki/digital.zip*

5.8.3. *Multicomponent solidification*

The following example relates to multicomponent alloy solidification in a Ni-based superalloy. A model alloy of the commercial alloy CMSX4 with four components, Ni, Al, Cr and Ta, is constructed so as to yield a comparable fraction of secondary γ' phases at the solvus temperature of this alloy. The solidification conditions are high heat extraction through the bottom of the calculation box. The heat conduction equation,

$$\rho C_p \dot{T} = \nabla \cdot (\lambda \nabla T) + L\dot{\phi}_{\text{solid}}, \qquad [5.27]$$

is solved implicitly. λ is the heat conductivity and L the latent heat of solidification. From the simulation results shown in Figure 5.8 (left), we can see the growth of dendrites in cellular form owing to the very high temperature gradient and cooling rate. The line scan of the temperature plot along the z axis highlights that the dendrite tip is the hottest region during solidification. This is due to the release of latent heat during phase transformation (right). In the middle of the figure, the temperature at the outside of the boundary of the calculation box is shown: it is not uniform, and also not in horizontal direction. A prescribed temperature distribution, as often applied in directional solidification, is not applicable here due to the strong transient nature of the process.

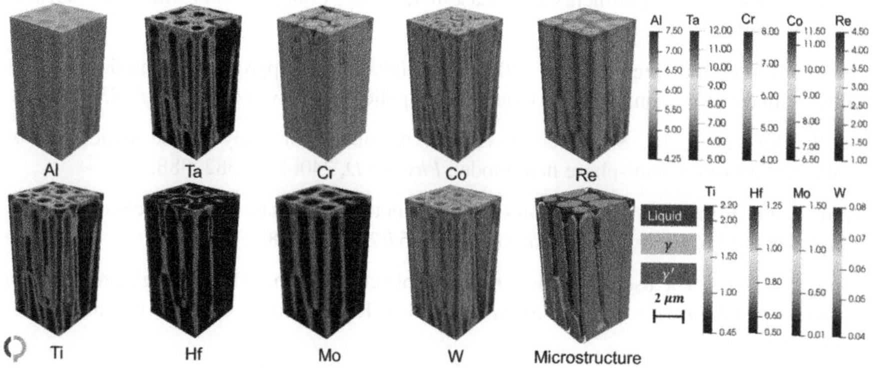

Figure 5.9. *Concentration distribution of all solute elements of CMSX-4 at the end of the solidification. The microstructure with phase information is included at the bottom right. Al and Cr, with equilibrium partition coefficient close to one, exhibit weaker segregation behavior. Ta and Ti enrich more into interdendritic region. Re, W and Co with partition coefficient greater than one segregate more into the dendritic core region. Cr is enriched in interdendritic liquid but depleted in the eutectic phase. For a color version of this figure, see www.iste.co.uk/bernacki/digital.zip*

Figure 5.9 shows the composition distribution and the microstructure at the end of the solidification of CMSX-4 full specification superalloy (Uddagiri et al. 2023). The microsegregation of individual alloying elements either to the trunk of the dendrites or to the interdendritic liquid, which finally transforms into $\gamma - \gamma'$ eutectic, can be seen. This result serves as a starting point for the simulation of "self-tempering" during the manufacturing process, when a new beam path heats up the sample to almost melting temperature, of the final solution heat treatment. Simulations were performed again with OpenPhase.

5.9. References

Beckermann, C., Diepers, H.-J., Steinbach, I., Karma, A., Tong, X. (1999). Modeling melt convection in phase-field simulations of solidification. *J. Comp. Phys*, 154, 468–496.

Borukhovich, E., Engels, P.S., Mosler, J., Shchyglo, O., Steinbach, I. (2015). Large deformation framework for phasefield simulations at the mesoscale. *Computational Materials Science*, 108, 367–373.

Cahn, J.E. and Hilliard, J.E. (1958). Free energy of a nonuniform system. I. Interfacial free energy. *Journal of Chemical Physics*, 28, 258–267.

Darvishi Kamachali, R., Kim, S., Steinbach, I. (2015). Texture evolution in deformed AZ31 magnesium sheets: Experiments and phase-field study. *Computational Materials Science*, 104, 193–199.

Eiken, J., Böttger, B., Steinbach, I. (2006). Multiphase-field approach for multicomponent alloys with extrapolation scheme for numerical application. *Physical Review E*, 73, 066122.

Guo, W., Spatschek, R., Steinbach, I. (2011). An analytical study of the static state of multi-junctions in a multi-phase field model. *Physica D*, 240(4–5), 382–388.

Hakim, V. and Karma, A. (2009). Laws of crack motion and phase-field models of fracture. *Journal of the Mechanics and Physics of Solids*, 57(2), 342–368.

Herring, C. (1952). The use of classical macroscopic concepts in surface-energy problems. In *Structure and Properties of Solid Surfaces*, Gomer, R. and Smith, C.S. (eds). University of Chicago, Chicago.

Hiebeler, J. (2017). Recovery and recrystallization during hot deformation in austenitic steel. Thesis, Ruhr-University, Bochum.

Hiebeler, J., Khlopkov, K., Shchyglo, O., Pretorius, T., Steinbach, I. (2017). Modelling of flow behaviour and dynamic recrystallization during hot deformation of MS-W 1200 using the phase field framework. In *MATEC Web of Conferences*, 80, 01003. doi: 10.1051/matecconf/20168001003.

Jantos, D.R., Junker, P., Hackl, K. (2016). An evolutionary topology optimization approach with variationally controlled growth. *Computer Methods in Applied Mechanics and Engineering*, 310, 780–801 [Online]. Available at: https://www.sciencedirect.com/science/article/pii/S0045782516307800.

Khachaturyan, A.G. (1983). *Theory of Structural Transformations in Solids*. Wiley, New York.

Kim, S.G., Kim, W.T., Suzuki, T. (1999). Phase-field model for binary alloys. *Physical Review E*, 60(6), 7186–7197.

Koumatos, K. and Muehlemann, A. (2017). A theoretical investigation of orientation relationships and transformation strains in steels. *Acta Crystallographica Section A: Foundations and Advances*, 73(2), 115–123.

Kundin, J. and Steinbach, I. (2020). Quantum-phase-field: From the Broglie–Bohm double-solution program to doublon networks. *Zeitschrift für Naturforschung*, 75(2a), 155–170.

Kundin, J., Steinbach, I., Abrahams, K., Divinski, S.V. (2021). Pair-exchange diffusion model for multicomponent alloys revisited. *Materialia*, 16, 101047.

Lázaro, G.R., Pagonabarraga, I., Hernández-Machado, A. (2015). Phase-field theories for mathematical modeling of biological membranes. *Chemistry and Physics of Lipids*, 185, 46–60.

Levitas, V.I. (1998). Thermomechanical theory of martensitic phase transformations in inelastic materials. *International Journal of Solids and Structures*, 35(9), 889–940 [Online]. Available at: https://www.sciencedirect.com/science/article/pii/S0020768397000899.

Levitas, V.I. (2013). Phase-field theory for martensitic phase transformations at large strains. *International Journal of Plasticity*, 49, 85–118 [Online]. Available at: https://www.sciencedirect.com/science/article/pii/S0749641913000727.

Miehe, C., Welschinger, F., Hofacker, M. (2010). Thermodynamically consistent phase-field models of fracture: Variational principles and multi-field FE implementations. *International Journal for Numerical Methods in Engineering*, 83(10), 1273–1311.

Miyoshi, E. and Takaki, T. (2016). Extended higher-order multi-phase-field model for three-dimensional anisotropic-grain-growth simulations. *Computational Materials Science*, 120, 77–83.

Nguyen, S., Folch, R., Verma, V.K., Henry, H., Plapp, M. (2010). Phase-field simulations of viscous fingering in shear-thinning fluids. *Phys. Fluids*, 22, 103102.

Plapp, M. (2011). Unified derivation of phase-field models for alloy solidification from a grand-potential functional. *Physical Review E*, 84, 031601.

Salama, H., Kundin, J., Shchyglo, O., Mohles, V., Marquardt, K., Steinbach, I. (2020). Role of inclination dependence of grain boundary energy on the microstructure evolution during grain growth. *Acta Materialia*, 188, 641–651.

Salama, H., Ali, M., Shchyglo, O., Steinbach, I. (2024). Phase-field simulation framework for modeling martensite and bainite formation in steel. *Computational Materials Science*, 241, 113033. doi: 10.1016/j.commatsci.2024.113033.

Scott, A., Chu, F., McLauchlin, D. (1973). The soliton: A new concept in applied science. *Proceedings IEEE*, 61, 1443–1483.

Shchyglo, O., Ali, M., Salama, H. (2024). Efficient finite strain elasticity solver for phase-field simulations. *NPJ Computational Materials*, 10, 52. doi: 10.1038/s41524-024-01235-4.

Steinbach, I. (2009). Phase-field models in materials science. *Modelling and Simulation in Materials Science and Engineering*, 17, 073001–073031.

Steinbach, I. (2013). Phase-field model for microstructure evolution at the mesoscopic scale. *Annual Review of Materials Research*, 43, 89–107.

Steinbach, I. (2017). Quantum-phase-field concept of matter: Emergent gravity in the dynamic universe. *Zeitschrift fur Naturforschung A*, 72(1), 51–58.

Steinbach, I. and Pezzola, F. (1999). A generalized field method for multiphase transformations using interface fields. *Physica D*, 134, 385–393.

Steinbach, I. and Salama, H. (2023). *Lectures on Phase Field*. Springer, Cham. doi: 10.1007/978-3-031-21171-3.

Steinbach, I. and Shchyglo, O. (2011). Phasefield modelling of microstructure evolution in solids: Perspectives and challenges. *Current Opinion in Solid State and Materials Science*, 15, 87–92.

Steinbach, I., Pezzolla, F., Nestler, B., Seeßelberg, M., Prieler, R., Schmitz, G., Rezende, J.L. (1996). A phase field concept for multiphase systems. *Physica D*, 94(3), 135–147.

Steinbach, I., Zhang, L., Plapp, M. (2012). Phase-field model with finite interface dissipation. *Acta Materialia*, 60, 2689–2701.

Steinbach, I., Uddagiri, M., Salama, H., Ali, M., Shchyglo, O. (2024). Highly complex materials processes as understood by phase-field simulations: Additive manufacturing, bainitic transformation in steel and high-temperature creep of superalloys. *MRS Bulletin*, 49, 1–11. doi: 10.1557/s43577-024-00703-y.

Tiaden, J., Nestler, B., Diepers, H.J., Steinbach, I. (1998). The multiphase-field model with an integrated concept for modeling solute diffusion. *Physica D*, 115, 73–86.

Uddagiri, M., Shchyglo, O., Steinbach, I., Tegeler, M. (2023). Solidification of the Ni-based superalloy CMSX-4 simulated with full complexity in 3-dimensions. *Progress in Additive Manufacturing*, 8, 1. doi: 10.1007/s40964-023-00513-9.

Wang, Y.U. and Khachaturyan, A. (1997). Three-dimensional field model and computer modeling of martensitic transformations. *Acta Materialia*, 45(2), 759–773 [Online]. Available at: https://www.sciencedirect.com/science/article/pii/S1359645496001802.

Wang, Y.U., Jin, Y.M., Khachaturyan, A.G. (2002). Phase field microelasticity theory and modeling of elastically and structurally inhomogeneous solid. *Journal of Applied Physics*, 92(3), 1351–1360. doi: 10.1063/1.1492859.

Zhang, L. and Steinbach, I. (2012). Phase-field model with finite interface dissipation: Extension to multi-component multi-phase alloys. *Acta Materialia*, 60, 2702–2710.

6

Level-Set Method for the Modeling of Microstructure Evolution

Marc BERNACKI
CNRS UMR 7635, Centre for Material Forming (CEMEF), MINES Paris, PSL Research University, Sophia Antipolis, France

6.1. Introduction

Metal forming modeling can be predictive only if the strain rate, strain and temperature dependency of the flow behavior are correctly described. The mechanical properties and behavior of metallic materials mainly depend on the density and structure of dislocation network, as illustrated in Chapter 1; this points out the need of incorporating microstructure concepts into the numerical models. On the other hand, mastering grain size, solid phase transformations (PTs) and dislocation density evolution during industrial processes become necessary to fulfill more and more severe microstructure and properties' specifications of the processed materials. This objective implies correctly describing the main physical mechanisms occurring in metals during thermomechanical processing, notably continuous (CDRX) or discontinuous (DDRX) dynamic recrystallization (DRX) and postdynamic recrystallization (PDRX) – possibly including metadynamic recrystallization (MDRX) and static recrystallization (SRX) mechanisms, solid-state phase transformations (SSPT), grain growth (GG), Ostwald ripening (OR), etc.

However, microstructure predictions on multi-pass processes are very challenging due to the strong evolution of the microstructure topology from the beginning to the

Digital Materials,
coordinated by Marc BERNACKI and Samuel FOREST. © ISTE Ltd 2024.

end of the process. The competition between boundary migration, recovery and nucleation leads to complex coupled effects. Minor variations of the process parameters (interpass and/or reheating times, temperature and strain rate values at each pass) may have huge effects on the way the previously cited mechanisms can take place.

Thanks to the explosion of computer capacities, mesoscopic modeling techniques for metallic materials at the solid state are now available. These models are based on a full description of the microstructure topology and have demonstrated an exciting potential for an extensive range of microstructure evolutions like the precise modeling of recrystallization (ReX) (Rollett et al. 2017) in DRX or PDRX conditions, GG (Rollett et al. 2017) or diffusive solid-state phase transformation (DSSPT) (Militzer 2011). The main numerical frameworks involved Monte Carlo Potts (MC) (Rollett et al. 1989; Rollett and Manohar 2004), cellular automata (CA) (Raabe 2002; Janssens 2010; Golab et al. 2014) (see Chapter 7), phase field (PF) or multiphase field (MPF) (Steinbach et al. 1996; Krill III and Chen 2002; Moelans et al. 2008) (see Chapter 5), front-tracking or Vertex (Barrales Mora et al. 2008) (see Chapter 4) or level set (LS) models (Merriman et al. 1994; Zhao et al. 1996; Bernacki et al. 2008; Hallberg and Bulatov 2019). These numerical methods are currently used and developed by many researchers (Rollett et al. 2017) and compared at regular intervals.

Each of the models mentioned has its unique strengths and limitations. Probabilistic voxel-based approaches, including MC and certain CA formulations, are widely recognized. These models utilize uniform grids of cells to represent microstructures and employ stochastic laws to predict interface movements. Such simulations are not only computationally efficient but also excel in scalability. In contrast, deterministic methods that solve partial differential equations (PDEs) offer precise descriptions of the physical mechanisms involved, albeit at a higher computational expense. For example, front-tracking or Vertex methods explicitly describe interfaces using vertices, with interface motion controlled incrementally by calculating the velocity of specific points. However, a significant challenge in these approaches lies in managing complex topological events, such as the emergence or disappearance of grains, which becomes particularly intricate in three dimensions. Other deterministic strategies, known as front-capturing approaches like the MPF and LS methods, circumvent these topological issues through an implicit interface representation. Nonetheless, a primary drawback of these methods, especially when integrated into a finite element (FE) framework, is their high computational demand. For a detailed exploration and illustration of the Vertex, MPF and CA methods, refer to Chapters 4, 5 and 7, respectively.

LS simulations in context of regular grids and Fourier transform resolution, with a very large number of grains, can be found for GG (Elsey et al. 2009; Mießen et al. 2015) and for static recrystallization (Elsey et al. 2011) modeling. When global or local meshing/remeshing operations have to be considered (large deformation, presence of second-phase particles (SPPs), etc.), LS approach in context of unstructured FE mesh and reasonable number of grains can be considered (Bernacki et al. 2008; Logé et al. 2008; Bernacki et al. 2011; Hallberg and Ristinmaa 2013; Scholtes et al. 2016a; Maire et al. 2017). This explains the limited number of studies dedicated to the full field modeling of ReX in context of industrial hot metal forming. Indeed, the full field modeling of real thermomechanical paths generally implies to take into account all the concomitant physical mechanisms in context of large or very large deformations.

In the following sections, we will provide a comprehensive summary of the current state of the art in using the finite element-level set (FE-LS) method for modeling microstructural evolution. Initially, we will revisit the kinetic equations related to microstructure evolution. This will be followed by a concise overview of the LS methodology in the context of polycrystalline material characterization. Subsequently, we will delve into algorithms designed for simulating ReX, GG and other diffusive phenomena within unstructured FE meshes. We will then engage in discussions about the anisotropy of grain boundary (GB) energy and the methods of incorporating static or evolving SPPs in these models. Following this, we will outline a typical procedure that transitions from an industrial thermomechanical path to employing the FE-LS method for predicting local GB migration. Finally, we will discuss some of the latest developments and advancements in this approach.

6.2. Kinetics equations of microstructure evolution at the mesoscopic scale

Numerous metallurgical phenomena have been mentioned. Consequently, it is essential to present a comprehensive evaluation of the kinetic equations currently established in the state of the art, particularly at the mesoscopic scale. This evaluation should include an analysis of their limitations and a revisitation of the pertinent physical mechanisms involved. When ReX and related phenomena or DSSPT are considered, the classical framework consists of defining the interface kinetics as the cross product between a mobility, μ [$m^4 \cdot s^{-1} \cdot J^{-1}$], generally viewed as an intrinsic scalar property of the considered interface and function of temperature, and the considered driving pressures, P [$J \cdot m^{-3}$]:

$$\vec{v} = \mu P \vec{n}, \qquad [6.1]$$

with $\vec{n}$ the interface unitary outside normal. In the context of ReX and related phenomena, P is classically defined as:

$$P = \tau [\![\rho]\!] - \gamma \kappa, \qquad [6.2]$$

where τ [J·m^{-1}] is the dislocation line energy, $[\![\rho]\!]$ [m^{-2}] is the homogenized dislocation density jump across interfaces estimated from the modeling of plastic deformation, γ [J·m^{-2}] is the GB energy and κ [m^{-1}] is the mean curvature (i.e. the trace of the curvature tensor). In the mesoscopic-scale analysis, it is commonly postulated that the dislocation density in each grain, subject to further deformation (ϵ), follows a specific evolution equation:

$$\partial_\epsilon \rho = K_1 \rho^\xi - K_2 \rho, \qquad [6.3]$$

Here, K_1 and K_2 signify two material constants that characterize strain hardening and recovery, respectively, while ξ is a parameter dependent on the chosen model. The renowned Yoshie–Laasraoui–Jonas (YLJ) model (Laasraoui and Jonas 1991) aligns with the case of $\xi = 0$, and the Kocks–Mecking model corresponds to $\xi = 1/2$ (Kocks 1976; Mecking and Kocks 1981). Furthermore, more intricate numerical frameworks incorporating crystal plasticity simulations, along with various models of intermediate complexity, are also viable. Particularly in the realm of hot metal forming and substantial plastic deformation, the crystal plasticity finite element method (CPFEM) is frequently employed. In CPFEM, the stress–strain response of each FE is governed by a single crystal model adhering to an elastoviscoplastic formulation (Cuitino and Ortiz 1993; Asaro and Needleman 1985; Marin and Dawson 1998), and a Lagrangian framework is typically used for updating the positions of mesh nodes. Within this framework, the local evolution of dislocation density is determined by the hardening rule, which is dependent on the slip rates $\dot{\gamma}_\alpha$ across all slip systems α, defined by the selected flow rule. Specifically, for FCC materials, an equivalent crystal plasticity YLJ law to equation [6.3] for describing geometrically dislocation density (GND) evolution is given by the following equation:

$$\partial_t \rho = \frac{1}{M}(K_1 - K_2\rho) \sum_{\alpha=1}^{n} | \partial_t \gamma_\alpha |, \qquad [6.4]$$

with M the Taylor factor.

In the context of DSSPT, P is generally defined through the well-known Gibbs–Thomson equation applying onto interphase boundaries (as already illustrated in the previous chapter by equation [5.9]):

$$P = [\![G]\!] - \sigma\kappa, \qquad [6.5]$$

where $[\![G]\!]$ [J·m^{-3}] is the PT Gibbs free energy jump at the phase interface between the involved phases, and σ [J·m^{-2}] is the phase boundary (PB) energy (totally equivalent to the γ definition for GB). Phase equilibrium and resulting $[\![G]\!]$ evaluation at the interface can be achieved by considering ortho-equilibrium (equilibrium for all components) or para-equilibrium (equilibrium only for the fast diffusional species). Para-equilibrium is often adopted in FF modeling.

In the context of spheroidization and coalescence of second-phase objects or participates, the kinetic equations can be summarized as (Wakai and Brakke 2011):

$$\vec{v} = -\Omega \vec{\nabla_s} \cdot \vec{j^s} \vec{n}, \quad [6.6]$$

with Ω the atomic volume and $\vec{j^s}$ the surface atom flux:

$$\vec{j^s} = -\frac{D_s \gamma_s}{kT} \overrightarrow{\nabla_s \kappa}, \quad [6.7]$$

with D_s the interfacial diffusivity, k the Boltzmann constant and T the absolute temperature. A common misconception in the literature often arises from interpreting observable metallurgical phenomena without acknowledging their underlying physical mechanisms. In simpler terms, the equations previously discussed quantitatively capture a wide array of observable phenomena, which, at first glance, may seem unrelated. To illustrate this point, consider the following examples: the OR phenomenon reflects the long-term evolution of phase equilibrium mechanisms, as defined by equations [6.1] and [6.5]. Similarly, the surface diffusion mechanism, outlined by equation [6.6] and equation [6.7], is crucial in explaining phenomena like sintering or the grooving of GBs/PBs on free surfaces. Abnormal grain growth (AGG) or critical grain growth (CGG) phenomena can be understood as local heterogeneities in terms defined by equation [6.2]. Smith–Zener pinning of GB or PB by SPPs can be comprehensively explained by the curvature term in equation [6.2] or [6.5], respectively, and so forth. A critical limitation of existing numerical frameworks is their tendency to treat each fundamental mechanism in isolation, disregarding their simultaneous nature and mutual influence in real materials. It is important to note that in all these mechanisms, plasticity resulting from hot deformation, as well as that potentially induced by the transformations themselves, can be precisely evaluated and predicted within FE frameworks. These effects can be incorporated into the aforementioned migration equations through the terms $[\![\rho]\!]$.

While equations [6.1]–[6.7] provide a classical depiction of a wide range of metallurgical phenomena at the mesoscopic scale, they are, in essence, approximations of mechanisms occurring at smaller scales. These kinetic equations and their corresponding physical mechanisms are subjects of ongoing evolution and refinement, as evidenced in recent literature (Chen et al. 2020; Bhattacharya et al. 2021; Florez et al. 2022). At the polycrystalline scale, particularly in the context of metal forming, they represent a thermodynamically and kinematically sound framework that has been both justified and validated. Furthermore, this framework offers the potential for enhancement through the integration of dynamic molecular simulation results, specifically in terms of refining the description of interface properties.

6.3. Level-set function, description of polycrystal and meshing adaptation

The LS method was firstly introduced by Dervieux and Thomasset (2006) under the terminology of "pseudo-density function" in 1979 before being extensively formalized and disseminated by the works of Osher and Sethian (1988) as a numerical tool to trace the spatial and temporal evolution of interfaces submitted to curvature flow problems. This approach was enhanced later for evolving multiple junctions (Merriman et al. 1994; Zhao et al. 1996) and applied to recrystallization and GG in Bernacki et al. (2008); Logé et al. (2008); Bernacki et al. (2011). The principle for modeling polycrystals is the following: grains are defined by LS functions ψ in the space Ω:

$$\begin{cases} \psi(x,t) = \pm d(x, \Gamma(t)), & x \in \Omega, \quad \Gamma(t) = \partial G \\ \psi(x, t \in \Omega) = 0 & \rightleftharpoons \quad x \in \Gamma(t), \end{cases} \qquad [6.8]$$

More precisely, each sub-domain G (grain) in a given domain Ω (polycrystal) is classically described implicitly by computing the signed distance function $\psi(x,t)$ representing the distance to the sub-domain boundaries $\Gamma(t) = \partial G$ (GBs). In a P1 formulation, the function $\psi(x,t)$ is calculated at each node on the FE mesh and is classically chosen, by convention, positive inside of the grain and negative outside as illustrated in Figure 6.1.

Using such an implicit description of the interfaces simplifies the definition of the FE mesh as it has not been correlated to the GB/PB interfaces. An experimental image can be easily immersed in the FE mesh by calculating the distance functions to the gray levels of each grain interface as illustrated in Figure 6.2.

Moreover, when only statistical data are known such as the mean grain size (MGS) or the grain size distribution, digital material must be built thanks to representative microstructures (Rollett et al. 2004). Voronoï or Laguerre–Voronoï tessellations can be used in order to respect the known data in the generated digital polycrystal made of polyhedral grains. Voronoï tessellation method (VTM) consists of generating randomly *N* Voronoï nuclei and defining each Voronoï cell as the region consisting of all locations in the space, which are closer to the considered nucleus than any other nucleus. Despite the good geometric correlation between Voronoï tessellations and many cellular structures, VTM presents some limitations due to the impossibility to respect, for example, a given statistical volume distribution of cells. Xu and Li (2009) described the divergence between statistical properties of grains classically observed in an equiaxial polycrystal and results obtained with the VTM. One way to improve the classical VTM consists of using the so-called Laguerre–Voronoï tessellation method (LVTM), by assigning a radius, or weight, to each nucleus and to respect this distribution in the Voronoï tessellation (Imai et al. 1985).

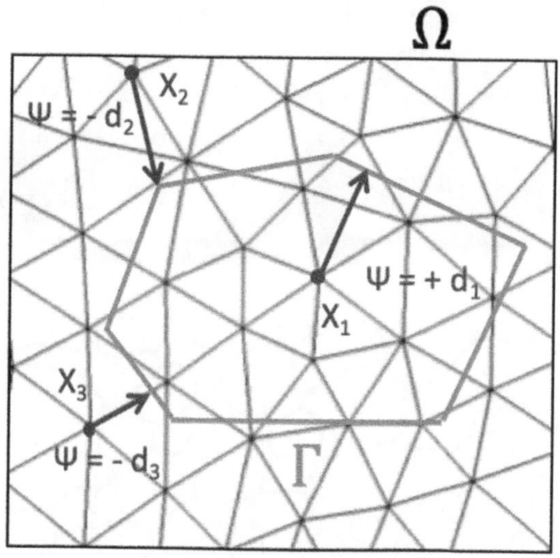

Figure 6.1. *LS description of a polyhedral grain (blue interface) in an unstructured FE mesh (in black). For a color version of this figure, see www.iste.co.uk/bernacki/digital.zip*

It is important to highlight that the principal difficulty of LVTM is common to the generation of polycrystal microstructure and powder RVEs: to respect a given statistical size of spheres with the highest possible density. If this criterion seems natural for powder RVEs, it is also crucial for LVTM in order to limit the heterogeneities between the weight imposed and the volume of Voronoï cells finally obtained. This discussion is classically unclear in the literature but put in another way, respecting the grain size distribution for the Laguerre–Voronoi sphere packing is not the same thing as respecting the grain size distribution for the resulting Laguerre–Voronoï cells. Different approaches, gathered in the so-called sphere packing methods (SPM), have been developed in the last 40 years (Jodrey and Tory 1985; Bagi 1993; He et al. 1999; Benabbou et al. 2009) and are classically divided into two families: the sequential addition models and the collective rearrangement ones. Both are used for polycrystal generation (Hitti et al. 2012; Ilin and Bernacki 2016a, 2016b; Quey and Renversade 2018; Depriester and Kubler 2019).

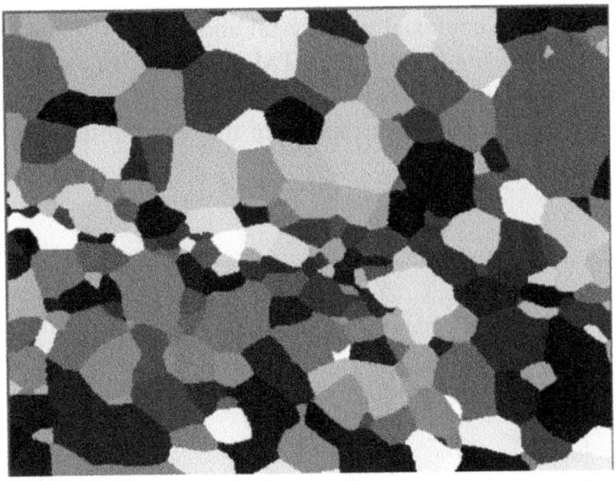

(a)

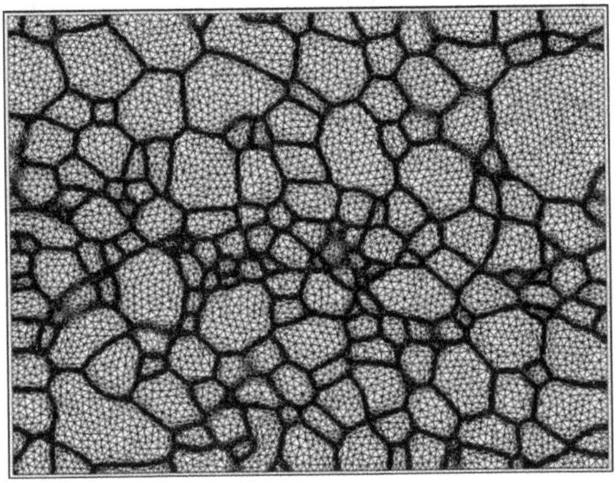

(b)

Figure 6.2. *(a) 2D micrograph obtained by EBSD (each grain being plotted with a different gray level) and (b) immersion of the experimental micrograph in an FE mesh thanks to an LS strategy and adaptive anisotropic remeshing around the grain boundaries*

In addition to the morphology of the grains and the phases composition, other attributes playing on the properties of the interfaces or the driving pressures in play can be sought to be respected in an exact or statistical way. At the mesoscopic scale, a GB, B_{ij}, between grains G_i and G_j, is characterized by its morphology and its crystallographic properties, which may be summarized by a tuple $B_{ij} = (M_{ij}, n_{ij})$ with two shape parameters describing the interfaces through the unitary–outward normal direction n_{ij}, and three crystallographic parameters describing the orientation relationship between the two adjacent grains known as the misorientation tensor M_{ij}. This GB space $\mathcal{B}$ parameterized by the misorientation and the GB inclination is illustrated in Figure 6.3. The misorientation is frequently defined with the axis-angle parameterization, that is, $M_{ij}(a_i, \theta)$, where a_i is the misorientation axis and θ is the misorientation angle leading to the disorientation angle when the symmetries of the crystal are taken into account (Morawiec 2003). The two quantities of interest, the GB energy γ and GB mobility μ, must then be seen as functions from the GB space $\mathcal{B}$ to $\mathbb{R}^+$. Thus, when EBSD maps are available, they can be used to exactly define the misorientation map; 3D data allowing us to add a complete determination of the n_{ij} parameter. When these information are missing, a classical strategy consists of using a random grain orientation, leading to a Mackenzie-like disorientation distribution function (DDF).

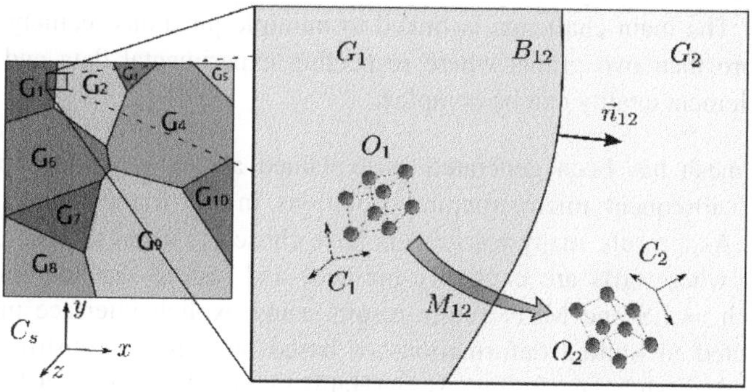

Figure 6.3. *Scheme depicting one GB and its parameters. Image available online at Flickr licensed under CC BY 2.0. Source: Brayan Murgas, 2021. For a color version of this figure, see www.iste.co.uk/bernacki/digital.zip*

Theoretically, each grain of a single-phase or multi-phase polycrystal must be represented by its own LS function. In order to reduce the computation time and memory storage, several non-neighboring grains in the initial microstructure (separated by a certain number of grains δ and/or by a minimal distance) can be grouped to form global level set (GLS) functions thanks to the Graph coloration technique. However, the grains belonging to the same function can no longer be distinguished. As a consequence, when two child grains of a GLS grow and meet each other, numerical coalescence occurs, that is, both grains merge to form a single grain. Thus, different strategies can be found in order to avoid or at least minimize numerical coalescence events: the initial separation δ and/or the minimal distance can be chosen small enough to limit the computation time and sufficiently high to avoid a significant amount of coalescence (Cruz-Fabiano et al. 2014); a complete optimal coloration can be performed at each time step, or re-coloration algorithms can be considered at each time step to treat the risky configurations (Krill III and Chen, 2002; Elsey et al. 2009; Scholtes et al. 2015).

In context of polyhedral cells, FE meshing of digital microstructures is quite straightforward, and consists, classically, of discretizing cells facets, and then the volume within each cell (Quey et al. 2011). This procedure is commonly used if the simulation domain is subjected to small deformations. Moreover, for real polycrystals, observed using scanning electron microscopy (SEM) or 3D X-ray imaging techniques (Ludwig et al. 2009; Proudhon et al. 2016), a Voronoï/Laguerre–Voronoï mesh is not accessible. While there has been much research done on meshing methods for real 3D microstructures (Zhang et al. 2005; Young et al. 2008), the generalization of these methods to massively multiphase materials such as polycrystals is not straightforward (Rollett et al. 2004; Brahme et al. 2006). The main challenge is linked to multiple junctions, namely, interfaces between more than two grains where respecting experimental data and achieving high mesh element quality can be complex.

Once a mesh has been generated as explained above, modeling large plastic strains and subsequent microstructure evolutions in FE framework is also very challenging. As a result, many researchers have chosen to avoid the use of conform meshes (i.e. where GBs are explicitly meshed) and instead use implicit interface methods such as LS and MPF. While results using explicit interface methods are either restricted to limited deformations or based on a full reconstruction of the computational mesh at each time step (Hallberg and Ristinmaa 2013), implicit interface methods allow modeling of a wider range of metallurgical phenomena where large deformations are considered (Bernacki et al. 2009; Scholtes et al. 2015b, 2016a; Maire et al. 2017). However, this numerical strategy generally requires fine FE meshes at the grain interfaces to reach acceptable accuracy with regard to ReX and GG modeling. The absence of a conform mesh at GBs generally requires a finer discretization. Of course, global isotropic refinement of the mesh or high-order interpolation of LS functions can be used to reach a desired accuracy in the interface description; however, these strategies lead to a significant increase in computation resources.

Adaptive local isotropic or anisotropic remeshing is therefore generally preferred. Different ways exist to generate locally adapted meshes to zero-isovalue of LS but, classically, metric field and topological meshers based on local mesh topology optimizations are used (Coupez et al. 2000; Gruau and Coupez 2005; Shakoor et al. 2015a). Concerning metric calculations, the most common approach consists of using a posteriori error analysis in order to obtain an optimal mesh for a given number of nodes (Almeida et al. 2000; Coupez 2011). However, this approach can be non-optimal in polycrystal cases when an important number of LS functions must be considered. When describing strictly disjoint objects by LS functions, a simple solution consists of adapting the mesh thanks to a posteriori error estimator to

$\psi_{max}(x,t) = \max_{i=1,\ldots,N_{LS}} \psi_i(x,t)$ with N_{LS} the number of GLS functions used to describe all the grains. However, such an approach, used in Scholtes et al. (2015b) and Maire et al. (2016), is not straightforward when the different objects considered are not strictly disjoint, which is the case in polycrystalline microstructures, as the ψ_{max} function is then not derivable at the GB network.

Alternatively, automatic geometrical methods can be used for the generation of local refined isotropic or anisotropic meshes adapted to polycrystals, based on the normal and/or mean curvature of the grain interfaces (Bernacki et al. 2009; Hitti et al. 2012). Figure 6.4(a) illustrates a 2D example where anisotropic metric is considered thanks to a geometric approach (Bernacki et al. 2009) for an Inconel 718 microstructure: the mesh is adapted to the grain interfaces (generated statistically with a Laguerre–Voronoï approach) and to SPPs interfaces immersed after thresholding of a SEM image; Figure 6.4(b) illustrates a 3D 304L case where a posteriori metric is adopted (Scholtes et al. 2015b). Figure 6.4(c) illustrates a 2D two-phase cases where an isotropic strategy is used with a mesh refinement in grain and phase interfaces but with an increased precision in the phase interface (Chandrappa and Bernacki 2023).

6.4. Isotropic framework for LS modeling of ReX and GG

In the LS method, the evolution of $\psi(x,t)$, submitted to a velocity field $\overrightarrow{v}(x,t)$, is then given by the following convective PDE (Osher and Sethian 1988):

$$\frac{\partial \psi(x,t)}{\partial t} + \overrightarrow{v}(x,t) \cdot \nabla \psi(x,t) = 0, \qquad [6.9]$$

$$\psi(x, t=0) = \psi^0(x), \qquad [6.10]$$

with $\psi^0(x)$ the initial description of the LS function. In the context of deterministic full field approaches and neglecting torque terms (Fausty et al. 2018), the velocity field can be defined using equation [6.1] and when ReX and GG are involved, the net pressure is classically defined using equation [6.2]. The isotropy hypothesis remains here, for the LS simulations, to consider γ as constant and M as only dependent of the temperature through an Arrhenius law $M = M_0 \exp(-Q/RT)$ where M_0 is a pre-exponential constant parameter, R is the gas constant and T is the absolute temperature. κ and $\overrightarrow{n}$ can be defined naturally by taking advantages of the possibilities offered by the LS framework. Indeed, considering that GLS functions remain distance functions all along the simulation (i.e. that $||\overrightarrow{\nabla\psi}|| = 1$), they can be defined as:

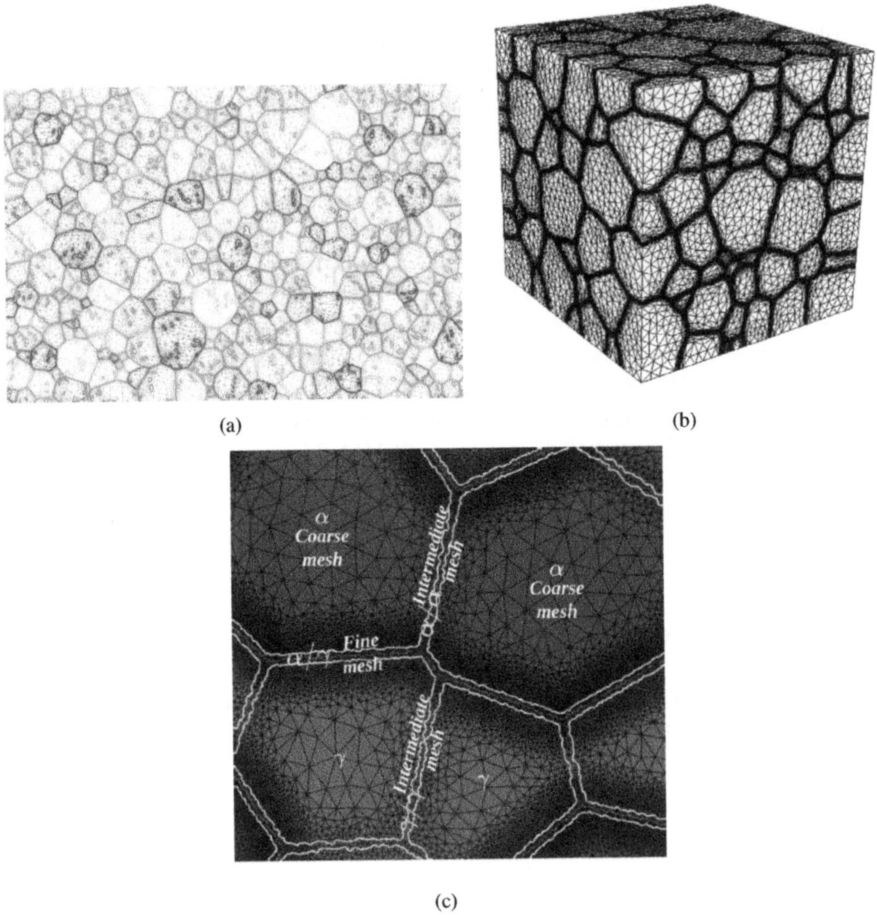

Figure 6.4. *(a) 2D Inconel 718 microstructure example where local anisotropic metric is considered for the FE mesh generation thanks to a geometric approach (Bernacki et al. 2009). The mesh is adapted to the grain interfaces (generated statistically with a Laguerre–Voronoï approach) and to δ phase interfaces immersed thanks to a SEM image. The color code corresponds to the GLS functions. (b) 3D equiaxed microstructure example where local anisotropic metric is considered for the FE mesh generation thanks to a posteriori approach (Scholtes et al. 2015b). The mesh is adapted to the grain interfaces (generated statistically with a Voronoï tessellation). (c) A 2D two-phase cases where an isotropic strategy is used with a mesh refinement in grain and phase interfaces but with an increased precision in the phase interface (Chandrappa and Bernacki 2023). For a color version of this figure, see www.iste.co.uk/bernacki/digital.zip*

$$\vec{n} = -\overrightarrow{\nabla\psi}, \qquad [6.11]$$

$$\kappa = -\Delta\psi. \qquad [6.12]$$

Finally, by using the coloring/recoloring algorithms (Scholtes et al. 2015b) and the described metric properties of the LS functions, we can solve after substituting equations [6.12] and [6.11] into equation [6.2] and equation [6.2] into equation [6.1], a set of N_{LS} convective-diffusive equations as detailed by equation [6.13], with $N_{LS} \ll N_g$ and $i \in \{1, 2, ..., N_{LS}\}$. The numerical strategy consisting of limiting the number of involved LS functions is then generally crucial in terms of numerical cost.

$$\frac{\partial \psi_i(x,t)}{\partial t} - \mu\gamma\Delta\psi_i(x,t) + \vec{v}_i^{[\![\rho]\!]} \cdot \nabla\psi_i(x,t) = 0, \qquad [6.13]$$

$$\vec{v}_i^{[\![\rho]\!]} = \mu\tau[\![\rho]\!]_i \vec{n}_i, \qquad [6.14]$$

$$\psi_i(x, t=0) = \psi_i^0(x). \qquad [6.15]$$

Generally, a particular numerical treatment must be imposed to avoid kinematic incompatibilities after the resolution of equation [6.13]. Indeed, as illustrated in equation [6.16], voids or overlaps can appear at multiple junctions and must be treated. A classical solution consists of correcting the GLS functions as follows (Merriman et al. 1994):

$$\psi_i(x,t) = \frac{1}{2}\left(\psi_i(x,t) - \max_{j \neq i} \psi_j(x,t)\right), \quad 1 \leq i \leq N_{LS}. \qquad [6.16]$$

An alternative could also consist (Zhao et al. 1996) of closing void regions thanks to an energy minimization principle enforced by a Lagrange multiplier related to a constraint added to equation [6.9].

A drawback of the LS formulation lies in the fact that after the resolution of equations [6.13] and [6.16], the GLS are no longer distance functions $\|\overrightarrow{\nabla\psi}\| \neq 1$. This is particularly problematic when a specific remeshing technique depending on the distance property is used at the interface, as described in the previous section. In addition, the convective–diffusive formulation proposed in equation [6.13] requires a distance function at least in a thin layer around the interface in order to properly compute the curvature driven mechanism. Finally, the conditioning of the transport problem also depends on the regularity of the LS function (Bernacki et al. 2009). For these reasons, the GLS functions need to be reinitialized. Restoring the metric property at the instant t is equivalent to solving the following eikonal equation for each GLS function:

$$\begin{cases} \|\overrightarrow{\nabla\psi}\| = 1, \ \forall x \in \Omega \\ \psi_i(x,t) = 0, \ \forall x \in \Gamma_i(t) \end{cases} \quad \forall i \in \{1, 2, ..., N_{LS}\}, \qquad [6.17]$$

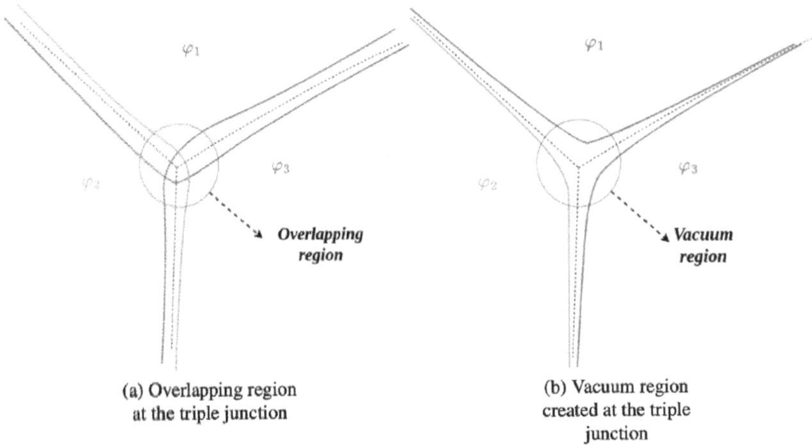

Figure 6.5. *Illustration of (a) overlapping of level sets at a triple junction, (b) decohesion of level sets at a triple junction. Adopted from Chandrappa and Bernacki (2024). For a color version of this figure, see www.iste.co.uk/bernacki/digital.zip*

Different approaches exist to solve this equation, including the well-known fast marching method introduced by Sethian (1996), which propagates a front from the interface and ensures directly a gradient equal to unity. Though this approach has been extended to unstructured meshes (Kimmel and Sethian 1998), its implementation becomes complicated when it comes to considering anisotropic triangulations (Sethian and Vladimirsky 2003) and parallel efficiency is poor.

In Sussman et al. (1994), a Hamilton–Jacobi (H–J) formulation equivalent to equation [6.17] was proposed in order to correct iteratively the LS values around the interface by solving a PDE. This method is massively used in the LS modeling of ReX and GG (Bernacki et al. 2011; Agnoli et al. 2014; Cruz-Fabiano et al. 2014; Hallberg and Bulatov 2019). It requires the definition of a purely numerical parameter known as a fictitious time step for reinitialization and the ratio between the desired reinitialized thickness and this parameter gives the number of required increments. Coupled convection–reinitialization (CR) methods emerged wherein the LS function is automatically reinitialized during the resolution of the transport equation (Coupez 2007; Bernacki et al. 2008, 2009). Their main advantage lies in the fact that only one solver is needed for the simulation instead of two for the classical H–J technique. The signed distance function can also be replaced by any smooth function which satisfies the metric property, at least in a thin layer around the interface.

In conclusion, a straightforward method to reinitialize gradient level-set (GLS) functions involves employing a brute force algorithm for a complete reconstruction of the distance function. This technique unfolds in two phases: first, the interface (zero-isovalue of the LS function) is discretized into a collection of simple elements, and then, for each node, the distance to every element in the collection is calculated. The minimum distance found is then stored as the updated value of the distance function. While this direct reinitialization (DR) technique ensures optimal accuracy, it is often noted for its significant computational demands (Sussman et al. 1994; Jones et al. 2006; Elias et al. 2007). However, it is important to note that these studies primarily focus on regular grids or hierarchical meshes (Fortmeier and Martin Bücker 2011). More recently, a novel direct approach that is both fast and accurate for use in unstructured FE meshes has been introduced (Shakoor et al. 2015b). This method leverages a space-partitioning technique utilizing a $k - d$ tree, coupled with an efficient bounding box strategy, to enhance numerical efficiency, especially in parallel computing environments. The discussion also extends to the residual errors inherent in this approach, as examined in Florez et al. (2020).

By extrapolating the shape of the velocity term, $\overrightarrow{v}_i^{[\![\rho]\!]}$ of equation [6.13], for the interface between grain G_i and G_j, the corresponding velocity $\overrightarrow{v}_{ij}^{[\![\rho]\!]}$ can be written as:

$$\overrightarrow{v}_{ij}^{[\![\rho]\!]} = \mu\tau[\![\rho]\!]_{ij}\overrightarrow{n}_{ij}. \qquad [6.18]$$

Classically in LS or MPF approaches, a constant stored energy is considered in each grain (Bernacki et al. 2008, 2009; Logé et al. 2008; Elsey et al. 2009; Hallberg and Ristinmaa 2013; Scholtes et al. 2016a; Maire et al. 2017). Then, for each interface between grain G_i and G_j, it is assumed that:

$$[\![E]\!]_{ij} = \tau[\![\rho]\!]_{ij} = e_j - e_i, \qquad [6.19]$$

where e_i and e_j are the mean plastic stored energies in the grains G_i and G_j, respectively. These averages can directly come from either constant approximative values where only a gradient of the stored dislocations between the nuclei and the non-recrystallized grains is considered, as in Bernacki et al. (2008, 2011) and Elsey et al. (2009), or simplified mechanical formulations as in Maire et al. (2017). They can also be evaluated thanks to an existing dislocation field in the FE mesh of the calculation domain Ω as in Logé et al. (2008); Bernacki et al. (2009); Hallberg and Ristinmaa (2013); Scholtes et al. (2016a); Ruiz et al. (2020a, 2020b) or come from a dislocation density field measured from experimental data, immersed in the FE mesh and averaged per grain as in Agnoli et al. (2015); Villaret et al. (2020); Grand et al. (2022). More precise approximations of the energetic field can also be considered where $[\![E]\!]_{ij} = E_j(x,t) - E_i(x,t)$ and $E_k(x,t)$ can be derived from averages at the interfaces scale (Ilin et al. 2018) or even from local node values (Grand et al. 2023).

Special attention has to be paid to the velocity field $\vec{v}_i^{[\![\rho]\!]}$ in the vicinity of multiple junctions as emphasized in Bernacki et al. (2008). In fact, rather than dealing with $\vec{v}_i^{[\![\rho]\!]}$ per grain as described in equation [6.13] and considering the contributions of each neighbor as in equation [6.18], a global common velocity can be built in the calculation domain and used for each convection–diffusion system. Generally, the following formulation is adopted:

$$\vec{v}^{[\![\rho]\!]}(x,t) = \sum_{i=1}^{N_{LS}} \sum_{\substack{j=1 \\ j\neq i}}^{N_{LS}} \chi_{G_i}(x,t)\, \mu_{ij} \exp\left(-\beta|\psi_j|\right)[\![E]\!]_{ij}\left(-\vec{n}_j\right), \qquad [6.20]$$

where χ_{G_i} is the characteristic function of the grain G_i, M_{ij} is the interface mobility between the neighboring grains i and j, the exponential term is a continuous decreasing function varying from 1 to 0 on either side of the interface and has the function of smoothening the velocity field across the interface, and β is a positive parameter that controls the degree of smoothness. 3D results obtained this way are summarized in Figure 6.6 for a complex thermomechanical path applied onto 304L stainless steel. The microstructure is initially generated by a Laguerre–Voronoï strategy as detailed in Hitti et al. (2012). The system of equations [6.13] is solved by considering equations [6.16], [6.19] and [6.20], an optimized graph recoloration technique (Scholtes et al. 2015b), a direct reinitialization technique (Shakoor et al. 2015b) and the simplified mechanical framework detailed in Maire et al. (2017) where equation [6.3] is considered and associated with a critical stored energy law, a nucleation rate law and a critical nucleus radius law to define the apparition of recrystallized grains.

It must be highlighted that this numerical framework is well adapted for configurations where the anisotropy of GB properties is not of prime importance, that is, where isotropic GG and DDRX can be representative of the considered material. This is often the case for low stacking fault energy materials. When texture aspects and subgrain definition, organization and evolution can be of prime importance for microstructure evolution prediction like for high stacking fault energy materials, the discussed framework can become insufficient. In this context, anisotropy of GB interface energy and mobility must be introduced into the LS formalism. This aspect is discussed in the following section.

6.5. Anisotropy of GB properties and CDRX modeling

6.5.1. *LS formulations in context of anisotropic GB properties*

GB energy and mobility have been widely studied since they were reported as being anisotropic by Smith (1948) and Kohara et al. (1958).

(a) (b) (c) (d) (e) (f)

Figure 6.6. *Complex thermomechanical path for a 304 L stainless steel. From left to right and top to bottom: time evolution of the polycrystal with the grain boundary network colored by the grain size. As illustrated, DDRX, MDRX, SRX and GG mechanisms occur. For a color version of this figure, see www.iste.co.uk/bernacki/digital.zip*

The simplest models use constant values for the GB energy γ and a temperature-dependent mobility, $M(T)$, referred to as isotropic approach as introduced in the previous sections (Anderson et al. 1984; Gao and Thompson 1996; Garcke et al. 1999; Bernacki et al. 2011; Lazar et al. 2011). Heterogeneous models were also proposed, in which each boundary has its own energy and mobility (Rollett et al. 1989; Hwang 1998; Holm et al. 2001; Upmanyu et al. 2002; Mießen et al. 2015; Miyoshi and Takaki 2016; Fausty et al. 2018; Chang and Chang 2019; Zöllner and Zlotnikov 2019; Fausty et al. 2020; Miyoshi et al. 2020) trying to reproduce more complex microstructures with local heterogeneity, such as twin boundaries. Each grain has its own crystal orientation, and GB energy and mobility depend on the disorientation angle between two grains (Miyoshi and Takaki 2017; Fausty et al. 2020), but the effect of the misorientation axis and GB inclination is frequently omitted. General frameworks were thus proposed including the GB properties dependence on misorientation and inclination (Kazaryan et al. 2002; Hallberg and Bulatov 2019; Salama et al. 2020; Fausty et al. 2021), categorized as anisotropic models. It must be highlighted that the difference between three-parameter (heterogeneous) and five-parameter (anisotropic) full-field formulations is often unclear in the literature, heterogeneous GB properties being often categorized as anisotropic.

The main reason why most of the studies are carried out using heterogeneous GB properties is the lack of data of GB properties. The early measurements of GB properties (mainly GB reduced mobility) were carried out on bicrystals (Viswanathan and Bauer 1973; Demianczuk and Aust 1975; Maksimova et al. 1988; Gottstein and Shvindlerman 1992; Winning et al. 2002; Ivanov 2006), leading to the well-known Sigmoidal model (Rollett et al. 2017). As experimental and computational technologies are improved, new experimental and computational 3D techniques allow us to study GG and recrystallization using, for instance, X-ray (Zhang et al. 2017a; Juul Jensen and Zhang 2020; Zhang et al. 2020; Fang et al. 2021) or molecular dynamics (Janssens et al. 2006; Olmsted et al. 2009a, 2009b). Hence, at the mesoscopic scale, few studies have been carried out in 2D using anisotropic GB properties designed by mathematical models (Kazaryan et al. 2002; Fausty et al. 2021) or by fitting data from molecular dynamics (Hallberg and Bulatov 2019). Nevertheless, these 2D models neglect a part of the 3D space, that is, the GB inclination is measured in the sample plane, and GB properties are simplified. Finally, regarding the study of GG in 3D, we frequently find heterogeneous GB properties based on mathematical descriptors of GB properties (Fjeldberg and Marthinsen 2010; Chang and Chang 2019; Song et al. 2020; Miyoshi et al. 2021) or based on databases of GB energy values (Kim et al. 2011, 2014). Based on this, two open questions arise frequently in the state of the art: can GB properties be described in 2D using the classical Read–Shockley (Read and Shockley 1950) and Sigmoidal (Rollett et al. 2017) model? Is the effect of anisotropy stronger in 3D? The latter implies carrying out 3D simulations instead of 2D, thus using a better description of GB properties in the 5D GB space.

In the state of art, four different formulations using an FE-LS approach can be reported. The first is the isotropic formulation described previously and used to model different annealing phenomena, such as GG, recrystallization and GG in presence of SPPs (Bernacki et al. 2008, 2009, 2011; Scholtes et al. 2015b; Maire et al. 2017). In the pure isotropic GG context, equation [6.13] can be reformulate as:

$$\frac{\partial \psi}{\partial t} - \mu\gamma\Delta\psi = 0. \tag{6.21}$$

This formulation has shown good agreement with experimental data regarding GG predictions concerning the MGS and even the grain size distribution (GSD). However, this approach is limited to reproduce complex grain morphology (non-equiaxed ones), described special GBs and to respect textures. The second is an extension of the isotropic formulation considering heterogeneous values of GB energy and mobility thanks to the axis-angle parameterization.

$$\frac{\partial \psi}{\partial t} - \mu(M)\gamma(M)\Delta\psi = 0. \tag{6.22}$$

With this formulation, it is expected to obtain more physical grain shapes. Indeed, some GBs can evolve faster thanks to higher GB mobility values, and triple junctions may have different dihedral angles thanks to different GB energy values. This strategy classically used in full-field formulations (not only in LS ones) can lead to confusion when it is named as "heterogeneous". Indeed, stricto sensu, the heterogeneity shape of μ and γ can lead to additional terms in the driving pressure of the kinetic equation (equation [6.2]) but also in the weak formulation derived to solve the GB motion. However, the term "heterogeneous" is classically used to distinguish this formulation from the purely isotropic model.

Such a discussion is described in Fausty et al. (2018) where an additional term capturing the local heterogeneity of the multiple junctions is added to the velocity equation such that:

$$\vec{v} = \mu\left(\vec{\nabla}\gamma \cdot \vec{\nabla}\psi - \gamma\Delta\psi\right)\vec{\nabla}\psi. \tag{6.23}$$

Inserting this term into the transport equation (equation [6.9]) leads to the, hereafter called, "heterogeneous with gradient" formulation (Fausty et al. 2018):

$$\frac{\partial \psi}{\partial t} + \mu\vec{\nabla}\gamma \cdot \vec{\nabla}\psi - \mu\gamma\Delta\psi = 0. \tag{6.24}$$

The introduction of the term $\vec{\nabla}\gamma \cdot \vec{n}$ only acts at multiple junctions because these are the only places where this term does not vanish. This formulation is equivalent to the isotropic one if no heterogeneity is added. This third formulation was proposed for triple junctions in Fausty et al. (2018) and extended to model GG using heterogeneous

GB energy in Fausty et al. (2020) and both heterogeneous GB energy and mobility in Murgas et al. (2021).

The last one is an anisotropic formulation which was initially developed using thermodynamics and differential geometry in Fausty et al. (2021) and was improved in Murgas et al. (2021) in order to consider heterogeneous GB mobility. Both the GB normal and misorientation are taken into account and an intrinsic torque term is present:

$$\vec{v} = \mu(M)\left(\mathbb{P}\vec{\nabla}\gamma(M,n)\cdot\vec{\nabla}\psi + \left(\vec{\nabla}_{\vec{n}}\vec{\nabla}_{\vec{n}}\gamma(M,n) + \gamma(M,n)\mathbb{I}\right) : \mathbb{K}\right)\vec{\nabla}\psi \quad [6.25]$$

where $\mathbb{I}$ is the unitary matrix, $\mathbb{P} = \mathbb{I} - \vec{n}\otimes\vec{n}$ is the tangential projection tensor, $\vec{\nabla}_{\vec{n}}$ is the surface gradient and $\mathbb{K} = \vec{\nabla}\vec{n} = -\vec{\nabla}\vec{\nabla}\psi$ is the curvature tensor. The term $\Gamma(M,n) = \vec{\nabla}_{\vec{n}}\vec{\nabla}_{\vec{n}}\gamma + \gamma\mathbb{I}$ is a tensorial diffusion term, known as GB stiffness tensor (Du et al. 2007; Abdeljawad et al. 2018). The term $\mathbb{P}\vec{\nabla}\gamma\cdot\vec{\nabla}\phi$ in equation [6.25] should be null in the grain interfaces. However, the front-capturing nature of the LS approach, which consists to solve equation [6.9] at the GB network and in its vicinity, requires us to consider this term which could be non-null around the interfaces near multiple junctions.

This stabilization term is then totally correlated to the front-capturing nature of the LS approach. The term $\Gamma(M,n)$ is the subject of recent studies for twin boundaries (TB) $\Sigma3$, $\Sigma5$, $\Sigma7$, $\Sigma9$ and $\Sigma11$ (Abdeljawad et al. 2018; Moore et al. 2021). In the state of the art, the use of these different formulations leads to the conclusion that the prediction of GG at the polycrystal scale can be ambiguous, depending the aimed attributes and the available data. If the isotropic formulation is able to reproduce MGS and grain size distribution evolutions when a moderated anisotropy is involved, for a textured configuration, the anisotropic formulation presents the best behavior in terms of grain morphology, DDF and interfacial energy evolution predictions while keeping a reasonable efficiency comparatively to the isotropic formulation. Moreover, it was also proved that 3D simulations should be considered. Of course, this aspect is, first, essential to improving the representativity of the considered polycrystals but is also essential to correctly describe the γ dependence to the inclination. Indeed, the proposed existing 2D model/data limit the actual use of the inclination as this parameter is described here with one degree of freedom and not in a 3D framework with 3D experimental data. Finally, this objective must also be correlated to the fact to integrate the torque effects, and so the GB stiffness tensor, in the simulations and analysis. It should be highlighted that this conclusion is common to the majority of existing works of the state of the art involving anisotropic 2D GG simulations and 3D simulations, where the inclination dependence, or torque terms, or both are not taken into account.

Figures 6.7–6.10 illustrate a 2D example discussed by Murgas et al. (2021): it consists of a square domain with length $L = 1.6\,\text{mm}$ and 5,000 grains generated using a Laguerre–Voronoi tessellation (Hitti et al. 2012). Figure 6.7 exhibits the initial configuration.

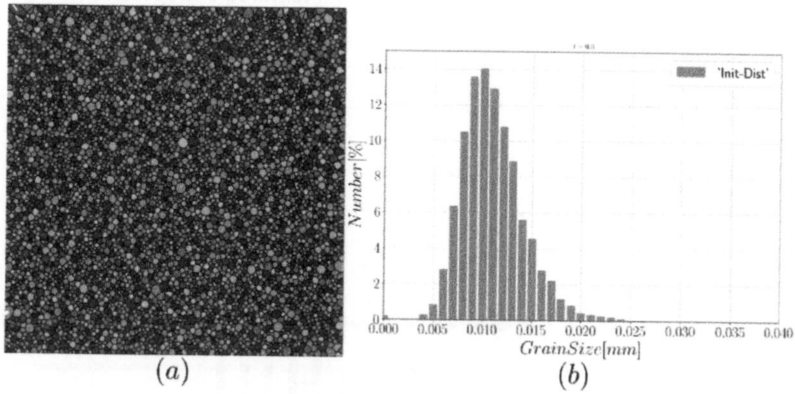

Figure 6.7. *Initial microstructure (a) with 5,000 grains and the grain size distribution in number (b). Adapted from Murgas et al. (2021)*

A misorientation dependent GB energy defined with a Read–Shockley (RS) function (Read and Shockley 1950) is considered:

$$\begin{cases} \gamma(\theta) = \gamma_{max} \frac{\theta}{\theta_{max}} \left(1 - ln\left(\frac{\theta}{\theta_{max}}\right)\right),\ \theta < \theta_{max} \\ \gamma_{max},\ \theta \geq \theta_{max} \end{cases} \qquad [6.26]$$

where γ_{max} is the maximal GB energy. θ_{max} =15° is the disorientation defining the transition from a low-angle grain boundary (LAGB) to a high-angle grain boundary (HAGB). μ =1.379 mm^4·J^{-1}·s^{-1}, and $\gamma_{max} = 6\times10^{-7}$ J·mm^{-2}, which are typical for a stainless steel (Cruz-Fabiano et al. 2014), are considered. Two ways are used to exacerbate the GB heterogeneties. In a first one (called Random), the Euler angles defining the crystallographic orientations are generated randomly, leading to a Mackenzie-like DDF (Mackenzie 1958). In a second one (called Uniform), one Euler angle is generated randomly with a uniform distribution function and the two others are constants. As a result, the final disorientation distribution is more uniform. Both considered orientation distribution and resulting GB energy distribution (GBED) are described in Figure 6.8 and the resulting orientation field using the vector magnitude $O_G = \sqrt{\varphi_1^2 + \phi^2 + \varphi_2^2}$, where $(\varphi_1, \phi, \varphi_2)$ are the three Euler angles, is described in Figure 6.9 for the Random configuration.

The time evolution is summarized in Figures 6.10 and 6.11. First, it is noticeable that all the formulations have a similar evolution concerning the total GB energy E_Γ, the number of grains N_g and the MGS weighted in number $\bar{R}_{Nb[\%]}$ or in surface $\bar{R}_{S[\%]}$ when the Random configuration is considered, whereas more stronger GB heterogeneities have a significant impact on the different attributes of the microstructure when the different formulations are adopted. This aspect corroborates the fact of avoiding an isotropic formulation when a textured configuration exists in the material.

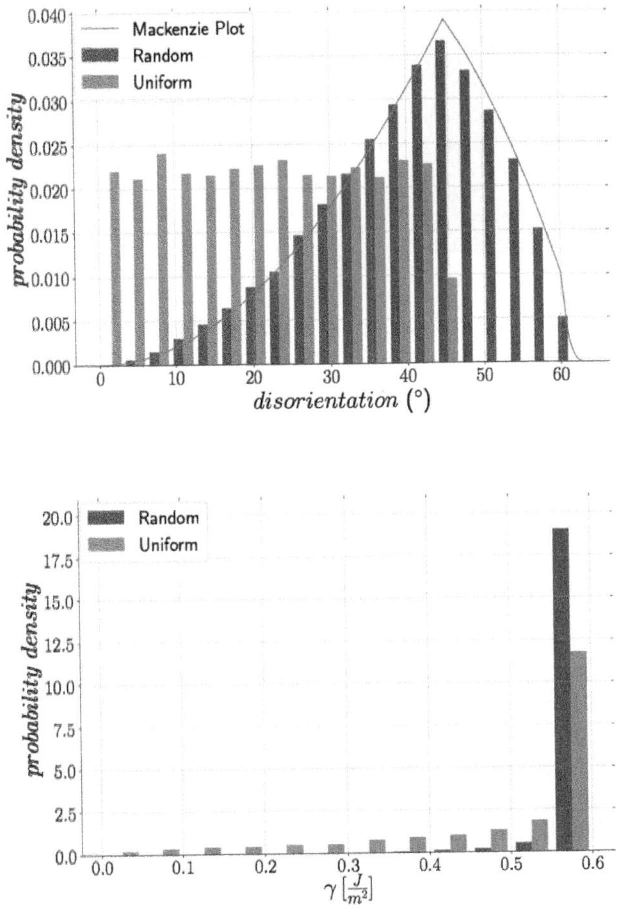

Figure 6.8. *Comparison of GB distribution properties: (a) DDF and (b) GBED. Adapted from Murgas et al. (2021). For a color version of this figure, see www.iste.co.uk/bernacki/digital.zip*

Another interesting aspect to deal with GB heteregoneities is to be able to consider materials where the substructure (LAGBs) are of prime importance in ReX mechanisms as illustrated in the next section.

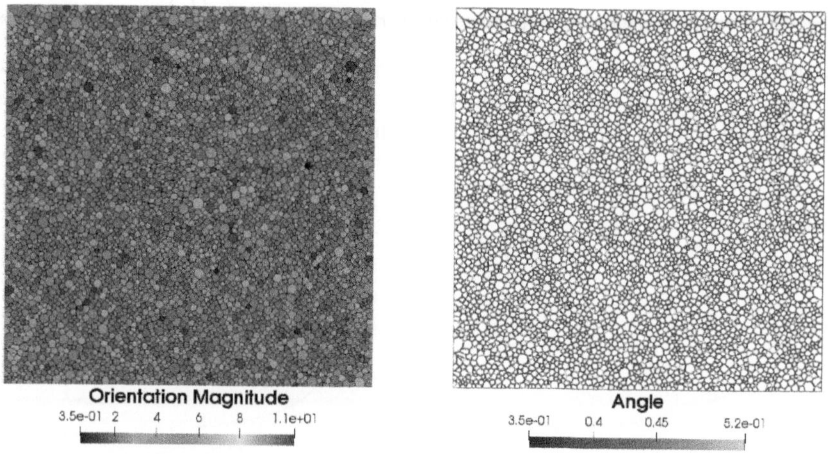

(a) Orientation (b) Disorientation

Figure 6.9. *Initial crystallographic characteristics. Adapted from Murgas et al. (2021). For a color version of this figure, see www.iste.co.uk/bernacki/digital.zip*

6.5.2. *CDRX modeling*

As already highlighted, the LS-FE approach was largely considered in context of DDRX for low stacking fault energy materials by associated equation [6.3] to a critical stored energy law, a nucleation rate law and a critical nucleus radius law. For high stacking fault energy materials, the progressive formation and evolution of subgrains must be taken into account. Dislocations could rearrange themselves to form new LAGB or accumulate into preexisting LAGB. This last phenomenon is responsible for a progressive increase of LAGB misorientation (Huang and Logé 2016; Rollett et al. 2017). During recrystallization, recrystallized grains form slowly and continuously during deformation. Indeed, in this CDRX context, grain formation is induced by the progressive reorganization of dislocations into cells or subgrains and the gradual increase of misorientation angle between those subgrains. Recently, a first LS-FE framework was proposed by Grand et al. (2022) and Gaillac et al. (2022) in order to deal with this mechanism. In these works, laws introduced by the Gourdet–Montheillet model were implemented (Gourdet and Montheillet 2003). While keeping the global dislocation density evolution law defined by equation [6.3], several mechanisms of evolution of dislocations are now taken into account:

– rearrangement into LAGB that bound new subgrains. Subgrain formation is described through following equation (Gourdet and Montheillet 2003):

$$dS^+ = \frac{\alpha b K_2 \rho d\varepsilon}{\eta \theta_0}, \qquad [6.27]$$

where dS^+ is the surface of LAGB created. $\alpha = 1 - \exp\left(\frac{D}{D_0}\right)^m$ is a coefficient describing the fraction of dislocations recovered to form new subgrains. D is the grain diameter, D_0 is a grain reference diameter and m is a fixed coefficient. η is the number of sets of dislocations and θ_0 the disorientation of newly formed subgrains;

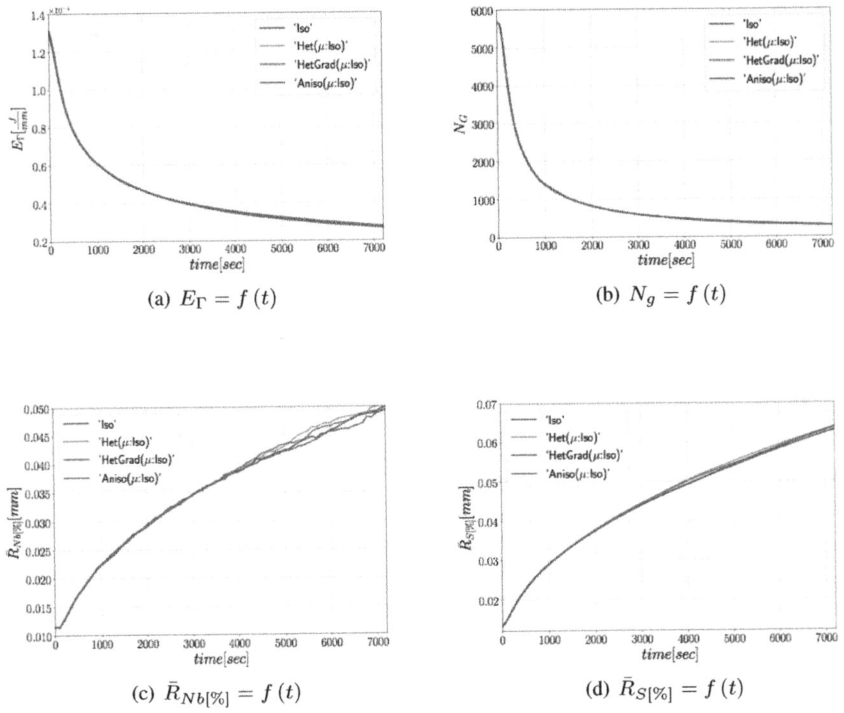

Figure 6.10. *Time evolution for the different formulations for the Random configuration: (a) the total GB energy; (b) the number of grains; (c) the mean radius weighted in number; and (d) the mean radius weighted in surface. From Murgas et al. (2021). For a color version of this figure, see www.iste.co.uk/bernacki/digital.zip*

– stacking into preexisting LAGB which is modeled according to the following equation (Gourdet and Montheillet 2003):

$$d\theta = \frac{b}{2\eta}(1-\alpha)DK_2\rho d\varepsilon. \quad [6.28]$$

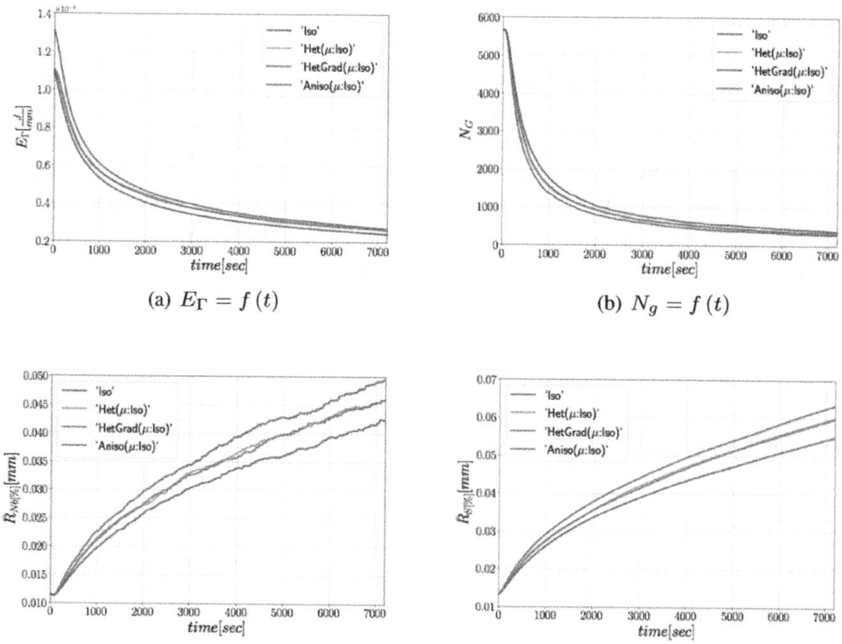

(a) $E_\Gamma = f(t)$ (b) $N_g = f(t)$

(c) $\bar{R}_{Nb[\%]} = f(t)$ (d) $\bar{R}_{S[\%]} = f(t)$

Figure 6.11. *Time evolution for the different formulations for the Uniform configuration: (a) the total GB energy; (b) the number of grains; (c) the mean radius weighted in number; and (d) the mean radius weighted in surface. For a color version of this figure, see www.iste.co.uk/bernacki/digital.zip*

– absorption during HAGB migration: this is naturally captured by affecting to the areas swept by moving boundaries a low dislocation density as described earlier.

At each time step, dislocation density of each grain is updated using equation [6.3], which impacts the computation of the velocity term related to stored energy differences. Then, the length of subgrain interfaces formed into each grain is computed using equation [6.27]. Depending on the simulation case, subgrains are added grain by grain based on the value of this grain property or globally, after having summed the length of subgrain interfaces for all grains. Subgrain orientation is initialized by applying a small misorientation to the parent grain orientation. The misorientation angle is selected to respect a distribution measured experimentally, whereas the misorientation axis satisfies a uniform distribution. The misorientation axis attributed to a subgrain at its formation is kept constant. Then, during the next increments, the misorientation increase described by equation [6.28] is realized by rotating of $d\theta$ around this axis.

An illustration of this framework in context of CDRX modeling for Zircaloy-4 (Zy-4) is depicted in Figure 6.12 and was proposed in De Micheli et al. (2024a) thanks to the DIGIMU software package. Material parameters have been estimated based on the experimental results obtained conducting a thermomechanical testing campaign associated with extensive EBSD characterization. The number of subgrains that are formed at each deformation increment is computed individually per grain/subgrain. GB energy is described by the RS equation (equation [6.26]) and GB mobility is assumed isotropic.

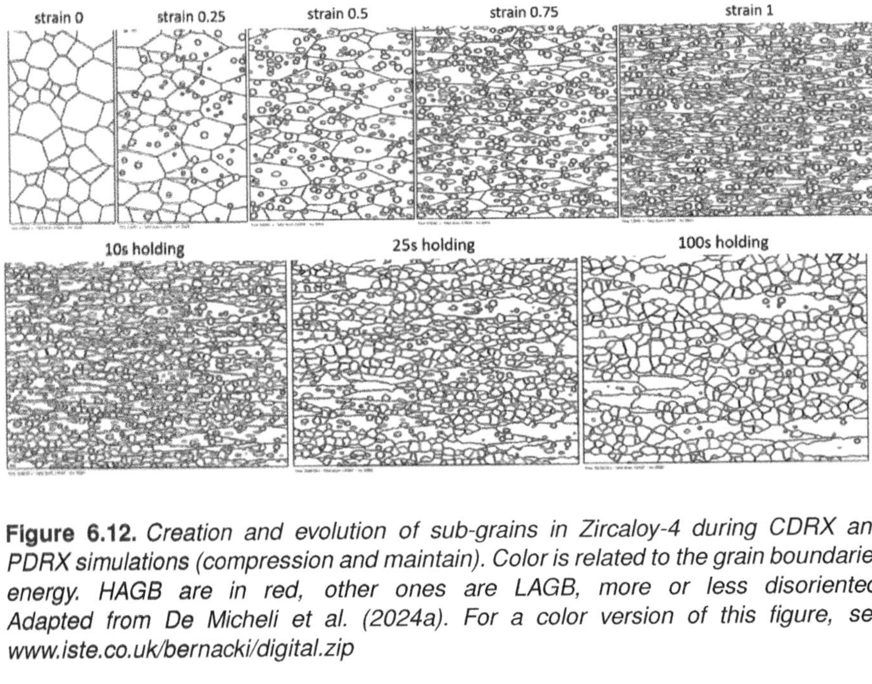

Figure 6.12. *Creation and evolution of sub-grains in Zircaloy-4 during CDRX and PDRX simulations (compression and maintain). Color is related to the grain boundaries energy. HAGB are in red, other ones are LAGB, more or less disoriented. Adapted from De Micheli et al. (2024a). For a color version of this figure, see www.iste.co.uk/bernacki/digital.zip*

6.6. Static/evolutive SPPs

This phenomenon was first discussed by Smith (1948) and then detailed by Zener one year after (Zener 1949); precipitates act as obstacles to the displacement of the GBs and may hinder GG. Under certain conditions, SPPs can pin the microstructure, leading eventually to a limiting MGS, which is characteristic of the Smith–Zener pinning. Since these first developments to equate this phenomenon, many variants have been developed in order to dispel some of the initial hypotheses (see Manohar et al. 1998 for a review). This phenomenon is widely used by metallurgists to control the grain size during the forming process of many alloys, including superalloys. Predictive tools are then needed to accurately model this phenomenon and thus optimize the final grain size and in-use properties of the materials. Classical laws predicting the limiting MGS (Manohar et al. 1998), denoted as $\overline{R}_\infty$, have the following form:

$$\overline{R}_\infty = K\frac{\overline{r}}{f^m}, \qquad [6.29]$$

where K and m are fitted parameters which can be assumed constant (Manohar et al. 1998) or dependent on the material and/or the characteristics of the SPP population (Manohar et al. 1998; Agnoli et al. 2014). In the last 30 years, numerous full field modeling of the Smith–Zener phenomenon have been proposed, including MC/CA framework (Srolovitz et al. 1984; Anderson et al. 1989; Hassold et al. 1990; Gao et al. 1997; Kad and Hazzledine 1997; Phaneesh et al. 2012), front tracking or Vertex (Weygand et al. 1999; Couturier et al. 2003), MPF (Chang et al. 2009; Moelans et al. 2006; Chang and Moelans 2014; Tonks et al. 2015) and LS (Agnoli et al. 2012, 2014, 2015; Scholtes et al. 2016b; Villaret et al. 2020; Alvarado et al. 2021a, 2021b).

Concerning the LS framework, the idea to consider inert SPP in an FE framework was first proposed to perform 2D GG (Agnoli et al. 2012) and 2D static recrystallization (Agnoli et al. 2014) simulations for Inconel 718. SPP are inserted in the FE mesh thanks to statistical description or experimental data and the local curvature of the GBs in contact with SPP is constrained. This approach enables us to consider SPP with no assumption on their size or morphology, isotropic and anisotropic particle/grain interface energies (incoherent or coherent interfaces). The dragging effect is naturally modeled by the modification of the local curvature when the GB encounters the particles as illustrated in Figure 6.13, which implies that no assumption is made on the dragging force exerted by the particles.

Indeed, the SZP effect generated by the presence of particles can be naturally taken into account by imposing the relevant boundary conditions related to the phenomenon at the interfaces between GBs and SPPs. More precisely, the influence of SPP on microstructure evolution can be taken into account by imposing a Neumann type limit condition on the GLS at the surface of the precipitates following the Young–Herring surface tension equilibrium:

$$\frac{\nabla\psi}{|\nabla\psi|}\cdot\vec{n} = \nabla\psi\cdot\vec{n} = sin(\alpha) = \frac{\gamma_p^2 - \gamma_p^1}{\gamma}, \qquad [6.30]$$

where $\vec{n}$ is the external unitary normal vector to the precipitate, α is the angle established by the balance of surface tensions at contact positions between the SPP and the GB. So, when the particle is assumed incoherent with the matrix, $\gamma_p^1 \simeq \gamma_p^2$, which leads to $\alpha \simeq 0$, a null Neumann boundary condition is applied at the precipitate/GB interface through the respect of equation [6.30].

Figure 6.13. *Illustration of the interaction between a particle (red surface) and a GB (green surface). For a color version of this figure, see www.iste.co.uk/bernacki/digital.zip*

This approach was used to support the idea that the phenomenon reported as abnormal GG in Inconel 718 could be explained by the growth of lower energy grains in a pinned microstructure and as such be a particular case of static recrystallization. Thanks to these simulations, the sensitivity of this phenomenon to the initial stored energy distribution could be studied (Agnoli et al. 2015). Optimization of parameters K and m of equation [6.29] thanks to a full-field simulations campaign and first 3D LS simulations were proposed in Scholtes et al. (2016a, 2016b), respectively. Comparisons with MC simulations and experimental data have been made for ODS steels (Villaret et al. 2020). Figure 6.14 illustrates a 3D GG LS simulation for Inconel 718 with an idealized spherical population of SPP (f=3%), until reaching a final stable microstructure.

However, this first LS approach concerning the SPP description presents different weaknesses. The main one concerns the fact that the evolution of SPPs, due to diffusive mechanisms, such as precipitation/dissolution, OR, agglomeration, spheroidization cannot be taken into account (as SPPs are considered static in the existing LS formulations). For these reasons, a new LS approach to model ReX and GG mechanisms in the presence of meshed SPPs and be able to reproduce evolving particles was developed by Alvarado et al. (2021a, 2021b).

In this LS formulation, the description of SPPs is made by a new LS function, ψ_{SPP}, over the domain calculation without considering holes in the FE mesh. The

GLS fields describing the grains are initially modified with simple topological operations (following by a reinitialization step) to introduce the presence of the SPPs, without modifying the particle interface:

$$\forall i \in [\![1, N_p]\!] \quad \hat{\psi}_i\left(t=0, x\right) = min\left(\psi_i\left(t=0, x\right), -\psi_{SPP}\left(t=0, x\right)\right). \qquad [6.31]$$

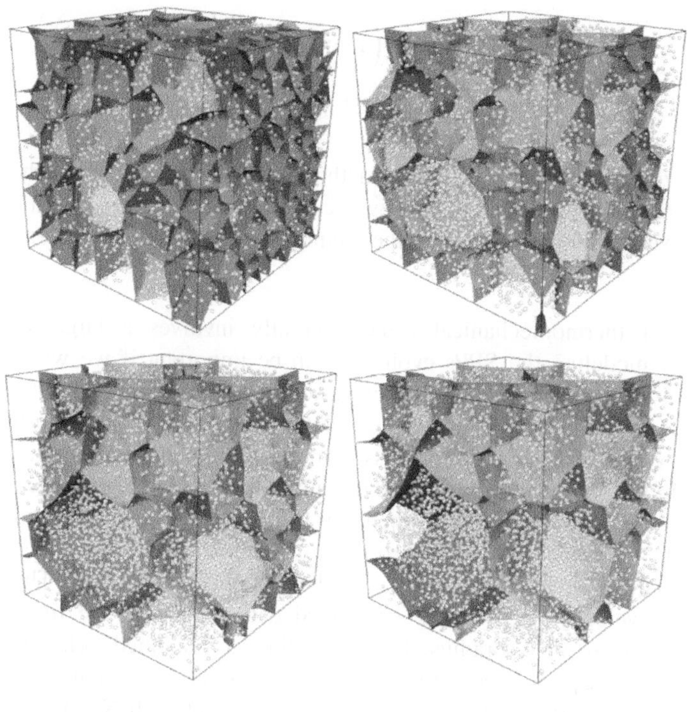

Figure 6.14. *3D GG LS simulations for Inconel 718 with an idealized spherical population of second-phase particles (f=3%). SPP are described in white, and the grain boundary network is described with a color code corresponding to the grain size until a stable configuration. From top to bottom and left to right: time evolution during the thermal treatment from the initial configuration to the stable grain boundary network. For a color version of this figure, see www.iste.co.uk/bernacki/digital.zip*

This operation is followed by a reinitialization procedure as the resulting LS functions are not (when the intersection is not empty) a distance function, even if GLS and ψ_{SPP} are. Of course, the function $\psi_{SPP}\left(t=0, x\right)$ can be easily estimated as the distance function to the union of simple objects (as circles for spherical particles) but also obtained through the FE-immersion of an experimental map (Agnoli et al. 2014; Scholtes et al. 2016a).

As already discussed (see equation [6.16]), the appearance of voids or overlaps especially at the multiple junctions after solving the convective–diffusive equations was first treated by Merriman et al. (1994) and implement in several cases, in 2D and 3D, using the LS method (Scholtes et al. 2016a; Bernacki et al. 2011; Ilin et al. 2018). In order to respect the Young–Herring equilibrium without hollowing out the SPPs, this classical treatment is extended by taking into account ψ_{SPP} in the procedure:

$$\hat{\psi}_i(t,x) = \frac{1}{2}\left(\psi_i(t,x) - \max\left(\max_{j\neq i}\left(\psi_j(t,x)\right), \psi_{SPP}(t,x)\right)\right), \quad \forall i = 1 \ldots N_{LS}, \forall x \in \Omega. \qquad [6.32]$$

In the zones without SPPs, this methodology is equivalent to the classical numerical treatment (equation [6.16]), whereas when SPPs are present, it enables us, by successive iterations, to impose the Young–Herring equilibrium for incoherent SPPs.

As a real thermomechanical route generally involves a large variation of temperature, modeling the SPPs evolution can be important if we want to predict quantitatively and qualitatively the microstructure evolution, notably the grain size distribution. Therefore, being able to model evolving particles during GG is necessary to treat real industrial processes and to accurately predict microstructural evolution. In this new LS context, once the initial mesh is generated, the polycrystal can be created statistically or experimentally from an EBDS map. Subsequently, the grains fields ψ_i are adjusted to incorporate the new particles field ψ_{SPP}. The convective equation, equation [6.9], is then applied to ψ_{SPP}. This involves computing a velocity, v_{spp}, using prescribed data related to the time-dependent evolution of the particle's radius. Following this, a smoothed velocity field v is computed using Laplace's equation (see equation [6.33]) with Dirichlet boundary conditions, v_{spp}, set at the particle interfaces, and null-Neumann boundary conditions applied to the boundaries of the domain. Finally, v is used to compute the velocity field $\vec{v}$ oriented toward the center of each precipitate to be applied to ψ_{SPP}:

$$\begin{cases} \Delta v = 0 \\ v = v_{spp} \quad at \quad \Gamma_{spp} \\ \nabla v \cdot \vec{n} = 0 \quad at \quad \partial\Omega \setminus \Gamma_{spp} \end{cases} \qquad [6.33]$$

and

$$\vec{v} = v\vec{n}_{SPP} = -v\nabla\psi_{SPP}, \qquad [6.34]$$

with $\vec{n}$ as the normal to the boundaries of the domain, $\vec{n}_{SPP}$ as the unitary inside normal vector to the SPP, v_{spp} as the velocity that we want to impose to the SPPs and

$\vec{v}$ as the resulting velocity field that we really impose through a convection equation to ψ_{SPP}.

This methodology is illustrated in Figure 6.15 where a simulation domain of 200 × 200 μm with an initial number of grains around 50,000 is considered in context of $AD730^{TM}$ nickel base superalloy. The mean grain radius (in number) $\bar{R} = 5\,\mu$m and a spherical particles population with an initial surface fraction $f_{spp} = 6\%$ divided in big SPPs with $f_{spp_{big}} = 4\%$ of radius $r_{spp_{big}} = 2\ \mu$m and small SPPs with $f_{spp_{small}} = 2\%$ of radius $r_{spp_{small}} = 1\ \mu$m, distributed in two bands in the domain.

Material properties and particle velocity v_{spp} for the chosen annealing temperature profile were obtained through experimental investigations (Alvarado et al. 2021b). The small particles will be dissolved before the large particles, thus the bands regions will present a classical GG mechanism sooner than the entire domain.

Figure 6.15 illustrates the evolution of the microstructure at different stages of the simulation (2580, 4950, 6880, and 10800 [s]) showing the particle and GG evolution during annealing.

For the initial isothermal treatment where the particles do not evolve, the grain size is smaller in the zones composed of small SPPs. This can be easily explained by a bigger resulting pinning pressure in this zone than in the zone with large SPPs (see Figure 6.15(a)). When the temperature increases, the particles begin to dissolve and the grains evolve, especially at the regions composed of small particles where the small grains of these regions begin to grow (see Figure 6.15(b)). Once the small SPPs are completely dissolved, the grains can grow freely and as represented in Figure 6.15(c) (white ellipses) some grains can grow more than others, thus a heterogeneous grain evolution begins to take place, where some grains are likely to grow more and more, leading potentially to abnormal GG.

The thermal treatment ends with a maintain of temperature at 1120°C, which is superior to the solvus temperature, so no particles remain and all the domains converge toward a classical GG mechanism as illustrated in Figure 6.15(d).

This recent LS formalism also opens the possibility to simulate material deformations more naturally by taking into account the mechanical behavior of SPPs. Validation and comparisons of this proposed LS-FE numerical framework comparatively to previous formulations (Agnoli et al. 2014; Scholtes et al. 2016a) and experimental data were proposed in Alvarado et al. (2021a, 2021b), respectively.

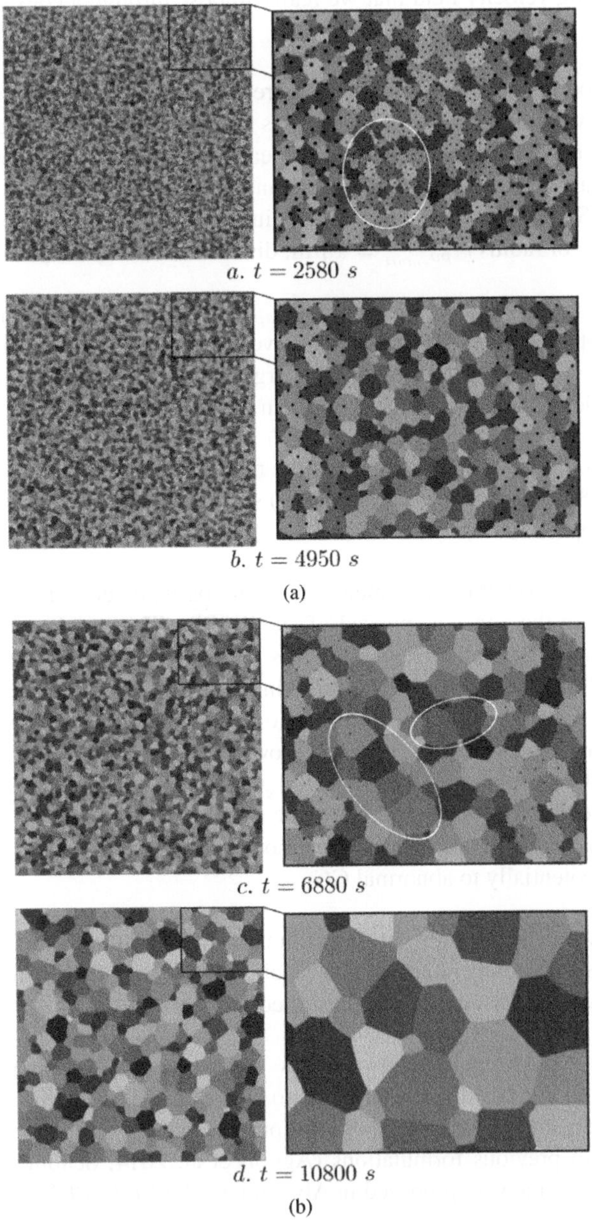

Figure 6.15. *Microstructure evolution at different times for a heterogeneous SPPs distribution (SPPs are in black). For a color version of this figure, see www.iste.co.uk/bernacki/digital.zip*

6.7. Modeling of diffusive solid-state phase transformation

Until a few years ago, the use of the LS approach in context of polycrystalline microstructure was mainly dedicated to the modeling of recrystallization and related phenomena for single-phase materials. However, PT at the solid-state involves crystallographic changes in the parent phase through rearrangement of the lattice structure to form a different, more stable product phase at the same solid state. PT can be either displacive or diffusive. Displacive transformation (James 1986) is characterized by the spontaneous, coherent, and cooperative movement of atoms across distances that are typically smaller than one nearest neighbor spacing. Diffusive transformation (Gamsjäger et al. 2002) involves gradual reorganization of the lattice through short- and long-range diffusion of atoms. Two basic mechanisms drive diffusive PT: (i) the diffusion of solutes across the phase interfaces and in the bulk of the grains, resulting in a change in chemical composition, and (ii) the interface migration resulting in the lattice rearrangement or structural changes. PT plays a critical role in producing diverse materials with varying microstructural features during TMTs. As illustrated in Chapter 5, in the context of DSSPT, phase-field methods (PFM) are popular and extensively used. The thermodynamic consistency and the ability to model arbitrary complex morphological changes without any presumption on their shape or mutual distribution make PFM a powerful and attractive tool. The early works of Wheeler et al. (1992), Steinbach et al. (1996) and Tiaden et al. (1998) on solidification using PFM provided some of the mathematical foundations of phase-field modeling for multi-component, multi-phase systems involving solute diffusion. However, globally, the existing numerical frameworks are finally dedicated to the modeling of recrystallization and GG for single-phase materials, or to the modeling of PTs without considering recrystallization and related phenomena. In this context, a recent LS approach (Chandrappa and Bernacki 2023) was proposed in order to overcome this limitation.

DSSPT modeling at the mesoscopic scale typically involves two governing equations: a diffusion equation that governs the partitioning of solute atoms (such as carbon) across different phases, and another governing equation that takes care of the resulting interface network migration. Classically, the diffusion equation could be resolved within an LS framework. However, due to the presence of material discontinuities across the phase interfaces, the sharp interface approach considered in the LS framework enforces the explicit consideration of interface jump conditions during the resolution of the diffusion equation. This demands explicit localization of the interface at each instant to numerically treat the necessary jump conditions. Thus, to avoid this cumbersome step, Chandrappa and Bernacki (2023, 2024) propose considering a diffuse interface hypothesis across the phase interfaces during the resolution of the diffusion equation. In other words, the multi-phase grain interface network is migrated using an LS description, while a global diffusion equation based on a diffusive interface assumption for the phase interfaces is considered. This enables us to resolve a single diffusion equation in the whole computational domain without the need for any interface jump conditions.

As already described in equation [5.4], the transition between a distance function to a diffuse interface description can be established, thanks to a hyperbolic tangent relation:

$$\phi = \frac{1}{2}\tanh\left(\frac{3\psi}{\eta}\right) + \frac{1}{2}, \qquad [6.35]$$

where η is a diffuse interface thickness parameter. In the following, we shall refer this function (ϕ) yielding the diffuse interface as the phase-field function. In the following case, the particular case of austenite decomposition to ferrite by carbon diffusion is illustrated even if the framework can be generalized (Chandrappa and Bernacki 2024). The total carbon concentration field (C) can be expressed as a continuous variable:

$$C = \phi C_\alpha + (1-\phi)C_\gamma. \qquad [6.36]$$

To improve clarity, vectors are written in bold in the following. The solute fluxes of each phase ($\boldsymbol{J}_\gamma, \boldsymbol{J}_\alpha$) weighted by the phase-field variable across the phase interface is assumed:

$$\boldsymbol{J} = \phi \boldsymbol{J}_\alpha + (1-\phi)\boldsymbol{J}_\gamma. \qquad [6.37]$$

The diffuse phase interface is assumed to be composed of a mixture of the two phases. A constant concentration ratio is imposed between the phases, such that the redistribution of the solute atoms between them at the interface respects a partitioning ratio (k) equal to that at the equilibrium:

$$k = \frac{C_\alpha}{C_\gamma} = \frac{C_\alpha^{eq}}{C_\gamma^{eq}}, \qquad [6.38]$$

where C_α^{eq} and C_γ^{eq} are the equilibrium concentrations of α and γ phases, respectively, at temperature T.

Following Fick's laws of diffusion, the diffusion equation for carbon partitioning can be expressed as:

$$\frac{\partial C}{\partial t} = -\boldsymbol{\nabla} \cdot \boldsymbol{J} = -\boldsymbol{\nabla} \cdot \left[\phi \boldsymbol{J}_\alpha + (1-\phi)\boldsymbol{J}_\gamma\right],$$

with,

$$\boldsymbol{J}_\alpha = -D_\alpha^C \boldsymbol{\nabla} C_\alpha; \quad \boldsymbol{J}_\gamma = -D_\gamma^C \boldsymbol{\nabla} C_\gamma.$$

We then obtain:

$$\frac{\partial C}{\partial t} = \boldsymbol{\nabla} \cdot \left[\phi D_\alpha^C \boldsymbol{\nabla} C_\alpha + (1-\phi) D_\gamma^C \boldsymbol{\nabla} C_\gamma\right], \qquad [6.39]$$

where D_α^C and D_γ^C represent the diffusivity of the carbon element in ferrite and austenite phases, respectively.

Invoking equations [6.36] and [6.38] in equation [6.39], a modified carbon diffusion equation (Tiaden et al. 1998) is obtained:

$$\frac{\partial C}{\partial t} = \boldsymbol{\nabla} \cdot \left\{ D^*(\phi) \left[\boldsymbol{\nabla} C - \frac{C(k-1)}{1+\phi(k-1)} \boldsymbol{\nabla} \phi \right] \right\}, \qquad [6.40]$$

where $D^*(\phi)$ is called "mixed diffusivity" and is defined as:

$$D^*(\phi) = \frac{D_\gamma^C + \phi(k D_\alpha^C - D_\gamma^C)}{1+\phi(k-1)}.$$

With further simplifications, the above equation [6.40] can be transformed into a convective–diffusive–reactive (CDR) form as follows:

$$\frac{\partial C}{\partial t} = \boldsymbol{\nabla} \cdot \left[D^*(\phi) \boldsymbol{\nabla} C - C \boldsymbol{A}(\phi) \right], \text{ that is,} \qquad [6.41]$$

$$\frac{\partial C}{\partial t} + (\boldsymbol{A} - \boldsymbol{\nabla} D^*) \cdot \boldsymbol{\nabla} C - D^* \Delta C + RC = 0, \qquad [6.42]$$

where,

$$\boldsymbol{A}(\phi) = \frac{D^*(\phi)(k-1)}{1+\phi(k-1)} \boldsymbol{\nabla} \phi, \quad \text{and} \quad R = \boldsymbol{\nabla} \cdot \boldsymbol{A}. \qquad [6.43]$$

Given the model focuses on pure solute diffusion within the computational domain without any influx or outflux of solute atoms, solute mass should remain conserved throughout. From a physical perspective, at first thought, we might consider employing pure Neumann-type boundary conditions on the concentration variable. However, from a mathematical perspective in equation [6.41], employing only Neumann-type boundary conditions on C conserves solely the diffusive flux $D^*(\phi)\boldsymbol{\nabla}C)$, while the advective flux $(C\boldsymbol{A})$ remains non-conserved. Thus, Robin-type boundary conditions are considered to close the problem (Chandrappa and Bernacki 2024):

$$[D^*(\phi)\boldsymbol{\nabla}C - C\boldsymbol{A}(\phi)] \cdot \boldsymbol{n}|_{\partial\Omega} = 0. \qquad [6.44]$$

Interestingly, when considering the weak form of the previous equation in the considered FE mesh, the gradient of the mixed diffusivity term $(\boldsymbol{\nabla}D^*)$ vanishes. In terms of numerical stability, this is of great interest considering the abrupt evolution of this term across a phase interface.

Concerning interface migration, by considering a combination between equations [6.5] and [6.2], P can be defined as:

$$P = \Delta G + [\![E]\!] - \kappa\sigma. \qquad [6.45]$$

The ΔG component acts only across the phase interfaces while vanishing across the grain interfaces of similar phases. Also, the sense and value of interface mobility, and interface energy could be different, depending on the type of interface (i.e. α/γ phase interface, α/α grain interface and the γ/γ grain interface). Equation [6.13] can then be generalized in the following form:

$$\frac{\partial\psi_i}{\partial t} + \left[\boldsymbol{v}_{\boldsymbol{\Delta G}} + \boldsymbol{v}_{[\![\boldsymbol{E}]\!]}\right]_i \cdot \boldsymbol{\nabla}\psi_i - \left[\sum_{l\in\mathcal{S}} \chi_l \mu_l \sigma_l\right] \Delta\psi_i = 0$$

$$\forall i \in \{1, 2, ..., N_{LS}\}, \qquad [6.46]$$

where $\mathcal{S} = \{\alpha\gamma, \alpha\alpha, \gamma\gamma\}$, $\boldsymbol{v}_{\boldsymbol{\Delta G}} = \chi_{\alpha\gamma}\mu_{\alpha\gamma}\Delta G_{\alpha\gamma}\boldsymbol{n}$ and $\boldsymbol{v}_{[\![\boldsymbol{E}]\!]} = \left[\sum_{l\in\mathcal{S}} \chi_l \mu_l [\![E]\!]_l\right] \boldsymbol{n}$.

As usual, these equations must be followed by a multiple-junction treatment and a reinitialization step. Moreover, as for $\boldsymbol{v}_{[\![\boldsymbol{E}]\!]}$ (see equation [6.20]), $\boldsymbol{v}_{\boldsymbol{\Delta G}}$ component can be built as an unique function in the FE mesh:

$$\boldsymbol{v}_{\boldsymbol{\Delta G}}(\boldsymbol{x}, t) = \sum_{i=1}^{N_{LS}} \sum_{\substack{j=1 \\ j\neq i}}^{N_{LS}} \chi_{G_i} \mu_{ij} \exp\left(-\beta|\psi_j|\right) \chi_{\alpha\gamma} \Delta G_{\alpha\gamma} \mathscr{F}_s(-\boldsymbol{n}_j), \qquad [6.47]$$

where $\chi_{\alpha\gamma}$, as seen earlier, helps to filter this component of velocity field only on the phase interfaces, and:

$$\mathscr{F}_s(\boldsymbol{x}, t) = \chi_\alpha(\boldsymbol{x}, t) - \chi_\gamma(\boldsymbol{x}, t) = 2\chi_\alpha(\boldsymbol{x}, t) - 1, \qquad [6.48]$$

with $\chi_\alpha(\boldsymbol{x},t)$ and $\chi_\gamma(\boldsymbol{x},t)$ the characteristic functions of α and γ phase, respectively. The last ingredient missing to completely prescribe the above kinetics is the change in Gibbs free energy between the two phases. $\Delta G_{\alpha\gamma}$ is typically dependent on the local composition of the solutes, temperature and the pressure. In many works, the description for $\Delta G_{\alpha\gamma}$ has been established by thermodynamic evaluations based on Calphad data (Steinbach et al. 2007) or ThermoCalc software (Thermo-Calc Software n.d.). For certain sharp interface descriptive models, the diffusion in the product phase is assumed to be instantaneous and so $\Delta G_{\alpha\gamma}$ is simply assumed to be proportional to the deviation in concentration at the interface in the parent phase ($C_{\gamma,eq}$) from the equilibrium concentration in this phase ($C_{\gamma,\gamma\alpha}$) (Mecozzi et al. 2011; Liu et al. 2018):

$$\Delta G_{\alpha\gamma} = \Upsilon \left(C_{\gamma,eq} - C_{\gamma,\gamma\alpha}\right),$$

where Υ is a proportionality factor that could be temperature-dependent and is derived from thermodynamic databases.

In Chandrappa and Bernacki (2023, 2024), as in the works of Mecozzi (2006), $\Delta G_{\alpha\gamma}$ is described based on a local linearization of the phase diagram. ΔG is basically assumed to be proportional to a small undercooling ($\Delta T = T^{eq} - T$). At low undercooling with the assumptions that the actual temperature T is close to the equilibrium temperature T^{eq} (corresponding to a local composition of C_α and C_γ), the variations of enthalpy (ΔH), and the entropy (ΔS) with temperature could be considered to be negligible ($\Delta S^{eq} \approx \Delta S$, $\Delta H^{eq} \approx \Delta H$) (Pariser et al. 2001; Hillert 2007). ΔG is thus given by:

$$\Delta G(T,C) = \frac{\Delta H^{eq}}{T^{eq}}(T^{eq} - T) = \Delta S^{eq}\Delta T. \qquad [6.49]$$

Linearizing at a reference temperature (T^R), and assuming only carbon element partitions, we can write:

$$\begin{aligned} T_\alpha^{eq} &= T^R + m_\alpha^R \left(C_\alpha - C_\alpha^R\right), \\ T_\gamma^{eq} &= T^R + m_\gamma^R \left(C_\gamma - C_\gamma^R\right), \end{aligned} \qquad [6.50]$$

where m_α^R and m_γ^R are the slopes of the boundary lines of the α and γ phases, respectively, linearized at T^R. C_α^R and C_γ^R are the equilibrium carbon concentrations at T^R of ferrite and austenite, respectively.

The undercooling is expressed as:

$$\Delta T = \frac{T_\alpha^{eq} + T_\gamma^{eq}}{2} - T \qquad [6.51]$$

So, if we substitute the two equations in [6.50] into equation [6.51], we can then write ΔT as:

$$\Delta T = T^R + 0.5m_\alpha^R \left(C_\alpha - C_\alpha^R\right) + 0.5m_\gamma^R \left(C_\gamma - C_\gamma^R\right) - T. \quad [6.52]$$

From equation [6.49] and equation [6.52], $\Delta G_{\alpha\gamma}$ component is then expressed as a function of the local concentrations and temperature as follows:

$$\Delta G_{\alpha\gamma}(T, C_\alpha, C_\gamma) = \Delta S \left[(T^R - T) + 0.5m_\alpha^R \left(C_\alpha - C_\alpha^R\right) + 0.5m_\gamma^R \left(C_\gamma - C_\gamma^R\right)\right]. \quad [6.53]$$

With the help of equations [6.36] and [6.38], the above description could be further expressed as a function of the total concentration variable, C, for each configuration of the phase-field function, $\phi(\boldsymbol{x}, t)$:

$$\Delta G = \Delta S \left[T^R - T + 0.5m_\gamma^R \left(\frac{C}{1 + \phi(k-1)} - C_\gamma^R\right) + 0.5m_\alpha^R \left(\frac{kC}{1 + \phi(k-1)} - C_\alpha^R\right)\right] \quad [6.54]$$

Based on the same linearization, the equilibrium carbon concentrations of each phase at temperature T can be estimated as follows:

$$C_i^{eq} = C_i^R + \frac{T - T^R}{m_i^R}, \quad with\ i = \{\alpha, \gamma\}. \quad [6.55]$$

Using equations [6.55], the equilibrium partitioning ratio (k) can be expressed at each temperature T as:

$$k(T) = \frac{C_\alpha^R + \frac{T - T^R}{m_\alpha^R}}{C_\gamma^R + \frac{T - T^R}{m_\gamma^R}}. \quad [6.56]$$

Figure 6.16 (from Chandrappa and Bernacki 2024) illustrates the described numerical framework in a 2D case, which corresponds to γ phase nucleation and growth in a α phase polycrystal. Moreover, Figure 6.17 also illustrates the capability of this approach to simulate OR mechanism.

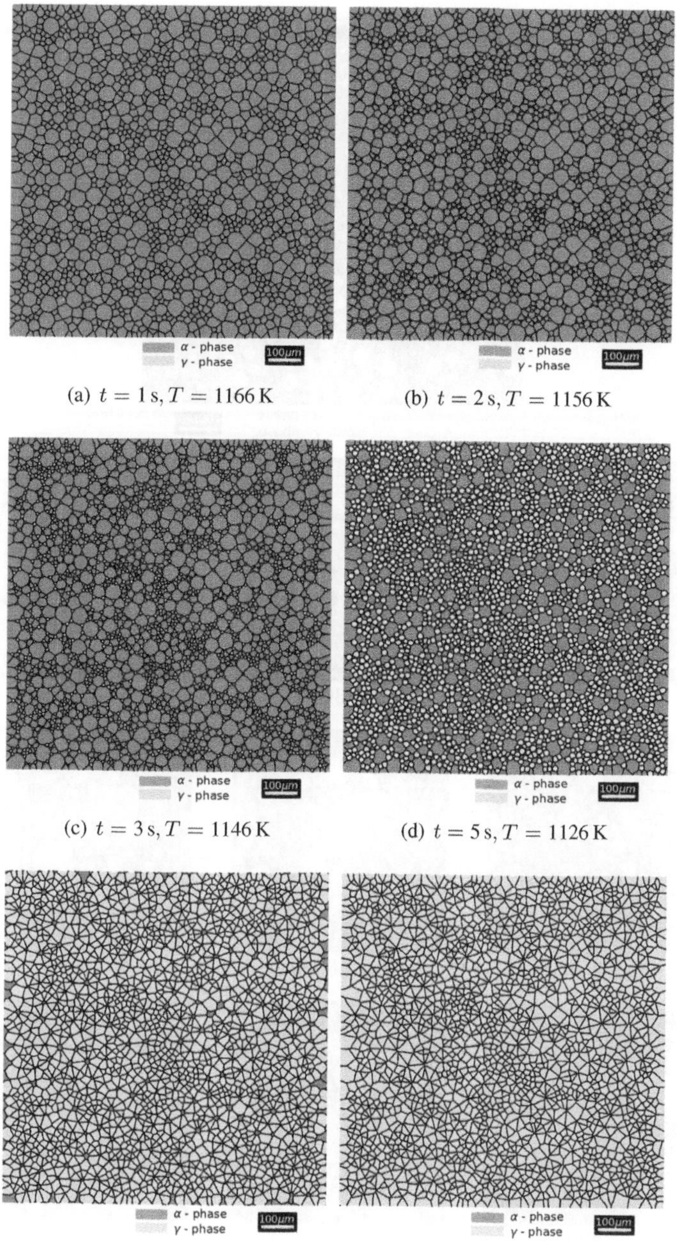

(a) $t = 1\,\mathrm{s}, T = 1166\,\mathrm{K}$

(b) $t = 2\,\mathrm{s}, T = 1156\,\mathrm{K}$

(c) $t = 3\,\mathrm{s}, T = 1146\,\mathrm{K}$

(d) $t = 5\,\mathrm{s}, T = 1126\,\mathrm{K}$

(e) $t = 10\,\mathrm{s}, T = 1076\,\mathrm{K}$

(f) $t = 20\,\mathrm{s}, T = 976\,\mathrm{K}$

Figure 6.16. *Snapshots of austenite decomposition into ferrite in a 2D microstructure, at different instants until the end of cooling. Adapted from Chandrappa and Bernacki (2024). For a color version of this figure, see www.iste.co.uk/bernacki/digital.zip*

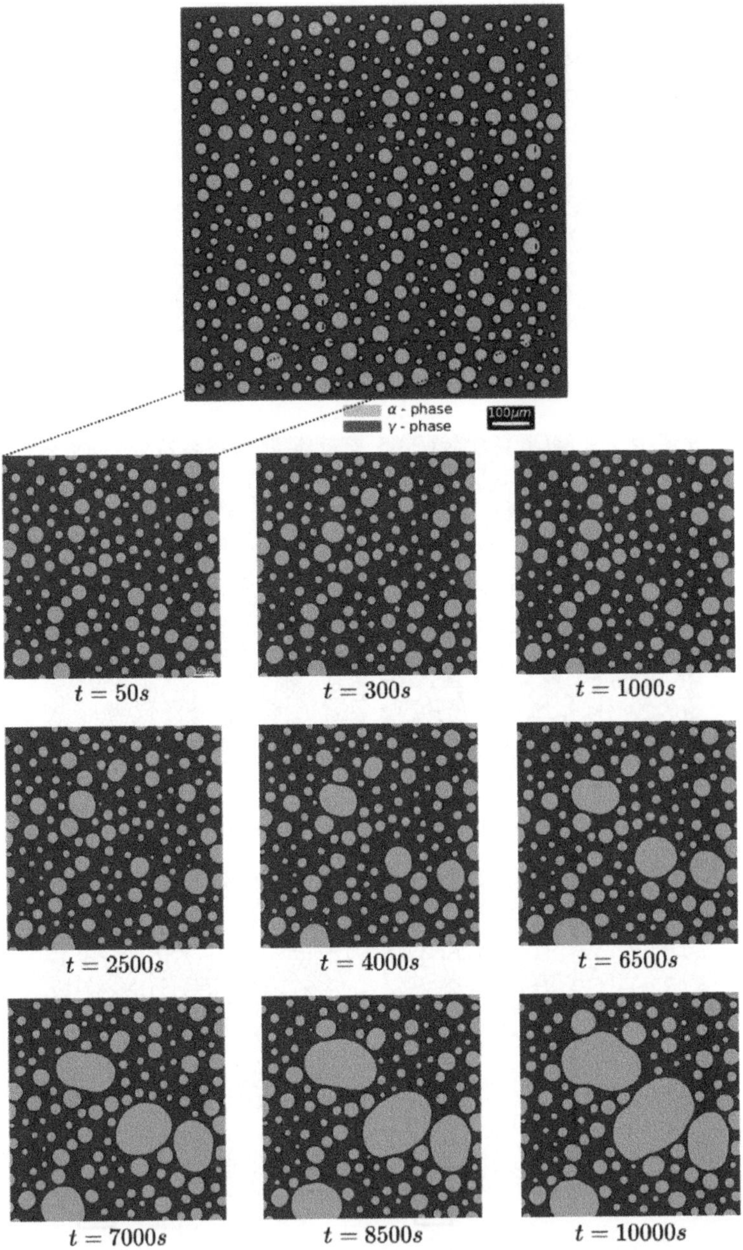

Figure 6.17. *Snapshots of coarsening and dissolution of second-phase particles at constant phases fraction (Ostwald ripening mechanism). Adapted from Chandrappa and Bernacki (2024). For a color version of this figure, see www.iste.co.uk/bernacki/digital.zip*

6.8. Solute drag aspect

Due to their low influence and involvement in solute diffusion, the contribution of the slow diffusing substitutional elements could sound insignificant toward the transformation kinetics in GG, recrystallization or SSPT. However, it has been experimentally observed (Hutchinson et al. 2006; Purdy et al. 2011) that the addition of these substitutional elements could induce additional dissipation effects as they interact with the migrating interface due to the solute segregation within the interface. Potentially, a solute concentration spike is developed in the vicinity of the interface. As the interface is driven, these solute atoms can either be dragged along with or pushed ahead of this migrating interface. In the process, a certain free energy is dissipated which could be attributed as a resistance or a drag pressure against the interface motion.

These effects have been classically qualified as "solute drag effects" that have the global effect of slowing down the rate of interface migration as a part of the driving pressure available for GB/PB migration is consumed to overcome the resistance. The solute drag effect can be described as a function of the interface velocity and impurities concentration and nature. The quantification of this drag pressure has been studied theoretically (Lücke and Detert 1957). Its intensity follows three main regimes (high, intermediate and low velocity), similarly to well-known dislocation impediment by Cottrell atmospheres (Cottrell and Bilby 1949). If a drag pressure (P_{SD}) is correlated to the solute drag effects, equation [6.45] could be modified with this additional contribution as:

$$P = \Delta G - \kappa\sigma + P_{SD}. \qquad [6.57]$$

For modeling the solute drag driving pressure (P_{SD}), Cahn's solute drag model (Cahn 1962) provides a simplified description, especially considering the nonlinearity of a solute drag pressure on the velocity field. Even though Cahn's model lacks physical sense across interphase boundaries and is more apt for GBs, the simplicity in its description is attractive for numerical implementation, especially in an FE framework. It was then used in the literature for GB (Furstoss et al. 2020) and PB (Chandrappa and Bernacki 2024). Thus, in this context, the magnitude of the solute drag pressure can be expressed as:

$$P_{SD} = \frac{\alpha x_X^0 \left\| \boldsymbol{v} \right\|}{1 + \beta^2 \left\| \boldsymbol{v} \right\|^2}, \qquad [6.58]$$

where $\boldsymbol{v}$ is the GB velocity, x_X^0 is the concentration of the element of interest, α and β are two parameters modulating the intensity of the drag. α controls the intensity of the solute drag pressure and β constrains the interface velocity which is the most impacted by the presence of the impurities.

The solute drag pressure renders the velocity field nonlinear. Indeed, if we consider a fully implicit time discretization scheme for the resolution of equation [6.46], we will obtain a nonlinear formulation. Considering Euler implicit time discretization, equation [6.46] could be rewritten as:

$$\frac{\partial \psi_i}{\partial t} \approx \frac{\psi_i^{k+1} - \psi_i^k}{\Delta t} = \mathcal{F}(\psi_i^{k+1}), \qquad \forall i \in \{1, 2, ..., N_{LS}\}, \tag{6.59}$$

where

$$\mathcal{F}(\psi_i^{k+1}) = \left[\sum_{l \in \mathcal{S}} \chi_l \mu_l \sigma_l\right] \Delta \psi_i^{k+1} + \sum_{l \in \mathcal{S}} \chi_l \mu_l \left[P_{SD}(\boldsymbol{v}^{k+1})\right]_l - \left[\boldsymbol{v}_{\Delta G} + \boldsymbol{v}_{[\![E]\!]}\right]_i \cdot \boldsymbol{\nabla} \psi_i^{k+1}, \tag{6.60}$$

is a time-dependent operator, with k being the index of time stepping.

If we consider P_{SD}^{k+1} in the second term in equation [6.60], we obtain from equation [6.58] and from the facts that the interface migration velocity is normal to the interface and that this drag pressure is always opposed to the direction of migration:

$$P_{SD}(\boldsymbol{v}^{k+1}) = -\frac{\alpha x_X^0 (\boldsymbol{v} \cdot \boldsymbol{n})^{k+1}}{1 + \beta^2 \left\|\boldsymbol{v}^{k+1}\right\|^2}. \tag{6.61}$$

Since the interface migration velocity is normal to the interface, at any position $\boldsymbol{x}$, for signed distance ith GLS function, such that $\psi_i^{k+1}(\boldsymbol{x}) \geq 0$, it is also possible to express:

$$(\boldsymbol{v} \cdot \boldsymbol{n})^{k+1} = \frac{\psi_i^{k+1} - \psi_i^k}{\Delta t} \approx \frac{\partial \psi_i}{\partial t}, \tag{6.62}$$

and,

$$\left\|\boldsymbol{v}^{k+1}\right\| = \left|(\boldsymbol{v} \cdot \boldsymbol{n})^{k+1}\right| = \frac{\left|\psi_i^{k+1} - \psi_i^k\right|}{\Delta t}. \tag{6.63}$$

So, P_{SD}^{k+1} would then become:

$$P_{SD}^{k+1} = -\frac{\alpha x_X^0 \left(\frac{\psi_i^{k+1} - \psi_i^k}{\Delta t}\right)}{1 + \beta^2 \left(\frac{\psi_i^{k+1} - \psi_i^k}{\Delta t}\right)^2}. \tag{6.64}$$

Clearly, the term in the denominator due to $\left\|\boldsymbol{v}^{k+1}\right\|$ manifests the operator $\mathcal{F}$ in equation [6.60] and hence the time discretized formulation in equation [6.59] nonlinear. To avoid this nonlinearity and simplify the resolution procedure, as seen in the works of Furstoss et al. (2020), the following explicit scheme based assumption is made:

$$\left\|\boldsymbol{v}^{k+1}(\boldsymbol{x})\right\|^2 \approx \left\|\boldsymbol{v}^{k}(\boldsymbol{x})\right\|^2 = \left(\frac{\psi_i^k(\boldsymbol{x}) - \psi_i^{k-1}(\boldsymbol{x})}{\Delta t}\right)^2$$
$$= \left\|\boldsymbol{v}^{old}(\boldsymbol{x})\right\|^2 \text{ for } i \mid \psi_i^k(\boldsymbol{x}) \geq 0, \qquad [6.65]$$

where $\boldsymbol{v}^k$ or $\boldsymbol{v}^{old}$ is the net migration velocity field prescribed to compute the GLS solution ψ_i^k at the current state. Even though this formulation deviates slightly from the fully implicit time scheme, for the small ranges of time steps expected to be considered in the FE resolution, the errors accumulated are expected to be of the small order while considerably simplifying the resolution procedure, thanks to the linearization. Taking this into account, equation [6.64] could be rewritten as:

$$P_{SD}^{k+1} \approx -\left(\frac{\alpha x_X^0}{1+\beta^2\left\|\boldsymbol{v}^{old}\right\|^2}\right)\left(\frac{\psi_i^{k+1} - \psi_i^k}{\Delta t}\right). \qquad [6.66]$$

This would then linearize the operator $\mathcal{F}$ in equation [6.60] and hence the time discretized formulation in equation [6.59]. Substituting equation [6.66] into the time discretized formulation and going back to the continuum description in equation [6.46], we obtain:

$$\frac{\partial \psi_i}{\partial t} + \left[\boldsymbol{v}_{\Delta G} + \boldsymbol{v}_{[\![E]\!]}\right]_i \cdot \nabla\psi_i + \sum_{l\in\mathcal{S}} \chi_l \mu_l \left(\frac{\alpha x_X^0}{1+\beta^2\left\|\boldsymbol{v}^{old}\right\|^2}\right)_l$$
$$\times \frac{\partial \psi_i}{\partial t} = \left[\sum_{l\in\mathcal{S}} \chi_l \mu_l \sigma_l\right] \Delta\psi_i. \qquad [6.67]$$

By imposing,

$$\mu_{SD} = 1 + \sum_{l\in\mathcal{S}} \chi_l \mu_l \left(\frac{\alpha x_X^0}{1+\beta^2\left\|\boldsymbol{v}^{old}\right\|^2}\right)_l, \qquad [6.68]$$

we obtain:

$$\mu_{SD}\frac{\partial \psi_i}{\partial t} + \left[\boldsymbol{v}_{\Delta G} + \boldsymbol{v}_{[\![E]\!]}\right]_i \cdot \nabla\psi_i = \left[\sum_{l\in\mathcal{S}} \chi_l \mu_l \sigma_l\right] \Delta\psi_i. \qquad [6.69]$$

Eventually, we can express the above equation into the following linearized convective–diffusive formulation:

$$\frac{\partial \psi_i}{\partial t} + \frac{1}{\mu_{SD}} \left[\boldsymbol{v}_{\boldsymbol{\Delta G}} + \boldsymbol{v}_{[\![\boldsymbol{E}]\!]} \right]_i \cdot \boldsymbol{\nabla} \psi_i = \frac{1}{\mu_{SD}} \left[\sum_{l \in \mathcal{S}} \chi_l \mu_l \sigma_l \right] \Delta \psi_i,$$

$$\forall i \in \{1, 2, ..., N_{LS}\}, \qquad [6.70]$$

with $\mathcal{S} = \{\gamma\gamma, \gamma\alpha, \alpha\alpha\}$. We can finally refer μ_{SD} as the solute drag pressure coefficient. When the solute drag effects are absent or negligible, $\mu_{SD} \to 1$, while when they are non-negligible, $\mu_{SD} > 1$, which has the effect of lowering the interface mobility (or increasing the drag resistance) for migration.

6.9. Conclusion

The primary appeal of the LS approach likely lies in its flexibility, enabling the inclusion of various mechanisms within a comprehensive numerical framework, even when substantial deformation occurs in the calculation domain. This flexibility likely contributes to its importance in simulating intricate thermomechanical paths (Maire et al. 2017), highlighting its distinctive ability to predict microstructures in hot metal forming processes (Scholtes et al. 2015; De Micheli et al. 2019, 2024a, 2024b). Regarding this aspect, it is clear that LS approaches have not yet fully realized their potential in modeling a wide range of mechanisms. As previously discussed, solid-state transformations represent a promising new area for these methods. Additionally, LS methods are poised to encompass other diffusion processes, such as surface diffusion (Polychronopoulou et al. 2019). Lastly, these approaches are proving invaluable in capturing all the discussed mechanisms in powder metallurgy processes (Bruchon et al. 2011, 2012; Zouaghi et al. 2012), effectively considering particles and their polycrystalline structures.

The primary limitation of the LS approach, as with other full-field methods, is its significant computational demand, especially in 3D. To provide an idea of the computation times for LS simulations on FE meshes with remeshing algorithms on the latest CPUs, the following estimates are relevant: in 2D, simulating a typical microstructure comprising 1,000 to 10,000 grains without SPPs can take a few hours on nine CPUs. When complex SPP populations are included, this time may increase to between one day and several days. In 3D, using 32 CPUs, these times can extend to several days for simulations without SPP and up to weeks for simulations with SPP. Such discussions are detailed in Scholtes (2016) and Florez et al. (2020). While these simulations are valuable for modeling and understanding mechanisms, their integration with component-scale calculations is currently quite limited due to the high computational cost and the need for large computing clusters for extensive calculation domains. The most sophisticated tools use approaches where representative volume element (RVE) simulations (see Chapter 2) are performed at a few integration points within component-scale simulations. Therefore, in research and development, the use of a low dimension/number of RVEs and indirect coupling with component-scale simulations represent the more advanced tools.

In this context, a prominent direction is the emerging trend of machine learning-trained surrogate models. These models, rooted entirely in artificial intelligence, function as reduced-order models derived from extensive full-field simulation databases. Their main benefit is their ability to deliver predictive models with considerably lower computational demands, as highlighted in studies like Hashemi and Kalidindi (2021); Montes de Oca Zapiain et al. (2021); Oommen et al. (2022); Yan et al. (2022). Despite the clear potential of these initial efforts, they face several significant challenges. First, the training datasets might be limited in representativeness due to potential inaccuracies in the underlying full-field models. Second, the relatively simple mechanisms primarily explored so far, such as GG and spinodal decomposition, fall short of capturing the complexity present in real metal forming scenarios. Third, the microstructures considered in these models lack the necessary complexity, possibly leading to an inaccurate depiction of actual materials. In the realm of hot metal forming, these early-stage approaches currently struggle to provide physically representative simulations at the mesoscopic scale. Nonetheless, this area holds considerable promise for the future, both in developing reduced-order models and, more intriguingly, in expediting high-fidelity LS simulations through mixed formulations, such as combining high-order time step resolutions with lower order ones, as suggested by Oommen et al. (2022).

The representativeness of equation [6.1] at the mesoscopic scale, especially when integrated with the conventional 5D-space and models for GB energy and mobility, is increasingly debated. Recent studies have yielded conflicting conclusions on this topic. The advancements in 3D GG and recrystallization experiments now enable a more precise examination of the accuracy of this model. Non-destructive in situ experiments minimize biases typically associated with 2D observations and pave the way for more accurate reverse engineering of the $\|\boldsymbol{v}\|/\kappa$ ratio. In this setting, numerical reverse engineering, particularly employing anisotropic LS formulations that incorporate torque terms, will be significantly valuable in contributing to this ongoing discussion.

The previous paragraph highlights that GB migration is a multifaceted process driven by atomic-scale mechanisms involving disconnections, which are line defects at GBs possessing both dislocation and step characteristics. To bridge the gap between existing GB migration models and the latest experimental observations, new models respecting crystallography for GB kinetics have also been proposed, rooted in the disconnection mechanisms that drive GB motion. These models initially started with a formulation using one reference plane and a single disconnection mode (Zhang et al.

2017b). Expanding these models to include multiple disconnection modes within a single-reference framework allowed for the simulation of shear-coupled GB migration (Wei et al. 2019). To address more complex phenomena like grain rotation, these models were further extended to consider two reference GB planes (Qiu et al. 2024). A further generalization to a multi-reference model made it possible to study GB faceting and defaceting processes (Qiu et al. 2023). It is important to note that these developments are just the initial forays in this area of research. Another intriguing prospect for the LS approach could be to enhance it with kinetic equations derived from a continuous model of disconnection-based kinetics.

6.10. References

Abdeljawad, F., Foiles, S.M., Moore, A.P., Hinkle, A.R., Barr, C.M., Heckman, N.M., Hattar, K., Boyce, B.L. (2018). The role of the interface stiffness tensor on grain boundary dynamics. *Acta Materialia*, 158, 440–453.

Agnoli, A., Bernacki, M., Logé, R., Franchet, J., Laigo, J., Bozzolo, N. (2012). Understanding and modeling of grain boundary pinning in Inconel 718. *Superalloys 2012: The 12th International Symposium on Superalloys*, Pennsylvania.

Agnoli, A., Bozzolo, N., Logé, R., Franchet, J.-M., Laigo, J., Bernacki, M. (2014). Development of a level set methodology to simulate grain growth in the presence of real secondary phase particles and stored energy – Application to a nickel-base superalloy. *Computational Materials Science*, 89, 233–241.

Agnoli, A., Bernacki, M., Logé, R., Franchet, J.-M., Laigo, J., Bozzolo, N. (2015). Selective growth of low stored energy grains during δ sub-solvus annealing in the Inconel 718 nickel base superalloy. *Metallurgical and Materials Transactions A*, 46(9), 4405–4421.

Almeida, R.C., Feijoo, R.A., Galeao, A.C., Padra, C., Silva, R.S. (2000). Adaptive finite element computational fluid dynamics using an anisotropic error estimator. *Computer Methods in Applied Mechanics and Engineering*, 182(3), 379–400.

Alvarado, K., Florez, S., Flipon, B., Bozzolo, N., Bernacki, M. (2021a). A level set approach to simulate grain growth with an evolving population of second phase particles. *Modelling and Simulation in Materials Science and Engineering*, 29(3), 035009.

Alvarado, K., Janeiro, I., Florez, S., Flipon, B., Franchet, J.-M., Locq, D., Dumont, C., Bozzolo, N., Bernacki, M. (2021b). Dissolution of the primary γ' precipitates and grain growth during solution treatment of three nickel base superalloys. *Metals*, 11(12).

Anderson, M., Srolovitz, D., Grest, G., Sahni, P. (1984). Computer simulation of grain growth–I. Kinetics, *Acta Metallurgica*, 32(5), 783–791.

Anderson, M., Grest, G., Doherty, R., Li, K., Srolovitz, D. (1989). Inhibition of grain growth by second phase particles: Three dimensional Monte Carlo computer simulations. *Scripta Metallurgica*, 23(5), 753–758.

Asaro, R. and Needleman, A. (1985). Overview no. 42 texture development and strain hardening in rate dependent polycrystals. *Acta Metallurgica*, 33(6), 923–953.

Bagi, K. (1993). A quasi-static numerical model for micro-level analysis of granular assemblies. *Mechanics of Materials*, 16(1), 101–110.

Barrales Mora, L., Gottstein, G., Shvindlerman, L. (2008). Three-dimensional grain growth: Analytical approaches and computer simulations. *Acta Materialia*, 56(20), 5915–5926.

Benabbou, A., Borouchaki, H., Laug, P., Lu, J. (2009). Geometrical modeling of granular structures in two and three dimensions. Application to nanostructures. *International Journal for Numerical Methods in Engineering*, 80(4), 425–454.

Bernacki, M., Chastel, Y., Coupez, T., Logé, R. (2008). Level set framework for the numerical modelling of primary recrystallization in polycrystalline materials. *Scripta Materialia*, 58(12), 1129–1132.

Bernacki, M., Resk, H., Coupez, T., Logé, R. (2009). Finite element model of primary recrystallization in polycrystalline aggregates using a level set framework. *Modelling and Simulation in Materials Science and Engineering*, 6, 064006.

Bernacki, M., Logé, R., Coupez, T. (2011). Level set framework for the finite-element modelling of recrystallization and grain growth in polycrystalline materials. *Scripta Materialia*, 64(6), 525–528.

Bhattacharya, A., Shen, Y.-F., Hefferan, C.M., Li, S.F., Lind, J., Suter, R.M., Krill, C.E., Rohrer, G.S. (2021). Grain boundary velocity and curvature are not correlated in polycrystals. *Science*, 374(6564), 189–193.

Brahme, A., Alvi, M.H., Saylor, D., Fridy, J., Rollett, A.D. (2006). 3D reconstruction of microstructure in a commercial purity aluminum. *Scripta Materialia*, 55(1), 75–80.

Bruchon, J., Drapier, S., Valdivieso, F. (2011). 3D finite element simulation of the matter flow by surface diffusion using a level set method. *International Journal for Numerical Methods in Engineering*, 86(7), 845–861.

Bruchon, J., Pino-Muñoz, D., Valdivieso, F., Drapier, S. (2012). Finite element simulation of mass transport during sintering of a granular packing. Part I. Surface and lattice diffusions. *Journal of the American Ceramic Society*, 95(8), 2398–2405.

Cahn, J. (1962). The impurity-drag effect in grain boundary motion. *Acta Metallurgica*, 10(9), 789–798.

Chandrappa, N. and Bernacki, M. (2023). A level-set formulation to simulate diffusive solid/solid phase transformation in polycrystalline metallic materials – Application to austenite decomposition in steels. *Computational Materials Science*, 216, 111840.

Chandrappa, N. and Bernacki, M. (2024). High-fidelity level-set modeling of diffusive solid-state phase transformations for polycrystalline materials. *Computational Materials Science*, 243, 113142.

Chang, K. and Chang, H. (2019). Effect of grain boundary energy anisotropy in 2D and 3D grain growth process. *Results in Physics*, 12, 1262–1268.

Chang, K. and Moelans, N. (2014). Effect of grain boundary energy anisotropy on highly textured grain structures studied by phase-field simulations. *Acta Materialia*, 64, 443–454.

Chang, K., Feng, W., Chen, L. (2009). Effect of second-phase particle morphology on grain growth kinetics. *Acta Materialia*, 57(17), 5229–5236.

Chen, K., Han, J., Pan, X., Srolovitz, D.J. (2020). The grain boundary mobility tensor. *Proceedings of the National Academy of Sciences*, 117(9), 4533–4538.

Cottrell, A.H. and Bilby, B.A. (1949). Dislocation theory of yielding and strain ageing of iron. *Proceedings of the Physical Society, Section A*, 62(1), 49.

Coupez, T. (2007). Convection of local level set function for moving surfaces and interfaces in forming flow. *AIP Conference Proceedings*, 908(1), 61–66.

Coupez, T. (2011). Metric construction by length distribution tensor and edge based error for anisotropic adaptive meshing. *Journal of Computational Physics*, 230(7), 2391–2405.

Coupez, T., Digonnet, H., Ducloux, R. (2000). Parallel meshing and remeshing. *Applied Mathematical Modelling*, 25(2), 153–175.

Couturier, G., Maurice, C., Fortunier, R. (2003). Three-dimensional finite-element simulation of Zener pinning dynamics. *Philosophical Magazine*, 83(30), 3387–3405.

Cruz-Fabiano, A., Logé, R., Bernacki, M. (2014). Assessment of simplified 2D grain growth models from numerical experiments based on a level set framework. *Computational Materials Science*, 92, 305–312.

Cuitino, A.M. and Ortiz, M. (1993). Computational modelling of single crystals. *Modelling and Simulation in Materials Science and Engineering*, 1(3), 225.

De Micheli, P., Maire, L., Cardinaux, D., Moussa, C., Bozzolo, N., Bernacki, M. (2019). DIGIMU®: Full field recrystallization simulations for optimization of multi-pass processes. *AIP Conference Proceedings*, 2113(1), 040014.

De Micheli, P., Alvarado, K., Grand, V., Bernacki, M. (2024a). Full field continuous dynamic recrystallization simulations considering precipitates evolutions with DIGIMU®. *Materials Research Proceedings*, 41, 2339–2346.

De Micheli, P., Alvarado, K., Grand, V., Bernacki, M. (2024b). Full field grain size prediction considering precipitates evolution and continuous dynamic recrystallization with DIGIMU® solution. In *Proceedings of the 14th International Conference on the Technology of Plasticity – Current Trends in the Technology of Plasticity*. Springer, Cham.

Demianczuk, D.W. and Aust, K.T. (1975). Effect of solute and orientation on the mobility of near-coincidence tilt boundaries in high-purity aluminum, *Acta Metallurgica*, 23(10), 1149–1162.

Depriester, D. and Kubler, R. (2019). Radical Voronoï tessellation from random pack of polydisperse spheres: Prediction of the cells' size distribution. *Computer-Aided Design*, 107, 37–49.

Dervieux, A. and Thomasset, F. (2006). A finite element method for the simulation of a Rayleigh-Taylor instability. In *Approximation Methods for Navier-Stokes Problems: Proceedings of the Symposium*. Springer, Berlin.

Du, D., Zhang, H., Srolovitz, D.J. (2007). Properties and determination of the interface stiffness. *Acta Materialia*, 55(2), 467–471.

Elias, R., Martins, M., Coutinho, A. (2007). Simple finite element-based computation of distance functions in unstructured grids. *Internation Journal for Numerical Methods in Engineering*, 72(May), 1095–1110.

Elsey, M., Esedoglu, S., Smereka, P. (2009). Diffusion generated motion for grain growth in two and three dimensions. *Journal of Computational Physics*, 21, 8015–8033.

Elsey, M., Esedoglu, S., Smereka, P. (2011). Large-scale simulation of normal grain growth via diffusion-generated motion. *Proc. R. Soc. A.*, 467, 381–401.

Fang, H., Juul Jensen, D., Zhang, Y. (2021). Improved grain mapping by laboratory X-ray diffraction contrast tomography. *IUCrJ*, 8(4), 559–573.

Fausty, J., Bozzolo, N., Muñoz, D.P., Bernacki, M. (2018). A novel level-set finite element formulation for grain growth with heterogeneous grain boundary energies. *Materials & Design*, 160, 578–590.

Fausty, J., Bozzolo, N., Bernacki, M. (2020). A 2D level set finite element grain coarsening study with heterogeneous grain boundary energies. *Applied Mathematical Modelling*, 78, 505–518.

Fausty, J., Murgas, B., Florez, S., Bozzolo, N., Bernacki, M. (2021). A new analytical test case for anisotropic grain growth problems. *Applied Mathematical Modelling*, 93, 28–52.

Fjeldberg, E. and Marthinsen, K. (2010). A 3D Monte Carlo study of the effect of grain boundary anisotropy and particles on the size distribution of grains after recrystallisation and grain growth. *Computational Materials Science*, 48(2), 267–281.

Florez, S., Shakoor, M., Toulorge, T., Bernacki, M. (2020). A new finite element strategy to simulate microstructural evolutions. *Computational Materials Science*, 172, 109335.

Florez, S., Alvarado, K., Murgas, B., Bozzolo, N., Chatain, D., Krill, C.E., Wang, M., Rohrer, G.S., Bernacki, M. (2022). Statistical behaviour of interfaces subjected to curvature flow and torque effects applied to microstructural evolutions. *Acta Materialia*, 222, 117459.

Fortmeier, O. and Martin Bücker, H. (2011). Parallel re-initialization of level set functions on distributed unstructured tetrahedral grids. *Journal of Computational Physics*, 230(12), 4437–4453.

Furstoss, J., Petit, C., Tommasi, A., Ganino, C., Muñoz, D.P., Bernacki, M. (2020). On the role of solute drag in reconciling laboratory and natural constraints on olivine grain growth kinetics. *Geophysical Journal International*, 224(2), 1360–1370.

Gaillac, A., Grand, V., Arsen, A., Gaillard, Q., Bernacki, M. (2022). Towards multi-scale modeling of zirconium alloys recrystallization and application to thermo-mechanical processes optimization. In *Key Engineering Materials*, 926. Trans Tech Publications, Stafa-Zurich.

Gamsjäger, E., Antretter, T., Schmaranzer, C., Preis, W., Chimani, C., Simha, N., Svoboda, J., Fischer, F. (2002). Diffusional phase transformation and deformation in steels. *Computational Materials Science*, 25(1), 92–99.

Gao, J. and Thompson, R. (1996). Real time-temperature models for Monte Carlo simulations of normal grain growth. *Acta Materialia*, 44(11), 4565–4570.

Gao, J., Thompson, R.G., Patterson, B.R. (1997). Computer simulation of grain growth with second phase particle pinning. *Acta Materialia*, 45(9), 3653–3658.

Garcke, H., Nestler, B., Stoth, B. (1999). A multiphase field concept: Numerical simulations of moving phase boundaries and multiple junctions. *SIAM Journal on Applied Mathematics*, 60(1), 295–315.

Golab, R., Sitko, M., Szyndler, J., Madej, L. (2014). Cellular automata finite element approach for modelling microstructure evolution under thermo-mechanical processing conditions. In *Cellular Automata*, 8751. Springer, Cham.

Gottstein, G. and Shvindlerman, L.S. (1992). On the true dependence of grain boundary migration rate on driving force. *Scripta Metallurgica et Materialia*, 27(11), 1521–1526.

Gourdet, S. and Montheillet, F. (2003). A model of continuous dynamic recrystallization. *Acta Materialia*, 51(9), 2685–2699.

Grand, V., Flipon, B., Gaillac, A., Bernacki, M. (2022). Characterization and modeling of the influence of initial microstructure on recrystallization of zircaloy-4 during hot forming. *arXiv*.

Grand, V., Flipon, B., Gaillac, A., Bernacki, M. (2023). Modeling CDRX and MDRX during hot forming of zircaloy-4. *Modelling and Simulation in Materials Science and Engineering*, 31(8), 085008.

Gruau, C. and Coupez, T. (2005). 3D tetrahedral, unstructured and anisotropic mesh generation with adaptation to natural and multidomain metric. *Computer Methods in Applied Mechanics and Engineering*, 194(48–49), 4951–4976.

Hallberg, H. and Bulatov, V.V. (2019). Modeling of grain growth under fully anisotropic grain boundary energy. *Modelling and Simulation in Materials Science and Engineering*, 27(4), 045002.

Hallberg, H. and Ristinmaa, M. (2013). Microstructure evolution influenced by dislocation density gradients modeled in a reaction-diffusion-system. *Computational Materials Science*, 67, 373–383.

Hashemi, S. and Kalidindi, S.R. (2021). A machine learning framework for the temporal evolution of microstructure during static recrystallization of polycrystalline materials simulated by cellular automaton. *Computational Materials Science*, 188, 110132.

Hassold, G., Holm, E., Srolovitz, D. (1990). Effects of particle size on inhibited grain growth. *Scripta Metallurgica et Materialia*, 24(1), 101–106.

He, D., Ekere, N.N., Cai, L. (1999). Computer simulation of random packing of unequal particles. *Phys. Rev. E*, 60, 7098–7104.

Hillert, M. (2007). *Phase Equilibria, Phase Diagrams and Phase Transformations: Their Thermodynamic Basis*. Cambridge University Press, Cambridge.

Hitti, K., Laure, P., Coupez, T., Silva, L., Bernacki, M. (2012). Precise generation of complex statistical representative volume elements (RVEs) in a finite element context. *Computational Materials Science*, 61, 224–238.

Holm, E.A., Hassold, G.N., Miodownik, M.A. (2001). On misorientation distribution evolution during anisotropic grain growth. *Acta Materialia*, 49(15), 2981–2991.

Huang, K. and Logé, R. (2016). A review of dynamic recrystallization phenomena in metallic materials. *Materials & Design*, 111, 548–574 [Online]. Available at: https://www.sciencedirect.com/science/article/pii/S0264127516311753.

Hutchinson, C., Zurob, H., Brechet, Y. (2006). The growth of ferrite in Fe-CX alloys: The role of thermodynamics, diffusion, and interfacial conditions. *Metallurgical and Materials Transactions A*, 37, 1711–1720.

Hwang, N.M. (1998). Simulation of the effect of anisotropic grain boundary mobility and energy on abnormal grain growth. *Journal of Materials Science*, 33(23), 5625–5629.

Ilin, D.N. and Bernacki, M. (2016a). Advancing layer algorithm of dense ellipse packing for generating statistically equivalent polygonal structures. *Granular Matter*, 18, 43. doi: 10.1007/s10035-016-0646-9.

Ilin, D.N. and Bernacki, M. (2016b). A new algorithm for dense ellipse packing and polygonal structures generation in context of FEM or DEM. *MATEC Web Conference*, 80, 02004.

Ilin, D.N., Bozzolo, N., Toulorge, T., Bernacki, M. (2018). Full field modeling of recrystallization: Effect of intragranular strain gradients on grain boundary shape and kinetics. *Computational Materials Science*, 150, 149–161.

Imai, H., Iri, M., Murota, K. (1985). Voronoi diagram in the Laguerre geometry and its applications. *SIAM Journal on Computing*, 14(1), 93–105.

Ivanov, V.A. (2006). On kinetics and thermodynamics of high angle grain boundaries in aluminum: Experimental study on grain boundary properties in bi-and tricrystals. Technical Report, Fakultät für Georessourcen und Materialtechnik.

James, R. (1986). Displacive phase transformations in solids. *Journal of the Mechanics and Physics of Solids*, 34(4), 359–394.

Janssens, K.G.F. (2010). An introductory review of cellular automata modeling of moving grain boundaries in polycrystalline materials. *Mathematics and Computers in Simulation*, 80(7), 1361–1381.

Janssens, K.G.F., Olmsted, D., Holm, E.A., Foiles, S.M., Plimpton, S.J., Derlet, P.M. (2006). Computing the mobility of grain boundaries. *Nature Materials*, 5(2), 124–127.

Jodrey, W.S. and Tory, E.M. (1985). Computer simulation of close random packing of equal spheres. *Phys. Rev. A*, 32, 2347–2351.

Jones, M., Baerentzen, J., Sramek, M. (2006). 3D distance fields: A survey of techniques and applications. *IEEE Transactions on Visualization and Computer Graphics*, 12(4), 581–599.

Juul Jensen, D. and Zhang, Y. (2020). Impact of 3D/4D methods on the understanding of recrystallization. *Current Opinion in Solid State and Materials Science*, 24(2), 100821.

Kad, B.K. and Hazzledine, P.M. (1997). Monte Carlo simulations of grain growth and Zener pinning. *Materials Science and Engineering: A*, 238(1), 70–77.

Kazaryan, A., Wang, Y., Dregia, S., Patton, B. (2002). Grain growth in anisotropic systems: Comparison of effects of energy and mobility. *Acta Materialia*, 50(10), 2491–2502.

Kim, H.-K., Ko, W.-S., Lee, H.-J., Kim, S.G., Lee, B.-J. (2011). An identification scheme of grain boundaries and construction of a grain boundary energy database. *Scripta Materialia*, 64(12), 1152–1155.

Kim, H.-K., Kim, S.G., Dong, W., Steinbach, I., Lee, B.-J. (2014). Phase-field modeling for 3D grain growth based on a grain boundary energy database. *Modelling and Simulation in Materials Science and Engineering*, 22(3), 034004.

Kimmel, R. and Sethian, J. (1998). Computing geodesic paths on manifolds. *Proc. Natl. Acad. Sci. USA*, 95(15), 8431–8435.

Kocks, U.F. (1976). Laws for work-hardening and low-temperature creep. *Journal of Engineering Materials and Technology*, 98(1), 76–85.

Kohara, S., Parthasarathi, M.N., Beck, P.A. (1958). Anisotropy of boundary mobility. *Journal of Applied Physics*, 29(7), 1125–1126.

Krill III, C. and Chen, L. (2002). Computer simulation of 3-D grain growth using a phase-field model. *Acta Materialia*, 50(12), 3059–3075.

Laasraoui, A. and Jonas, J. (1991). Prediction of steel flow stresses at high temperatures and strain rates. *Metallurgical Transactions A*, 22(7), 1545–1558.

Lazar, E.A., Mason, J.K., MacPherson, R.D., Srolovitz, D.J. (2011). A more accurate three-dimensional grain growth algorithm. *Acta Materialia*, 59(17), 6837–6847.

Liu, X., Li, H., Zhan, M. (2018). A review on the modeling and simulations of solid-state diffusional phase transformations in metals and alloys. *Manufacturing Review*, 5, 10.

Logé, R., Bernacki, M., Resk, H., Delannay, L., Digonnet, H., Chastel, Y., Coupez, T. (2008). Linking plastic deformation to recrystallization in metals using digital microstructures. *Philosophical Magazine*, 30–32, 3691–3712.

Lücke, K. and Detert, K. (1957). A quantitative theory of grain-boundary motion and recrystallization in metals in the presence of impurities. *Acta Metallurgica*, 5(11), 628–637.

Ludwig, W., King, A., Reischig, P., Herbig, M., Lauridsen, E.M., Schmidt, S., Proudhon, H., Forest, S., Cloetens, P., du Roscoat, S.R. et al. (2009). New opportunities for 3D materials science of polycrystalline materials at the micrometre lengthscale by combined use of x-ray diffraction and x-ray imaging. *Materials Science and Engineering: A*, 524(1), 69–76.

Mackenzie, J. (1958). Second paper on statistics associated with the random disorientation of cubes. *Biometrika*, 45(1–2), 229–240.

Maire, L., Scholtes, B., Moussa, C., Bozzolo, N., Pino Muñoz, D., Bernacki, M. (2016). Improvement of 3D mean field models for capillarity-driven grain growth based on full field simulations. *Journal of Materials Science*, 51(24), 10970–10981.

Maire, L., Scholtes, B., Moussa, C., Bozzolo, N., Muñoz, D.P., Settefrati, A., Bernacki, M. (2017). Modeling of dynamic and post-dynamic recrystallization by coupling a full field approach to phenomenological laws. *Materials & Design*, 133, 498–519.

Maksimova, E.L., Shvindlerman, L.S., Straumal, B.B. (1988). Transformation of σ17 special tilt boundaries to general boundaries in tin. *Acta Metallurgica*, 36(6), 1573–1583.

Manohar, P.A., Ferry, M., Chandra, T. (1998). Five decades of the Zener equation. *ISIJ International*, 38, 913–924.

Marin, E. and Dawson, P. (1998). On modelling the elasto-viscoplastic response of metals using polycrystal plasticity. *Computer Methods in Applied Mechanics and Engineering*, 165(1), 1–21.

Mecking, H. and Kocks, U.F. (1981). Kinetics of flow and strain-hardening. *Acta Metallurgica*, 29(11), 1865–1875.

Mecozzi, M.G. (2006). Phase field modelling of the austenite to ferrite transformation in steels. PhD Thesis, Delf University of Technology.

Mecozzi, M.G., Eiken, J., Apel, M., Sietsma, J. (2011). Quantitative comparison of the phase-transformation kinetics in a sharp-interface and a phase-field model. *Computational Materials Science*, 50(6), 1846–1853.

Merriman, B., Bence, J., Osher, S. (1994). Motion of multiple junctions: A level set approach. *Journal of Computational Physics*, 112, 334–363.

Mießen, C., Liesenjohann, M., Barrales-Mora, L., Shvindlerman, L., Gottstein, G. (2015). An advanced level set approach to grain growth – Accounting for grain boundary anisotropy and finite triple junction mobility. *Acta Materialia*, 99, 39–48.

Militzer, M. (2011). Phase field modeling of microstructure evolution in steels. *Current Opinion in Solid State and Materials Science*, 15(3), 106–115.

Miyoshi, E. and Takaki, T. (2016). Validation of a novel higher-order multi-phase-field model for grain-growth simulations using anisotropic grain-boundary properties. *Computational Materials Science*, 112, 44–51.

Miyoshi, E. and Takaki, T. (2017). Multi-phase-field study of the effects of anisotropic grain-boundary properties on polycrystalline grain growth. *Journal of Crystal Growth*, 474, 160–165.

Miyoshi, E., Takaki, T., Ohno, M., Shibuta, Y. (2020). Accuracy evaluation of phasefield models for grain growth simulation with anisotropic grain boundary properties. *ISIJ International*, 60(1), 160–167.

Miyoshi, E., Takaki, T., Sakane, S., Ohno, M., Shibuta, Y., Aoki, T. (2021). Large-scale phase-field study of anisotropic grain growth: Effects of misorientation-dependent grain boundary energy and mobility. *Computational Materials Science*, 186, 109992.

Moelans, N., Blanpain, B., Wollants, P. (2006). Phase field simulations of grain growth in two-dimensional systems containing finely dispersed second-phase particles. *Acta Materialia*, 54(4), 1175–1184.

Moelans, N., Blanpain, B., Wollants, P. (2008). Quantitative analysis of grain boundary properties in a generalized phase field model for grain growth in anisotropic systems. *Physical Review B*, 78(2), 024113.

Montes de Oca Zapiain, D., Stewart, J.A., Dingreville, R. (2021). Accelerating phase-field-based microstructure evolution predictions via surrogate models trained by machine learning methods. *NPJ Computational Materials*, 7(1), 3.

Moore, R.D., Beecroft, T., Rohrer, G.S., Barr, C.M., Homer, E.R., Hattar, K., Boyce, B.L., Abdeljawad, F. (2021). The grain boundary stiffness and its impact on equilibrium shapes and boundary migration: Analysis of the $\sigma 5$, 7, 9, and 11 boundaries in Ni. *Acta Materialia*, 218, 117220.

Morawiec, A. (2003). *Orientations and Rotations*. Springer, New York.

Murgas, B., Florez, S., Bozzolo, N., Fausty, J., Bernacki, M. (2021). Comparative study and limits of different level-set formulations for the modeling of anisotropic grain growth. *Materials*, 14(14).

Olmsted, D.L., Foiles, S.M., Holm, E.A. (2009a). Survey of computed grain boundary properties in face-centered cubic metals: I. Grain boundary energy. *Acta Materialia*, 57(13), 3694–3703.

Olmsted, D.L., Holm, E.A., Foiles, S.M. (2009b). Survey of computed grain boundary properties in face-centered cubic metals-II. Grain boundary mobility. *Acta Materialia*, 57(13), 3704–3713.

Oommen, V., Shukla, K., Goswami, S., Dingreville, R., Karniadakis, G.E. (2022). Learning two-phase microstructure evolution using neural operators and autoencoder architectures. *NPJ Computational Materials*, 8(1), 190.

Osher, S. and Sethian, J. (1988). Fronts propagating with curvature-dependent speed: Algorithms based on Hamilton-Jacobi formulations. *Journal of Computational Physics*, 79(1), 12–49.

Pariser, G., Schaffnit, P., Steinbach, I., Bleck, W. (2001). Simulation of the $\gamma - \alpha$ transformation using the phase-field method. *Steel Research*, 72, 354–360.

Phaneesh, K.R., Bhat, A., Mukherjee, P., Kashyap, K.T. (2012). On the Zener limit of grain growth through 2D Monte Carlo simulation. *Computational Materials Science*, 58, 188–191.

Polychronopoulou, D., Bozzolo, N., Muñoz, D.P., Bruchon, J., Shakoor, M., Millet, Y., Dumont, C., von Thüngen, I.F., Besnard, R., Bernacki, M. (2019). Introduction to the level-set full field modeling of laths spheroidization phenomenon in α/β titanium alloys. *International Journal of Material Forming*, 12(2), 173–183.

Proudhon, H., Li, J., Reischig, P., Gueninchault, N., Forest, S., Ludwig, W. (2016). Coupling diffraction contrast tomography with the finite element method. *Advanced Engineering Materials*, 18(6), 903–912.

Purdy, G., Ågren, J., Borgenstam, A., Bréchet, Y., Enomoto, M., Furuhara, T., Gamsjager, E., Gouné, M., Hillert, M., Hutchinson, C. et al. (2011). Alemi: A ten-year history of discussions of alloying-element interactions with migrating interfaces. *Metallurgical and Materials Transactions A*, 42, 3703–3718.

Qiu, C., Salvalaglio, M., Srolovitz, D.J., Han, J. (2023). Interface faceting-defaceting mediated by disconnections. *Acta Materialia*, 251, 118880.

Qiu, C., Salvalaglio, M., Srolovitz, D.J., Han, J. (2024). Disconnection flow-mediated grain rotation. *Proceedings of the National Academy of Sciences*, 121(1), e2310302121.

Quey, R. and Renversade, L. (2018). Optimal polyhedral description of 3D polycrystals: Method and application to statistical and synchrotron X-ray diffraction data. *Computer Methods in Applied Mechanics and Engineering*, 330, 308–333.

Quey, R., Dawson, P., Barbe, F. (2011). Large-scale 3D random polycrystals for the finite element method: Generation, meshing and remeshing. *Computer Methods in Applied Mechanics and Engineering*, 200(17–20), 1729–1745.

Raabe, D. (2002). Cellular automata in materials science with particular reference to recrystallization simulation. *Annual Review of Materials Research*, 32(1), 53–76.

Read, W.T. and Shockley, W. (1950). Dislocation models of crystal grain boundaries. *Physical Review*, 78(3), 275–289.

Rollett, A.D. and Manohar, P. (2004). Continuum scale simulation of engineering materials. In *The Monte Carlo Method*, Raabe, D., Roters, F., Barlat, F., Chen, L.-Q. (eds). John Wiley & Sons, New Jersey.

Rollett, A.D., Srolovitz, D.J., Anderson, M. (1989). Simulation and theory of abnormal grain growth–anisotropic grain boundary energies and mobilities. *Acta Metallurgica*, 37(4), 1227–1240.

Rollett, A.D., Saylor, D., Fridy, J., El-Dasher, B.S., Brahme, A., Lee, S., Cornwell, C., Noack, R. (2004). Modeling polycrystalline microstructures in 3D. *AIP Conference Proceedings*, 712(1), 71–77.

Rollett, A., Rohrer, G.S., Humphreys, J. (2017). *Recrystallization and Related Annealing Phenomena*. Elsevier, Amsterdam.

Ruiz, D., Bozzolo, N., Moussa, C., Maire, L., Pino Muñoz, D., Bernacki, M. (2020a). Full field modeling of dynamic recrystallization in a CPFEM context – Application to 304l steel. *Computational Materials Science*, 184, 109892.

Ruiz, D., Pino Muñoz, D., Bernacki, M. (2020b). A new numerical framework for the full field modeling of dynamic recrystallization in a CPFEM context. *Computational Materials Science*, 179, 109645.

Salama, H., Kundin, J., Shchyglo, O., Mohles, V., Marquardt, K., Steinbach, I. (2020). Role of inclination dependence of grain boundary energy on the microstructure evolution during grain growth. *Acta Materialia*, 188, 641–651.

Scholtes, B. (2016). Development of an efficient level set framework for the full field modeling recrystallization in 3D. PhD Thesis, MINES Paris PSL.

Scholtes, B., Shakoor, M., Bozzolo, N., Bouchard, P.-O., Settefrati, A., Bernacki, M. (2015a). Advances in level-set modeling of recrystallization at the polycrystal scale – Development of the Digi-μ software. In *Material Forming ESAFORM 2015*, 651. Trans Tech Publications Ltd, Stafa-Zurich.

Scholtes, B., Shakoor, M., Settefrati, A., Bouchard, P.-O., Bozzolo, N., Bernacki, M. (2015b). New finite element developments for the full field modeling of microstructural evolutions using the level-set method. *Computational Materials Science*, 109, 388–398.

Scholtes, B., Boulais-Sinou, R., Settefrati, A., Muñoz, D.P., Poitrault, I., Montouchet, A., Bozzolo, N., Bernacki, M. (2016a). 3D level set modeling of static recrystallization considering stored energy fields. *Computational Materials Science*, 122, 57–71.

Scholtes, B., Ilin, D., Settefrati, A., Bozzolo, N., Agnoli, A., Bernacki, M. (2016b). Full field modeling of the Zener pinning phenomenon in a level set framework – Discussion of classical limiting mean grain size equation. *Proceedings of the 13th International Symposium on Superalloys*, Seven Springs.

Sethian, J.A. (1996). Theory, algorithms, and applications of level set methods for propagating interfaces. *Acta Numerica*, 5, 309–395.

Sethian, J.A. and Vladimirsky, A. (2003). Ordered upwind methods for static Hamilton-Jacobi equations: Theory and algorithms. *Proceedings of the National Academy of Sciences of the United States of America*, 98(20), 11069–11074.

Shakoor, M., Bernacki, M., Bouchard, P.-O. (2015a). A new body-fitted immersed volume method for the modeling of ductile fracture at the microscale: Analysis of void clusters and stress state effects on coalescence. *Engineering Fracture Mechanics*, 147, 398–417.

Shakoor, M., Scholtes, B., Bouchard, P.-O., Bernacki, M. (2015b). An efficient and parallel level set reinitialization method–Application to micromechanics and microstructural evolutions. *Applied Mathematical Modelling*, 39(23), 7291–7302.

Smith, C.S. (1948). Introduction to grains, phases, and interfaces–An interpretation of microstructure. *Transactions of the American Institute of Mining and Metallurgical Engineers*, 175(2), 15–51.

Song, Y.-H., Wang, M.-T., Ni, J., Jin, J.-F., Zong, Y.-P. (2020). Effect of grain boundary energy anisotropy on grain growth in zk60 alloy using a 3D phase-field modeling. *Chinese Physics B*, 29(12), 128201.

Srolovitz, D., Anderson, M., Grest, G., Sahni, P. (1984). Computer simulation of grain growth-III. Influence of a particle dispersion. *Acta Metallurgica*, 32(9), 1429–1438.

Steinbach, I., Pezzolla, F., Nestler, B., Seesselberg, M., Prieler, R., Rezende, J. (1996). A phase field concept for multiphase systems. *Physica D: Nonlinear Phenomena*, 94, 135–147.

Steinbach, I., Böttger, B., Eiken, J., Warnken, N., Fries, S. (2007). Calphad and phase-field modeling: A successful liaison. *Journal of Phase Equilibria and Diffusion*, 28(1), 101–106.

Sussman, M., Smereka, P., Osher, S. (1994). A level set approach for computing solutions to incompressible two-phase flow. *Journal of Computational Physics*, 114, 146–159.

Thermo-Calc Software (n.d.). [Online]. Available at: http://www.thermocalc.se/.

Tiaden, J., Nestler, B., Diepers, H., Steinbach, I. (1998). The multiphase-field model with an integrated concept for modelling solute diffusion. *Physica D: Nonlinear Phenomena*, 115(1), 73–86.

Tonks, M., Zhang, Y., Butterfield, A., Bai, X. (2015). Development of a grain boundary pinning model that considers particle size distribution using the phase field method. *Modelling and Simulation in Materials Science and Engineering*, 23(4), 045009.

Upmanyu, M., Hassold, G.N., Kazaryan, A., Holm, E.A., Wang, Y., Patton, B., Srolovitz, D.J. (2002). Boundary mobility and energy anisotropy effects on microstructural evolution during grain growth. *Interface Science*, 10(2–3), 201–216.

Villaret, F., Hary, B., de Carlan, Y., Baudin, T., Logé, R., Maire, L., Bernacki, M. (2020). Probabilistic and deterministic full field approaches to simulate recrystallization in ODS steels. *Computational Materials Science*, 179, 109646.

Viswanathan, R. and Bauer, C.L. (1973). Kinetics of grain boundary migration in copper bicrystals with [001] rotation axes. *Acta Metallurgica*, 21(8), 1099–1109.

Wakai, F. and Brakke, K. (2011). Mechanics of sintering for coupled grain boundary and surface diffusion. *Acta Materialia*, 59(14), 5379–5387.

Wei, C., Thomas, S.L., Han, J., Srolovitz, D.J., Xiang, Y. (2019). A continuum multi-disconnection-mode model for grain boundary migration. *Journal of the Mechanics and Physics of Solids*, 133, 103731.

Weygand, S., Brechet, Y., Lépinoux, J. (1999). Zener pinning and grain growth: A two-dimensional vertex computer simulation. *Acta Materialia*, 47(3), 961–970.

Wheeler, A.A., Boettinger, W.J., McFadden, G.B. (1992). Phase-field model for isothermal phase transitions in binary alloys. *Phys. Rev. A*, 45, 7424–7439.

Winning, M., Gottstein, G., Shvindlerman, L.S. (2002). On the mechanisms of grain boundary migration. *Acta Materialia*, 50(2), 353–363.

Xu, T. and Li, M. (2009). Topological and statistical properties of a constrained Voronoi tessellation. *Philosophical Magazine*, 89(4), 349–374.

Yan, W., Melville, J., Yadav, V., Everett, K., Yang, L., Kesler, M.S., Krause, A.R., Tonks, M.R., Harley, J.B. (2022). A novel physics-regularized interpretable machine learning model for grain growth. *Materials & Design*, 222, 111032.

Young, P., Beresford-West, T., Coward, S., Notarberardino, B., Walker, B., Abdul-Aziz, A. (2008). An efficient approach to converting three-dimensional image data into highly accurate computational models. *Philosophical Transactions of the Royal Society A: Mathematical, Physical and Engineering Sciences*, 366(1878), 3155–3173.

Zener, C. (1949). Theory of growth of spherical precipitates from solid solution. *Journal of Applied Physics*, 20(10), 950–953.

Zhang, Y., Bajaj, C., Sohn, B.-S. (2005). 3D finite element meshing from imaging data. *Computer Methods in Applied Mechanics and Engineering*, 194(48), 5083–5106.

Zhang, J., Poulsen, S.O., Gibbs, J.W., Voorhees, P.W., Poulsen, H.F. (2017a). Determining material parameters using phase-field simulations and experiments. *Acta Materialia*, 129, 229–238.

Zhang, L., Han, J., Xiang, Y., Srolovitz, D.J. (2017b). Equation of motion for a grain boundary. *Physical Review Letters*, 119(24), 246101.

Zhang, J., Ludwig, W., Zhang, Y., Sørensen, H.H.B., Rowenhorst, D.J., Yamanaka, A., Voorhees, P.W., Poulsen, H.F. (2020). Grain boundary mobilities in polycrystals. *Acta Materialia*, 191, 211–220.

Zhao, H., Chan, T., Merriman, B., Osher, S. (1996). A variational level set approach to multiphase motion. *Journal of Computational Physics*, 127, 179–195.

Zöllner, D. and Zlotnikov, I. (2019). Texture controlled grain growth in thin films studied by 3D Potts model. *Advanced Theory and Simulations*, 2(8), 1900064.

Zouaghi, A., Bellet, M., Bienvenu, Y., Perrin, G., Cédat, D., Bernacki, M. (2012). Modelling of the compaction phase during hot isostatic pressing process at the mesoscopic scale. In *Advances in Powder Metallurgy and Particulate Materials*, Nashville, 3117–3125.

7

Resolution Methods for Digital Materials – Recent Developments of Cellular Automaton Method

Lukasz MADEJ and Mateusz SITKO
AGH University of Krakow, Poland

7.1. Introduction

Thermomechanical treatment is a possibility in order to provide the required range of options to control the final microstructure morphology and in-use properties of metallic products. A combination of deformation routes and conditions, as well as varying process temperatures, can be used to initiate and control three main groups of phenomena responsible for the microstructure evolution. The texture evolution is the first one (Kowalczyk-Gajewska et al. 2015; Kim et al. 2017; Frydrych and Kowalczyk-Gajewska 2018). The second involves phase transformations in metallic materials during both the heating and cooling stages (Schoof et al. 2018; Ollat et al. 2018; Song et al. 2022), and the last group is represented by thermally activated phenomena of recrystallization (static, dynamic and metadynamic) (Goetz and Seetharaman 1998; Li et al. 2010; Cho and Yoo 2001). The latter has been experimentally and numerically investigated for many decades in various laboratories around the world, increasing the level of understanding of interactions between

For a color version of all figures in this chapter, see www.iste.co.uk/bernacki/digital.zip.

Digital Materials,
coordinated by Marc BERNACKI and Samuel FOREST. © ISTE Ltd 2024.

processing conditions and microstructural changes (Chun et al. 2006; Salehi and Serajzadeh 2012; Ali et al. 2017; Luan et al. 2020). The modeling of recrystallization in metal forming is historically related to the initial works by Jonas (1994) in Canada, Sellars and Zhu (2000) in Europe, Sakai et al. (1988) in Japan and Hodgson in Australia (Wong et al. 1996). These works can be considered to be the most important contributions to the development of microstructure evolution models. Today, such models are generally classified as conventional phenomenological approaches and are based on closed-form or differential equations. They provide valuable results in the form of average data on flow stress values and process kinetics. Presently, they are still commonly used in practical industrial applications as they are fast and do not require extensive computing facilities. At the same time, the number of assumptions and calibration parameters that must be identified can be significant. A group of more advanced approaches to simulate recrystallization that attempt to consider the underlying microstructure morphology in an implicit manner are mean field models (Beltran et al. 2015; Després et al. 2020). The grains morphology is simplified in these approaches and reduced to interacting spherical objects or objects within the homogenous matrix, called the homogeneous equivalent medium.

Finally, even more advanced approaches, called full-field models, have been intensively developed in recent years. In contrast to mean field models, they can more accurately predict the evolution of microstructure morphology and grain interactions as they are considered in an explicit manner. Such an explicit consideration of the microstructure with its heterogeneities significantly increases the amount of information obtained from numerical modeling. Examples of explicit full-field models that are addressed within the chapter are Monte Carlo (MC) (Xu et al. 2022; Maazi and Lezzar 2020) and cellular automaton (CA) (Teferra and Rowenhorst 2021; Alone et al. 2020) techniques. Both approaches can be classified as discrete modeling techniques that can also take stochastic aspects of deformation into account. Conceptually different full-field approaches are the phase field (PF) (Takaki et al. 2007; Xiao et al. 2018) (see Chapter 5) and the level-set (LS) methods (Bernacki et al. 2008; Osher and Fedkiw 2001) (see Chapter 6), which are based on computational continuum mechanics principles. A commonly used approach in these applications is also the Vertex (Mellbin et al. 2017; Florez et al. 2020) method (see Chapter 4).

7.2. Cellular automaton and applications

The CA method is designed to perform calculations within computational spaces with a finite number of cells. One of the first mentions of the method can be found in the work of (Neumann 1966). In his research, he constructed a deterministic cellular automaton model that was based on three fundamental elements:

– Space: this consists of finite set of cells, where each cell is represented by a set of internal and state variables describing the evolutions of the cell state during the simulations.

– Neighborhood: a geometrical arrangement that describes the closest neighbors of a particular cell. Depending on the computational domain dimensions, the neighborhood can be considered in 1D, 2D and 3D.

– Transition rules: these are is a set of definitions controlling the evolution of the cell states. The state of each cell in the CA space is determined by the previous states of its neighbors and the cell itself by these transition rules. Generally, a transition rule can be used to define the new state of the CA cell Y according to the following relation:

$$Y_i^{t+dt} = f\left(Y_j^t\right), j \in N(i), \quad [7.1]$$

with $N(i)$ the surrounding of the ith cell. From the model point of view, the CA cell state can remain the same as it was in the previous time-step state, or it can change to a new state if a certain logistic function Λ is true:

$$Y_i^{t+dt} = \begin{cases} \text{new\_state if } \Lambda \\ Y_i^t \qquad \text{otherwise} \end{cases}, \quad \text{where } j \in N(i) \quad [7.2]$$

with Y_i the state of the ith cell, t is the time, dt is the time step and $\Lambda = \Lambda(p, q, Y_i, Y_j)$ the logical function, based on the transition rules, which depends on the internal variables q and external variables p and describes a condition for the cell to change its state. The above-presented definitions are often extended by a fourth component.

– Cell variables: these consider both state variables used in the transition rules and internal variables used to store information about required physical quantities. As a result, the CA model is defined as a quadrupole:

$$CA = (\omega, \mathbf{Y}, \mathbf{N}, f), \quad [7.3]$$

where ω is the cellular automata space, Y is the set of cell variables, N is the set of neighbors and f is the transition rules.

Therefore, the main idea of the CA technique is to divide an investigated computational domain into one-, two-, or three-dimensional lattices of finite cells. The 2D and 3D spaces are mainly investigated when microstructure evolution models are considered. Each CA cell in this space is called a cellular automaton, while the lattice of the cells is known as the CA space ω. In 2D, the CA space can be based on hexagons (circles), squares, triangles or pentagons (regular pentagons are not suitable to represent CA space, but a combination of specific angles and side lengths can generate different types of spaces (Bays 2018)), representing the CA cell geometry, as seen in Figure 7.1. Similar concepts to the CA space division exist in the 3D space.

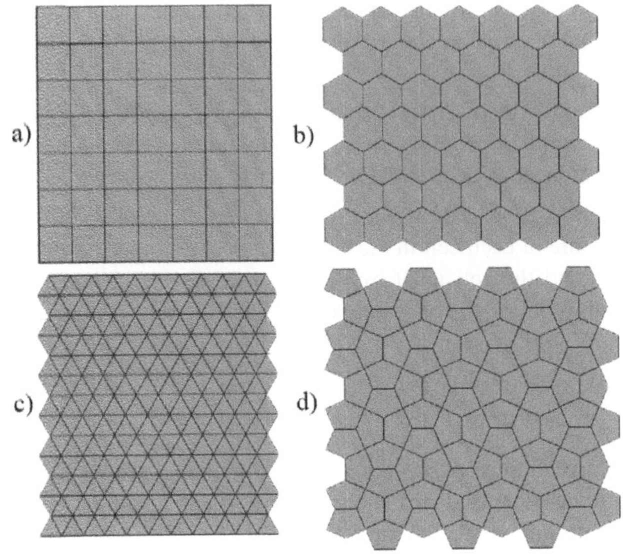

Figure 7.1. *CA spaces with different cell geometries (a) square, (b) hexagonal (circular), (c) triangular or (d) pentagonal (Bays 2018)*

The definition of an appropriate CA neighborhood is another crucial component in model development. The most commonly used CA neighborhoods are the von Neumann and Moore ones where in the 2D case, each CA cell is surrounded by four or eight neighboring cells, respectively. However, during the last several years, different CA neighborhoods were proposed to find a more suitable solution for particular investigated physical phenomena, for example, random pentagonal or random hexagonal. Graphical representation of commonly used types of neighborhood in 2D and 3D is presented in Figures 7.2 and 7.3, respectively.

It should also be pointed out that other nonstandard CA neighborhoods can be found in scientific literature (Bretin et al. 2020). A fully random neighborhood operating on sparsely distributed cellular automatons is especially interesting (Figure 7.4). These approaches are used more often in fully coupled cellular automata and finite element (CAFE) models of large microstructure deformation simulations (Madej et al. 2018).

The CA method's main assumption is that the computing space size is fixed during the simulation, and cells change their states synchronously in discretely defined time steps. Another assumption concerns cells' long-range interactions, which are generally neglected, and only the nearest neighbors are considered in a particular time step.

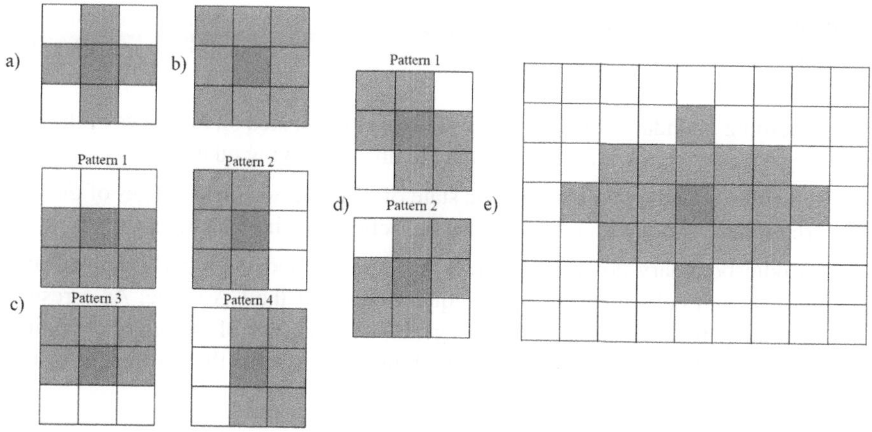

Figure 7.2. *Different neighborhood definitions in CA space (a) von Neumann; (b) Moore; (c) pentagonal random (template generated randomly from four candidates); (d) hexagonal random (template generated randomly from two candidates); and (e) elliptical (different values of rotation angle can be applied)*

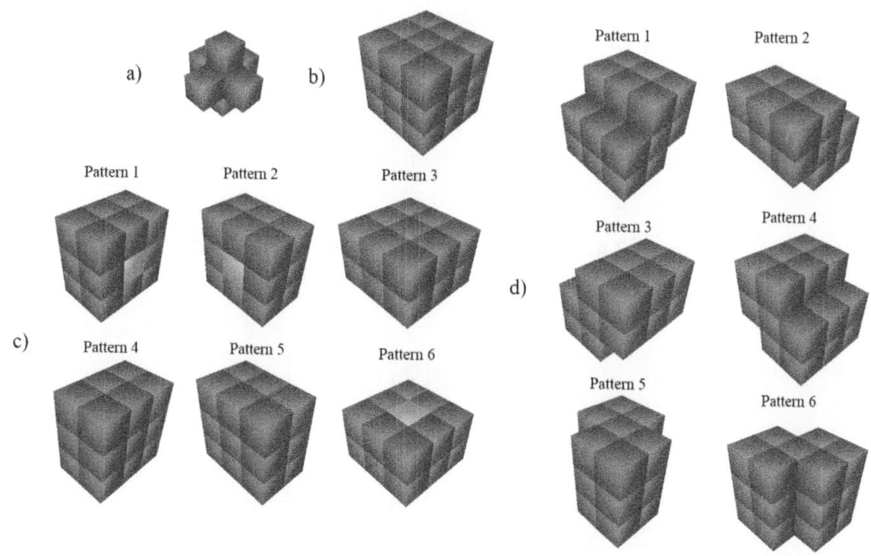

Figure 7.3. *Different neighborhood definitions in the 3D CA space (a) von Neumann; (b) Moore; (c) pentagonal random (template generated randomly from six candidates); and (d) hexagonal random (template generated randomly from six candidates)*

Finally, to deal with different kinds of physical phenomena, appropriate boundary conditions have also to be created. Three types of boundary conditions are generally considered:

– Absorbing boundary conditions: the state of cells located on the edges of the CA space are properly fixed with a specific state to absorb moving quantities.

– Reflective boundary conditions: the state of cells located on the edges of the CA space are properly fixed with a specific state to reflect moving quantities.

– Periodic boundary conditions: the CA neighborhood is properly defined and takes into account cells located on subsequent edges of the CA space. As a result, space continuity is obtained, as seen in Figure 7.5. This type of boundary condition is commonly used in many physical simulations, as it assures the fulfillment of the principle of conservation.

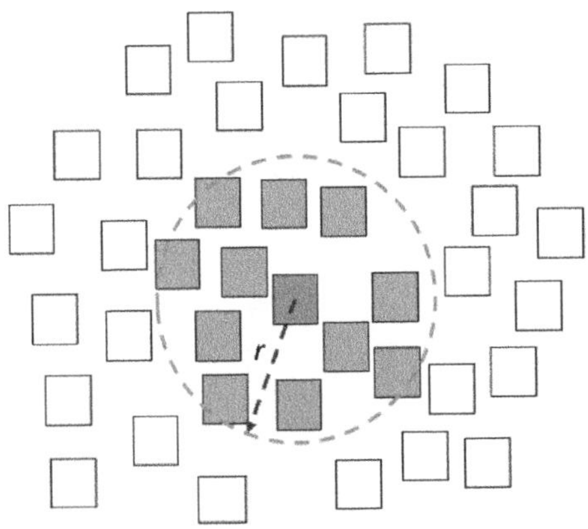

Figure 7.4. *The basic concept of the neighborhood selection algorithm in the random CA algorithm*

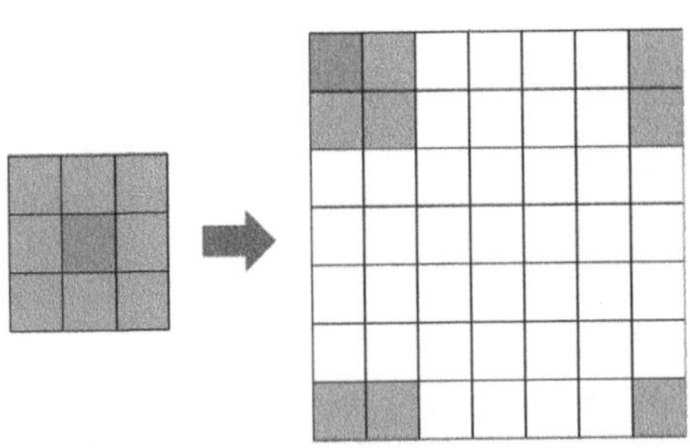

Figure 7.5. *Illustration of the periodic boundary condition concept in the 2D case*

The presented classical CA method found a wide range of applications in the material science area. The approach was used for simulation of solidification (Reuther and Rettenmayr 2014), oxidation (Jahns et al. 2017), crack propagation (Rauch et al. 2016), solid-state phase transformation (Halder et al. 2014) or dynamic (Teferra and Rowenhorst 2021) and static recrystallization phenomena (Sieradzki and Madej 2012), among many other applications. In addition to the classical approach to the CA algorithm, an interesting solution that can reduce computational cost is the concept of frontal cellular automata (FCA) (Łach et al. 2018; Svyetlichnyy 2010, 2012). The FCA are based on the same CA space and neighborhood definition as the classical method, but the direction of the information flow is different. Models based on the classical CA method iterate through a CA space cell by cell, applying transition rules to each of them. In this case, the cell changes its state based on the information from neighboring CA cells, as presented in Figure 7.6(a). In the FCA method, three different groups of cells with different interaction mechanisms (Figure 7.6(b)) can be considered:

– cells belonging to grain boundaries, all of these cells participate in the simulation and evaluation of transition rules;

– cells inside the grain waiting for a state change;

– cells, which have already changed their state and further changes are not expected.

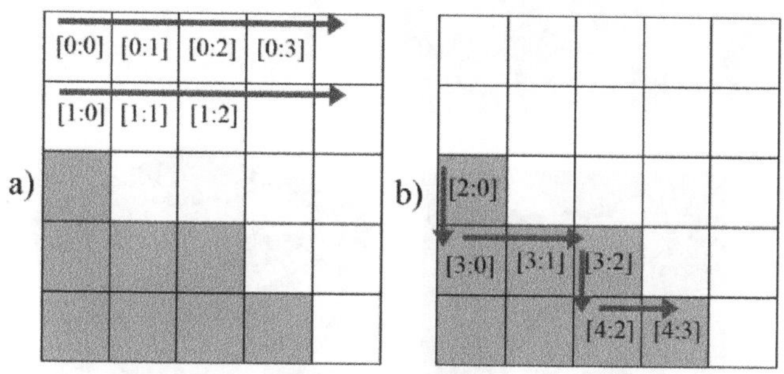

Figure 7.6. *Concept of a space iteration direction in the (a) classical CA and (b) frontal CA*

The FCA approach was successfully used for numerical modelling of various microstructural phenomena (Svyetlichnyy 2010; Svyetlichnyy et al. 2015), as it provides possibilities to reduce computational time, especially for 3D computational domains. However, restricting calculations only to CA cells located at the grain boundary may not replicate all the important phenomena also occurring in the CA cells within the grains, for example, recovery or diffusion.

7.3. CA front tracking model within the static recrystallization case study

One of the first used of the cellular automaton method for the modeling of the static recrystallization (SRX) mechanism was proposed by (Hesselbarth and Göbel 1991). This approach assumed that cells are in two states: recrystallized and non-recrystallized. The authors also adopted the approach where the recrystallization process is composed of two main steps, nucleation and grain growth. The model could predict the kinetics of the SRX and provide data on microstructure evolution during subsequent steps. This was a precursor of a series of CA SRX models that are still being developed and published in the scientific literature (Shelton and Dunand 1996; Raabe and Hantcherli 2005; Sieradzki and Madej 2013; Kühbach et al. 2016; Liu et al. 2017b; Shen et al. 2018; Davies 1997; Raabe 1999). The SRX models are characterized by different levels of complexity related to definitions of dimensions of the computational domain (Figure 7.7) (Han et al. 2014; Zheng et al. 2009), CA cell shapes (Janssens et al. 2010) or neighborhood types (Pietrzyk et al. 2015; Davies 1995).

Figure 7.7. *Different computational domains hexagonal cell shape in (a) 2D, (b) 3D spaces and square cell shape in (c) 2D and (d) 3D spaces*

However, the most important component of the CA method that diversifies the complexity and predictive capabilities of developed models is related to the physical mechanisms considered during the definition of transition rules. The stochastic properties of the system can be incorporated into CA microstructure evolution models mainly through the introduction of nondeterministic transition rules or by the definition of random-type neighborhood (Janssens 2003). The complexity of the CA SRX model is also diversified through the definition of the initial state of the microstructure morphology, accumulated energy and its distribution, nucleation type, or phenomena influencing grain growth. Different approaches have been proposed to provide initial microstructure morphology for SRX calculations throughout the years. The most commonly used are approaches based on Voronoi tessellation (Falco et al. 2017), statistically representative models (Mandal et al. 2018), MC (Mason et al. 2015) (Figure 7.8) or CA (Bakhtiari and Salehi 2018) algorithms (Figure 7.9).

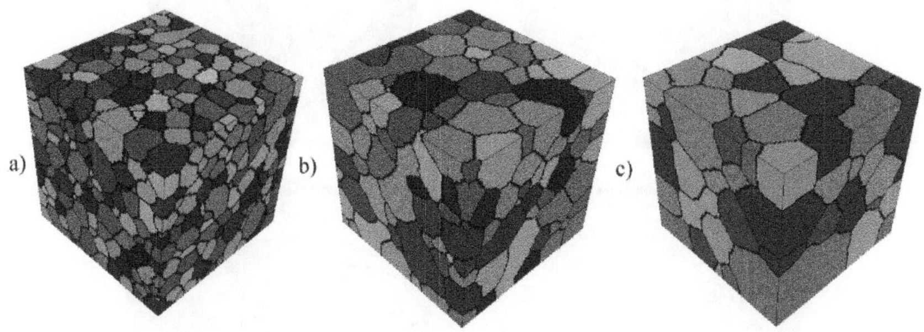

Figure 7.8. *Initial 200 × 200 × 200 μm microstructure obtained by the Monte Carlo grain growth algorithm: (a) 10 Monte Carlo steps (MCS); (b) 100 MCS and (c) 1000 MCS*

Image analysis methods were also used to eliminate unphysical artifacts in the digital microstructure morphologies (Liu et al. 2017a). This approach accurately represents the investigated microstructures, however, most of the time is still limited only to the 2D computational domain (Figure 7.10). It can operate on images acquired from light and scanning electron microscopy. In the latter case, not only the grain morphology but also their crystallographic orientations in the form of, for example, Euler angles (Φ_1, θ, Φ_2) can be easily reconstructed.

The main advantage of image processing approaches is that even very complex microstructures of multiphase materials can be digitalized for further SRX calculations. An example of ferritic-pearlitic microstructure investigation is presented in Madej et al. (2013). This approach is more demanding in the case of the generation of 3D input data, as large data sets from, for example, serial sectioning procedures or 3D computed tomography results based on, for example, synchrotron radiation techniques have to be used (Salvo et al. 2010; Patterson et al. 2014; Pokharel et al. 2015). Therefore, as mentioned, dedicated numerical algorithms are more often used for the 3D generation of microstructure morphology mentioned (Bargmann et al. 2018). Digital microstructures generated with these approaches statistically represent the grain size distribution, grain morphology, phase fractions, or crystallographic texture that are observed during standard metallographic investigations.

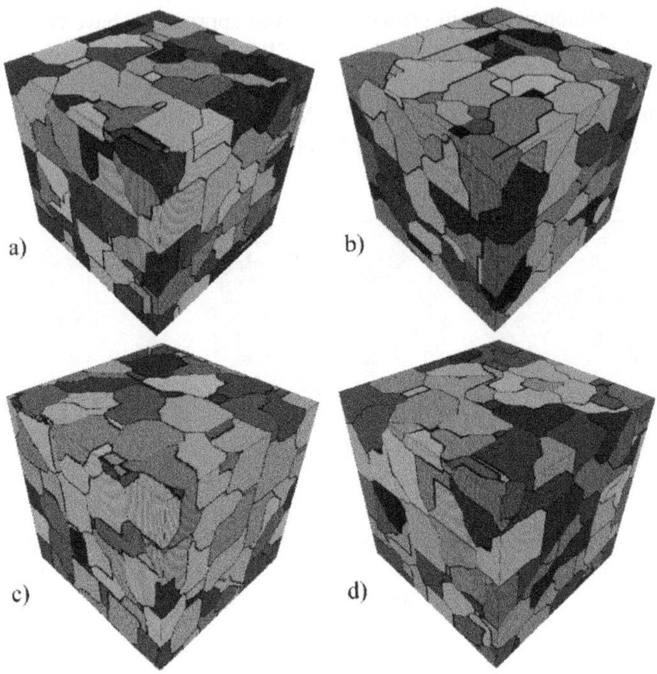

Figure 7.9. *Initial 200 × 200 × 200 μm microstructure obtained by the cellular automata grain growth algorithm with different neighborhood definitions: (a) Moore, (b) von Neumann, (c) random pentagonal and (d) random hexagonal*

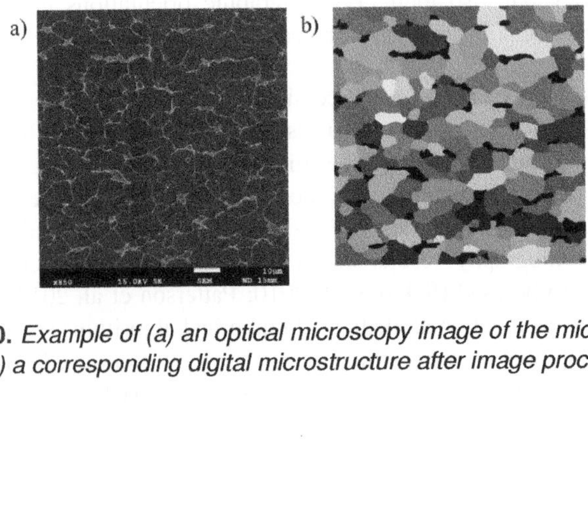

Figure 7.10. *Example of (a) an optical microscopy image of the microstructure and (b) a corresponding digital microstructure after image processing*

After an initial microstructure morphology is generated to simulate the SRX evolution, appropriate driving pressures must be considered. The primary is the plastic-stored energy. Therefore, the energy distribution across the entire sample has to be addressed. This energy field is behind one of the key driving pressures for grain growth during SRX. Again, there is a wide variety of approaches with different levels of complexity to handle this task. The simplest and most commonly used to introduce energy into the CA computational domain are based on a homogeneous or heterogeneous distribution of artificial energy (Kim et al. 2012).

Some approaches assume a homogeneous energy distribution after deformation, neglecting any influence of microstructure heterogeneities (Figure 7.11(a)) (Kugler and Turk 2006). As a result, these models can introduce inaccuracies in the calculations. The solution to this consists of considering an artificial heterogeneous distribution of accumulated energy with respect to grain boundaries and grain interior (Figures 7.11(b) and 7.11(c)). A comparison of results obtained from both approaches was studied in Sieradzki and Madej (2012) and proved the major influence of the energy distribution on SRX progress, as presented in Figure 7.12.

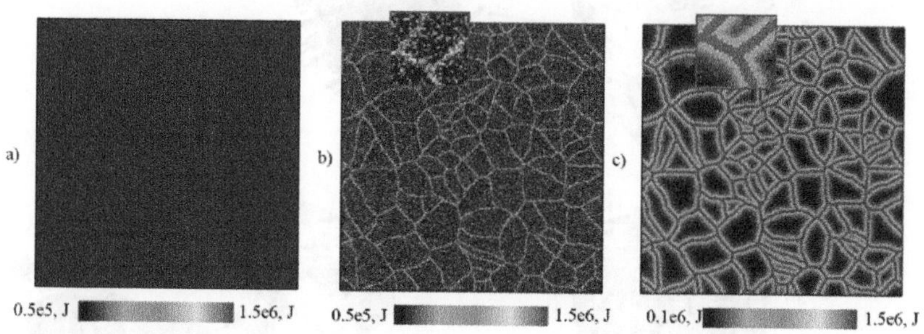

Figure 7.11. *Artificial energy distributed (a) homogeneously, (b) heterogeneously along the grain boundaries and (c) heterogeneously with the gradient distribution*

Thus, to properly reproduce the distribution of energy accumulated during deformation prior to CA SRX calculations, a combination of digital material representation (DMR) concept and finite element simulation can be used (Sieradzki and Madej 2012). In this case, models of underformed digital microstructures obtained from experimental investigation or numerically generated are discretized with finite elements by dedicated mesh generator DMRmesh (Kruzel et al. 2014; Madej et al. 2012, 2021). Then they are incorporated into the finite element software to perform a numerical simulation of material behavior under required loading conditions. The concept of the approach is presented in Figure 7.13.

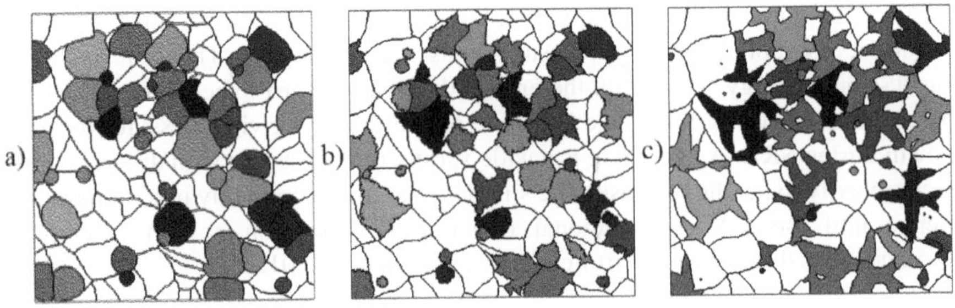

Figure 7.12. *Shape of growing recrystallized grains (white grains – deformed matrix, colur grains – recrystallized) for the energy distributed (a) homogeneously, (b) heterogeneously and (c) with gradient (site saturated nucleation, the same initial crystallographic orientation during all scenarios)*

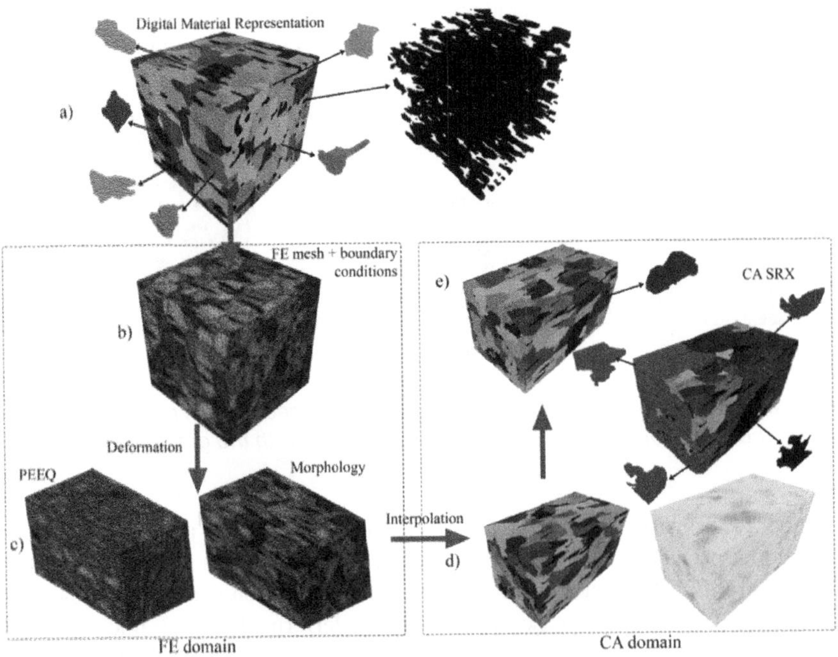

Figure 7.13. *Generation of the CA SRX model input data based on the DMR FE calculations: (a) artificially generated ferrite-perlite microstructure with a combination of CA and MC approaches (Boguń et al. 2021), (b) assembly of the DMR model, (c) results from FE tensile test simulation, (d) interpolation of results from FE mesh to 3D CA space and (e) CA simulation of SRX*

It should be pointed out that this approach is especially useful for the 3D simulation case studies, where the capabilities of experimental analysis are particularly limited. In this way, detailed information on the accumulated energy during deformation as a function of local microstructural heterogeneities can be obtained without extensive experimental effort. In this approach, the hardening behavior of the investigated microstructure features in the FE model can be described by conventional flow stress models or by more advanced crystal plasticity approaches. In the latter case, it is also possible to predict the local evolution of texture and provide data on the crystallographic orientation of each grain. However, in such a case, not only a microstructure morphology and a strain field but also crystallographic orientations have to be transferred between the FE and CA computational domains. Considering that calculations are performed in the FE and CA computational domains with incompatible meshes as seen in Figure 7.14, a proper data transfer procedure has to be used in this approach (Madej et al. 2013).

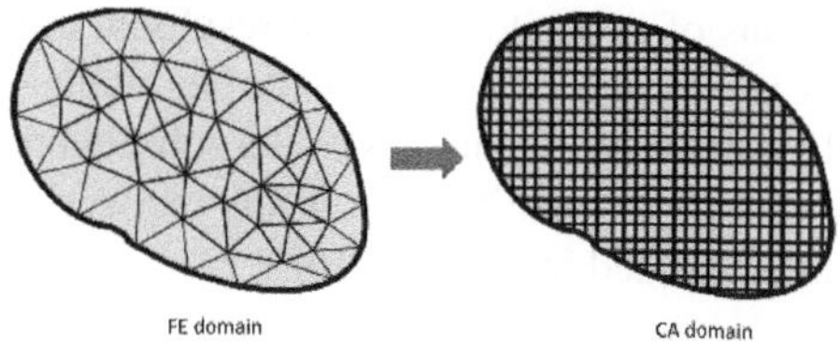

Figure 7.14. *Concept of data transfer between two computational domains, FE and CA, respectively (Madej et al. 2016b)*

This concept was successfully used in the literature to provide input data for single-phase metallic microstructures (Salehi and Serajzadeh 2012) as well as multi-phase microstructures (Madej et al. 2016a). As presented, this approach can provide detailed information not only on 2D but also on the 3D state of material prior to heat treatment. However, it requires a set of additional finite element numerical simulations and data interpolation techniques. Therefore, the alternative solution is again based on experimental data but this time obtained from metallographic analysis of the already deformed sample. The electron backscattered diffraction (EBSD) detector within the scanning electron microscope can provide information on microstructure morphology, crystallographic orientation and accumulated energy. These data can be used directly as input for the CA SRX model without the need for any additional numerical simulations. However, in most existing works, this approach is still limited only to 2D investigations due to difficulties in obtaining data in the 3D space that most of the time also requires the application of sophisticated 3D reconstruction algorithms (Sitko et al. 2020b) (Figure 7.15).

Finally, with the two possible approaches based on numerical or experimental investigations, the obtained input data are used for the CA modeling of microstructure evolution under the SRX conditions. Recovery is the first phenomenon that directly affects the kinetics of SRX and should be taken into account in the model (Rollett et al. 2017). Recovery can be simulated implicitly, as seen in (Shen et al. 2018). At the beginning of each iteration, its influence is evaluated and affects the plastic stored energy value obtained from the preprocessing module (Shen et al. 2018):

$$H_i = \left[H_{i0}^{1/2} - C_1\mu^{-1/2}k_B T \ln\left(1 + \frac{t}{\tau_0}\right)\right]^2, \qquad [7.4]$$

where H_{i0} is the initial plastic stored energy in the CA cell, C_1 is the recovery parameter, k_B – is the Boltzmann constant, τ_0 is the initial time of recovery, t is the current time and μ is the shear modulus of the material. Recovery is based on the annihilation of dislocations of opposite signs and results in a decrease in the plastic stored energy. Recovery is assumed to occur from the beginning of the heating process, which is why it directly influences the kinetics of the subsequent nucleation and grain growth stages. Therefore, proper identification of the recovery module parameters is of prime importance, as illustrated in Figure 7.16.

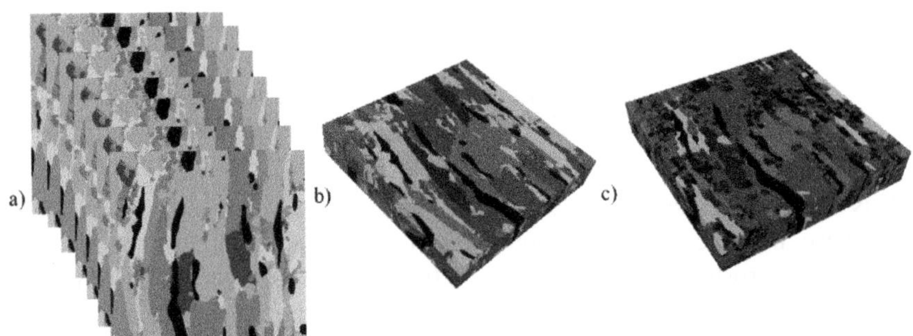

Figure 7.15. *Information on initial data for the CA SRX model after deformation from (a) the EBSD images from serial sectioning; (b) final input data for the CA simulation after 3D reconstruction procedure; and (c) CA SRX simulation results*

The second phenomenon that is of importance in SRX modeling is nucleation. This stage is one of the most critical elements of the developed SRX model. The number of nuclei that can appear in the material volume per unit of time is controlled by the following equation:

$$N_{NUC} = C_0\left(H_i - H^C\right)\exp\left(-\frac{Q_N}{RT}\right) \qquad [7.5]$$

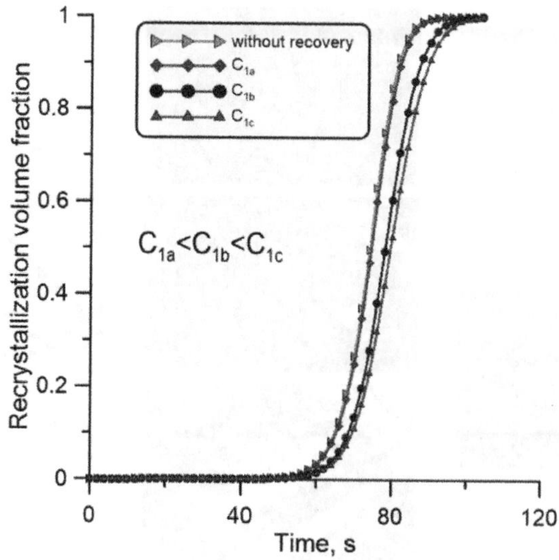

Figure 7.16. *Influence of recovery module on the SRX kinetics during heating to 775°C for different values of C_1 parameter*

where Q_N is the activation energy for nucleation, R is the perfect gas constant, T is the absolute temperature, C_0 is the nucleation parameter, H_i is the plastic stored energy in the CA cell and H^C is the critical amount of energy required for nucleation. H^C is calculated from the following formula (Kazeminezhad et al. 2007):

$$H^C = \frac{\varepsilon_c}{a\varepsilon_c + b}\gamma_{LAGB} \quad [7.6]$$

where ε_c is the critical equivalent strain value for nucleation, a, b are two constant parameters which are fitted during the experimental analysis and γ_{LAGB} is the low angle grain boundary energy. Finally, the probability of nucleation is evaluated as follows:

$$P_{NUC} = N_{NUC} S_N dt, \quad [7.7]$$

with S_N the volume in which the nucleus can appear. A schematic illustration of the nucleation stage with respect to accumulated energy and different equivalent critical strain values is presented in Figure 7.17.

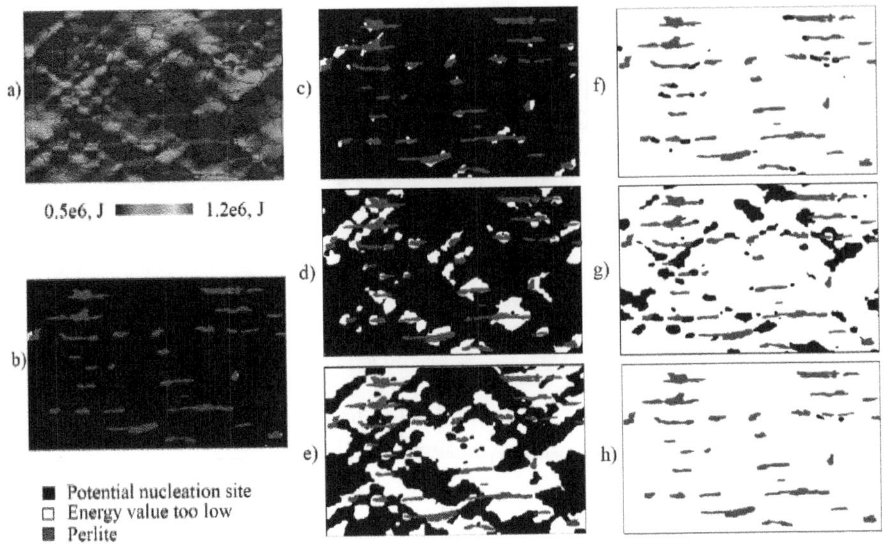

Figure 7.17. *Area with nucleation potential sites for different critical equivalent strain values: (a) energy distribution after deformation, (b) $\varepsilon_c = 0.1$, (c) $\varepsilon_c = 0.15$, (d)$\varepsilon_c = 0.2$, (e) $\varepsilon_c = 0.25$, (f) $\varepsilon_c = 0.3$, (g) $\varepsilon_c = 0.35$ and (h) $\varepsilon_c = 0.4$ (red regions represent pearlite, white regions represent an area that does not fulfill nucleation condition and black regions represent an area with potential nucleation sites)*

An example of the defined transition rule for the nucleation stage can be represented by:

$$Y_i^{t+dt} = \begin{cases} SRX \Leftrightarrow P_{NUC} > x, \text{ where } x \subset < 0:1 > \\ Y_i^t \end{cases} \quad [7.8]$$

The CA cell, representing a nucleus, changes its state from un-recrystallized ($uSRX$) into recrystallized (SRX) when the calculated P_{NUC} is higher than the randomly generated x value. At the same time, the unique internal variable, the id_i number, is assigned to the investigated CA cell. The id_i is used to identify CA cells associated with the particular grain in the microstructure investigated. Two different nucleation schemes are usually considered: site saturated (Raabe and Hantcherli 2005) and continuous nucleation (Zheng et al. 2009). The grain growth model follows the nucleation step, leading to grain growth and SRX progress. To recreate the complexity of grain growth, SRX models have to take into account the influence of the two main driving pressures for microstructure evolutions. The first controls the velocity norm as a function of accumulated deformation energy, while the second is associated with the influence of grain boundary curvature, both during and after the

SRX. The constitutive equation that relates the grain boundary velocity with the mobility and the net driving pressure is expressed as follows:

$$v_i = M_G P, \tag{7.9}$$

where M_G is the grain boundary mobility and P is the net driving pressure. The value of M_G is computed as:

$$M_G = M_0 M_\theta \frac{D_0 b_B^2}{k_B T} \exp\left(-\frac{Q_G}{RT}\right), \tag{7.10}$$

where D_0 is a diffusion constant, b_B is the Burgers vector, Q_G is the activation energy for grain boundary motion, M_0 is a mobility parameter, and M_θ is the mobility related with crystallographic orientation, computed through a sigmoidal equation:

$$M_\theta = \left(1 - \exp\left(-m_1 \left(\frac{\theta}{\theta_m}\right)^{m_2}\right)\right), \tag{7.11}$$

where m_1, m_2 are two mobility function parameters and θ_m is the high angle grain boundary threshold and θ – is grains misorientation angle calculated based on Euler angles (ϕ_1, Φ, ϕ_2) (Zhu et al. 1999). It should also be noted that equation 7.11 is most commonly used in the scientific literature (see, for example, Lin et al. (2016)). However, many studies clearly point out a more complex relation between mobility and misorientation value for different materials (Singh and Parashar 2018). Thus, to capture the behavior of a wide range of materials, an alternative equation based on the polynomial of fourth degree can also be proposed, for example:

$$M_\theta = \begin{cases} \left(1 - \exp\left(-m_1 \left(\frac{\theta}{\theta_m}\right)^{m_2}\right)\right) \text{ for } \theta < \theta_1 \\ a_5\theta^4 + a_4\theta^3 + a_3\theta^2 + a_2\theta + a_1 \text{ for } \theta_1 \le \theta \le \theta_2 \\ a_6 \text{ for } \theta > \theta_2 \end{cases}, \tag{7.12}$$

where $a_1 : a_5$ are polynomial coefficients and a_6 is the maximum mobility for misorientation above θ_2. Differences in mobility functions based on the two approaches are presented in Figure 7.18.

It is well known that recrystallized grains grow faster into the neighboring regions with higher misorientation values. That is why it was essential to consider this phenomenon in the SRX models. A set of simple case studies, where nuclei with different crystallographic orientations are placed into the microstructure, was selected to show differences in final recrystallized grain shapes. A comparison of the final grain shapes can be seen in Figure 7.19.

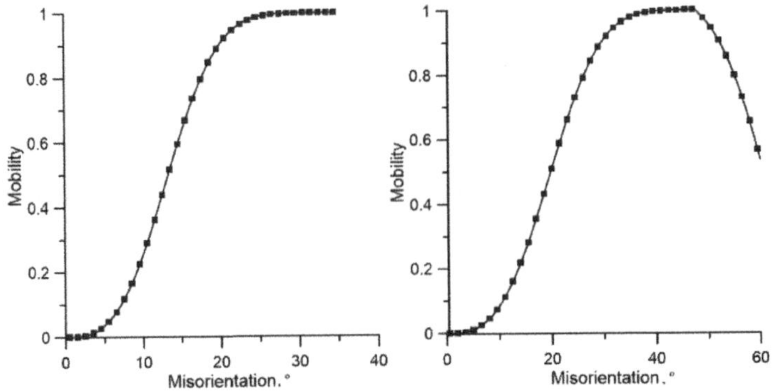

Figure 7.18. *Grain boundary mobility as a function of misorientation: (a) exponential function (equation [7.11]) with* $m_1 = 1$, $m_2 = 3$, $\theta_m = 15°$*; (b) polynomial function (equation [7.12])* $a_6 = 0.55$, $a_5 = 0$, $a_4 = 0$, $a_3 = -0.0021$, $a_2 = 0.1889$, $a_1 = -3.239$, $m_1 = 0.3$, $m_2 = 3$, $\theta_m = 15°$

Despite the mobility term in equation [7.11], the net pressure P is of particular importance for the SRX predictions. Often, the net pressure on the grain boundaries is based on the summation of the two main driving pressures associated with stored energy P_E as well as a grain boundary curvature P_{GB} and can be additionally complemented with the possible influence of precipitates P_Z (Smith–Zener pinning pressure):

$$P = P_E + P_Z + P_{GB}, \qquad [7.13]$$

The first component based on the accumulated deformation energy value is considered the major driving pressure controlling the grain boundary migration during SRX. As mentioned, the energy value H_i can be obtained directly from the EBSD investigation of a deformed sample or is calculated using the DMR-based FE model of a deformation process according to:

$$P_E = H_i = \frac{\varepsilon_i}{a\varepsilon_i + b}\gamma_{LAGB} \qquad [7.14]$$

where ε_i – is the equivalent plastic strain, a, b – are parameters and γ_{LAGB} – is the low-angle grain boundary energy. The second component is responsible for a reduction in the net pressure due to the presence of precipitates dispersed in the microstructure. This Smith–Zener pinning phenomenon (Raabe and Hantcherli 2005) is considered by the following formula:

$$P_Z = -\frac{3}{2}\gamma\frac{f_p}{r_p} \qquad [7.15]$$

where γ is the grain boundary energy, r_p is the mean precipitate radius and f_p is the volume fraction of precipitates. An example of simulation kinetic slowdown based on the impact of $0.01\mu m$ and $0.05\mu m$ precipitates is shown in Figure 7.20.

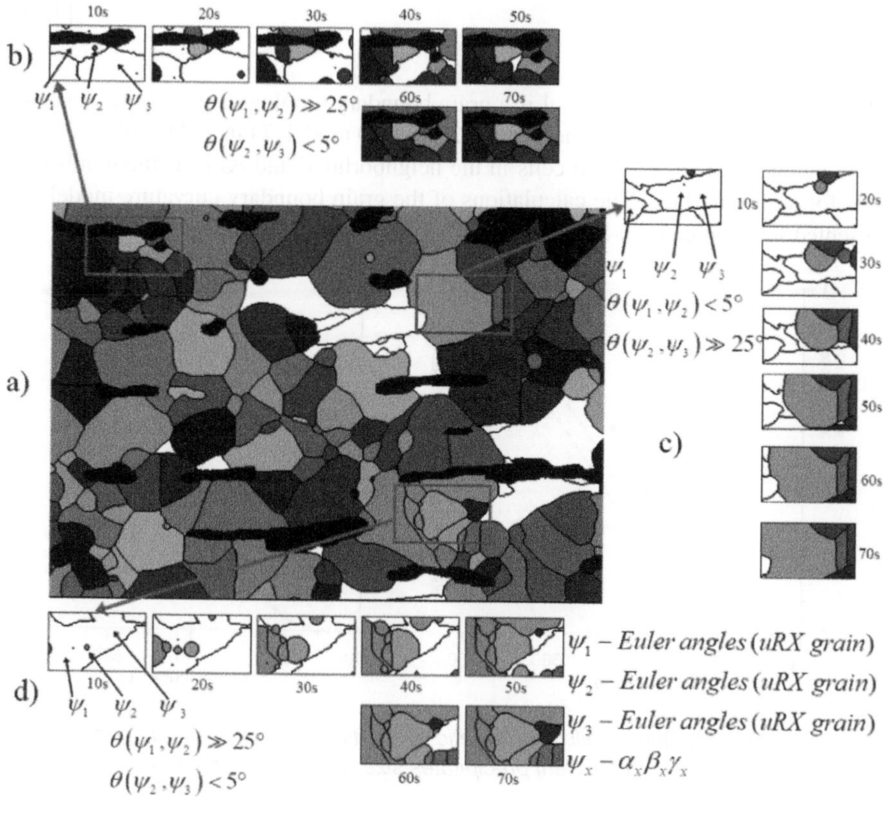

Figure 7.19. *Results of grain growth with different values of misorientations between grains and the nuclei: (a) final recrystallized microstructure; (b–d) progress of growth with different misorientation values*

In this case, the model is sensitive to the number of precipitates in the material. Additionally, having the r_p and f_p parameters stored in an internal cell variable allows the precipitations to be distributed, for example, in a highly nonhomogeneous manner across the CA space. Finally, the last part of equation [7.13] includes effect of grain boundary curvature, which is calculated by:

$$P_{GB} = -\kappa\gamma, \quad [7.16]$$

where κ – is the grain boundary curvature identified by a template approach (Lan et al. 2006):

$$\kappa = \frac{A_C}{C_S} \frac{\text{Kink} - c_{iSRX}}{N'}, \qquad [7.17]$$

where C_S – is the CA cell size, A is a grain boundary topological parameter (usually 1.28), N' is the number of CA neighbors (24 in a 2D model, 74 in a 3D model), c_{iSRX} is the number of recrystallized cells in the neighborhood and Kink is the number of cells for a flat interface. The calculations of the grain boundary curvature model are presented schematically in Figure 7.21.

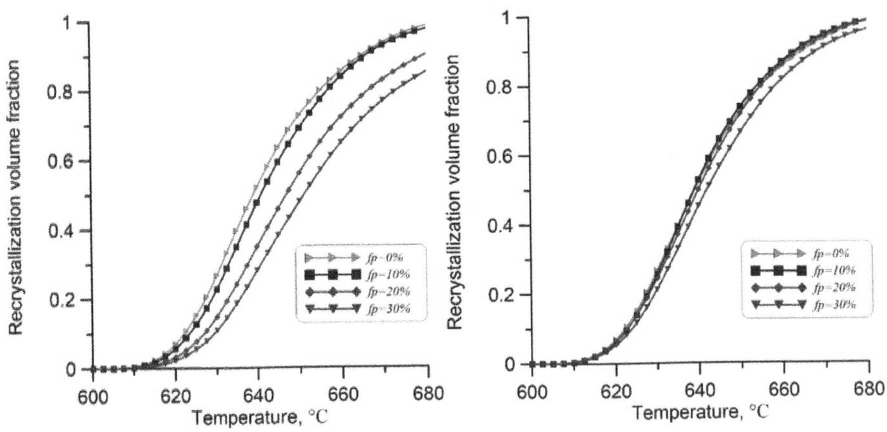

Figure 7.20. *Recrystallization kinetic for heating simulation (heating rate of* $1°\text{C/s}$*) with different precipitates size: (a)* $0.1\mu m$*; (b)* $0.05\mu m$

As seen in Figure 7.21, the grain boundary curvature can have a positive or negative contribution to the net driving pressure depending on the geometry of the boundary. More advanced models for capturing grain boundary curvatures are presented in Li et al. (2019) and Niewczas et al. (2022). Based on the above equations, equation [7.9] is solved in each CA cell located at the recrystallization front to evaluate the recrystallized area fraction in the subsequent tth time steps:

$$RX_{i,t}^{\text{fraction}} = RX_{i,t-dt}^{\text{fraction}} + \sum_{j=1}^{rx} \left(\frac{v_j t_{\text{step}}}{C_S} \right), \qquad [7.18]$$

where $RX_{i,t-dt}^{\text{fraction}}$ is the recrystallization volume fraction in the ith CA cell in the $t - dt$ time, rx is the number of recrystallized neighbors and v_j is the velocity of the recrystallization front in the jth cell from equation [7.9]. All of the presented equations are crucial for a definition of the transition rule that controls grain boundary migration during the SRX. The transition rule describes the change of the CA cell state from non-recrystallized ($uSRX$) to recrystallized (SRX) one and is defined as:

$$Y_i^{t+dt} = \begin{cases} SRX \Leftrightarrow RX_{i,t}^{\text{fraction}} > 1.0 \\ Y_i^t \end{cases} \qquad [7.19]$$

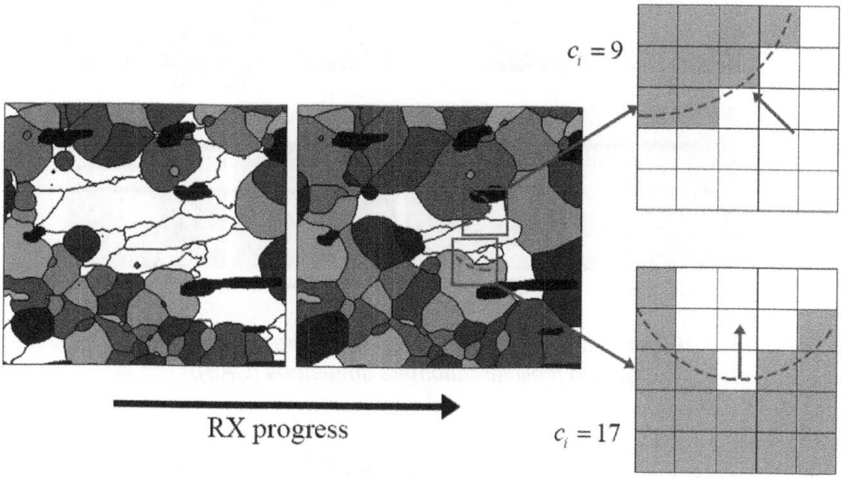

Figure 7.21. *Evaluation of a grain boundary curvature during recrystallization*

In the end of the process, after complete recrystallization of all CA cells, subsequent grain boundary migration is controlled only by the grain boundary curvature model, that is, the grain growth mechanism. An illustration of the curvature-driven grain growth model is presented in Figure 7.22.

Finally, examples of SRX results obtained from a case study simulation realized with the described workflow are presented in Figure 7.23. The initial microstructure of dimensions 930 × 570 CA cells (772 × 473μm) was deformed during the DMR FE channel die compression simulation and then annealed at the temperature of 1000°C for 50s.

7.4. Calculation times and perspectives

A lot of research dedicated to the simulation of material morphology evolution during SRX by the CA method can be found in the literature (Shen et al. 2018; Alavi and Serajzadeh 2020).

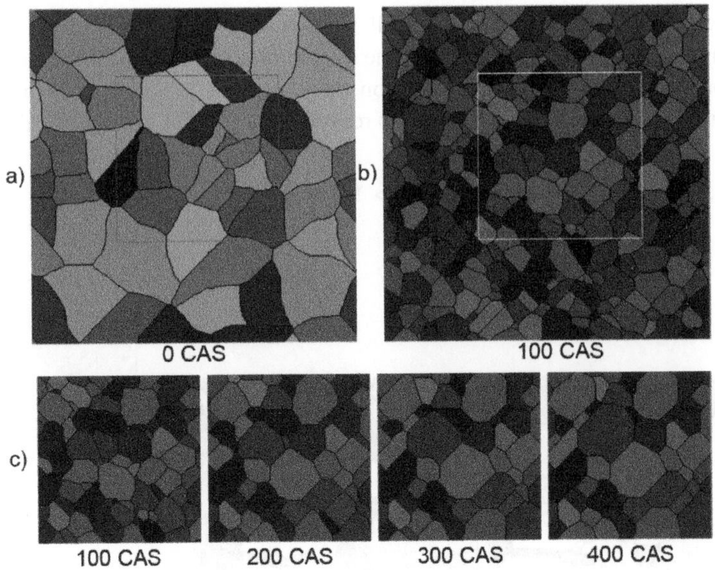

Figure 7.22. *Illustration of the curvature-driven grain growth (cellular automata time steps (CAS))*

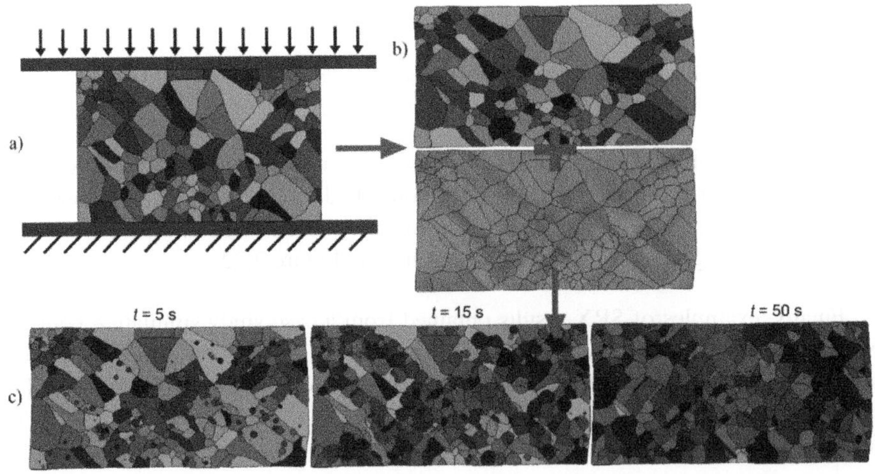

Figure 7.23. *Example of results obtained from CA SRX simulation: (a) initial microstructure morphology before deformation; (b) material morphology and accumulated energy values after deformation; (c) material morphology after* 5, 15 *and* 50s *of SRX*

Nevertheless, simulations based on the CA method are still strongly limited in potential applications under industrial conditions due to the excessive computational times. However, in recent years, an important increase in the computational power of single computational units and powerful servers has been observed. This situation allows for the possibility of making simulations based on the CA method much faster and, on the other hand, allows for the simulation of large microstructures in 3D. However, to properly take advantage of novel computer architectures, different implementation concepts have been introduced into the standard code for its parallel execution. In general, parallelization can be realized with the use of:

– CPU (central processing unit)-based computations (Gibson et al. 2015);

– GPGPU (general purpose graphical processor unit)-based computations (Kolnoochenko and Menshutina 2015);

– heterogeneous platform computations (Rauch and Bachniak 2017).

An example of the CPU parallelization concept developed in Sitko et al. (2020a) is presented in Figure 7.24.

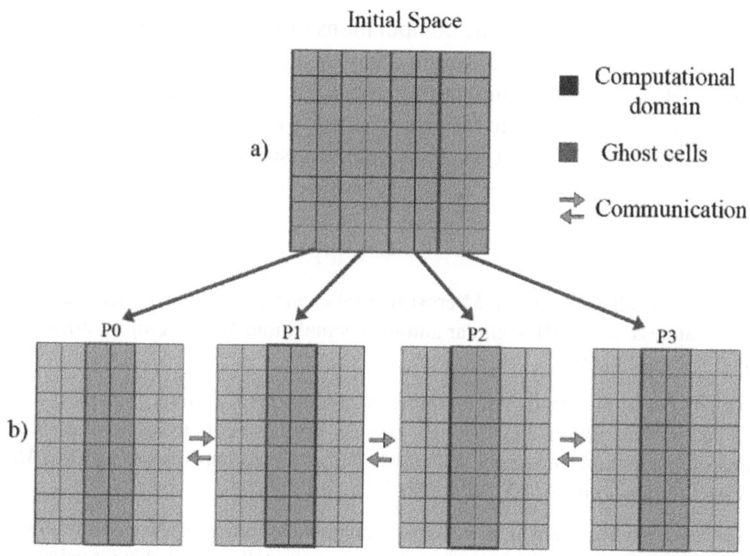

Figure 7.24. *CPU parallelization concept applied to the SRX model*

These kinds of parallelization approaches are the only possible solutions for practical applications of the CA microstructure evolution models, as the obtained speedup is significant (Figure 7.25).

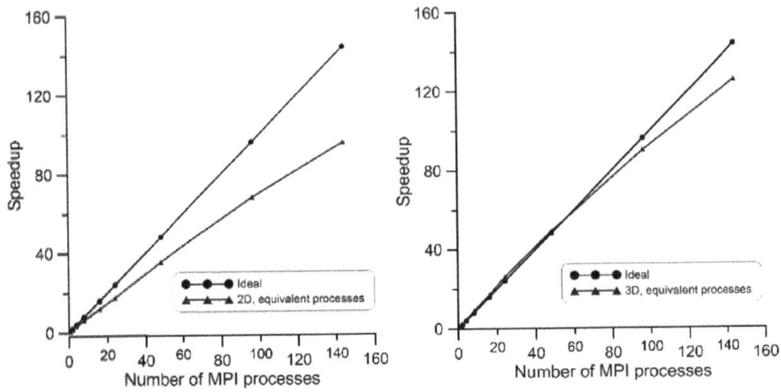

Figure 7.25. *Speedup from (a) 2D and (b) 3D investigations for a different number of MPI*

With such parallel solutions, the computations of more than 1 billion CA cells are possible with reasonable time, which provides a possibility to simulate even 3D DMR models with a large number of grains with varying sizes (Sitko et al. 2020a, 2021). An alternative solution to the presented CA method simulations is using the MC concept, which is somewhat based on similar assumptions.

7.5. References

Alavi, P. and Serajzadeh, S. (2020). Microstructural changes during static recrystallization of austenitic stainless steel 304l: Cellular automata simulation. *Metallography, Microstructure, and Analysis*, 9, 223–238.

Ali, U., Odoh, D., Muhammad, W., Brahme, A., Mishra, R.K., Wells, M., Inal, K. (2017). Experimental investigation and through process crystal plasticity-static recrystallization modeling of temperature and strain rate effects during hot compression of aa6063. *Materials Science and Engineering: A*, 700, 374–386.

Alone, A., Chatterjee, R., Alankar, A. (2020). A comparative study of the effect of random and preferred crystallographic orientations on dynamic recrystallization behavior using a cellular automata model. *Materials Today Communications*, 25, 101200.

Bakhtiari, M. and Salehi, M.S. (2018). Reconstruction of deformed microstructure using cellular automata method. *Computational Materials Science*, 149, 1–13.

Bargmann, S., Klusemann, B., Markmann, J., Schnabel, J.E., Schneider, K., Soyarslan, C., Wilmers, J. (2018). Generation of 3D representative volume elements for heterogeneous materials: A review. *Progress in Materials Science*, 96, 322–384.

Bays, C. (2018). *Cellular Automata in Triangular, Pentagonal, and Hexagonal Tessellations*. Springer, New York.

Beltran, O., Huang, K., Logé, R.E. (2015). A mean field model of dynamic and post-dynamic recrystallization predicting kinetics, grain size and flow stress. *Computational Materials Science*, 102, 293–303.

Bernacki, M., Chastel, Y., Coupez, T., Logé, R.E. (2008). Level set framework for the numerical modelling of primary recrystallization in polycrystalline materials. *Scripta Materialia*, 58(12), 1129–1132.

Boguń, K., Sitko, M., Mojżeszko, M., Madej, Ł. (2021). Cellular automata-based computational library for development of digital material representation models of heterogenous microstructures. *Archives of Civil and Mechanical Engineering*, 21, 1–15.

Bretin, R., Levesque, M., Pilvin, P., Bocher, P. (2020). Neighborhood effect within polycrystalline materials: Elastoplastic micromechanical analysis by cellular automaton and finite element. *International Journal of Fatigue*, 137, 105634.

Cho, S.-H. and Yoo, Y.-C. (2001). Metadynamic recrystallization of austenitic stainless steel. *Journal of Materials Science*, 36(17), 4279–4284.

Chun, Y.B., Semiatin, S., Hwang, S.-K. (2006). Monte Carlo modeling of microstructure evolution during the static recrystallization of cold-rolled, commercial-purity titanium. *Acta Materialia*, 54(14), 3673–3689.

Davies, C. (1995). The effect of neighbourhood on the kinetics of a cellular automaton recrystallisation model. *Scripta Metallurgica et Materialia*, 33(7), 1139–1143.

Davies, C. (1997). Growth of nuclei in a cellular automaton simulation of recrystallisation. *Scripta Materialia*, 36(1), 35–40.

Després, A., Greenwood, M., Sinclair, C. (2020). A mean-field model of static recrystallization considering orientation spreads and their time-evolution. *Acta Materialia*, 199, 116–128.

Falco, S., Jiang, J., De Cola, F., Petrinic, N. (2017). Generation of 3D polycrystalline microstructures with a conditioned Laguerre-Voronoi tessellation technique. *Computational Materials Science*, 136, 20–28.

Florez, S., Alvarado, K., Muñoz, D.P., Bernacki, M. (2020). A novel highly efficient Lagrangian model for massively multidomain simulation applied to microstructural evolutions. *Computer Methods in Applied Mechanics and Engineering*, 367, 113107.

Frydrych, K. and Kowalczyk-Gajewska, K. (2018). Microstructure evolution in cold-rolled pure titanium: Modeling by the three-scale crystal plasticity approach accounting for twinning. *Metallurgical and Materials Transactions A*, 49(8), 3610–3623.

Gibson, M.J., Keedwell, E.C., Savić, D.A. (2015). An investigation of the efficient implementation of cellular automata on multi-core CPU and GPU hardware. *Journal of Parallel and Distributed Computing*, 77, 11–25.

Goetz, R. and Seetharaman, V. (1998). Static recrystallization kinetics with homogeneous and heterogeneous nucleation using a cellular automata model. *Metallurgical and Materials Transactions A*, 29, 2307–2321.

Halder, C., Madej, L., Pietrzyk, M. (2014). Discrete micro-scale cellular automata model for modelling phase transformation during heating of dual phase steels. *Archives of Civil and Mechanical Engineering*, 14(1), 96–103.

Han, F., Tang, B., Kou, H., Cheng, L., Li, J., Feng, Y. (2014). Static recrystallization simulations by coupling cellular automata and crystal plasticity finite element method using a physically based model for nucleation. *Journal of Materials Science*, 49, 3253–3267.

Hesselbarth, H. and Göbel, I. (1991). Simulation of recrystallization by cellular automata. *Acta Metallurgica et Materialia*, 39(9), 2135–2143.

Jahns, K., Balinski, K., Landwehr, M., Trindade, V.B., Wübbelmann, J., Krupp, U. (2017). Modeling of intergranular oxidation by the cellular automata approach. *Oxidation of Metals*, 87, 285–295.

Janssens, K.G.F. (2003). Random grid, three-dimensional, space-time coupled cellular automata for the simulation of recrystallization and grain growth. *Modelling and Simulation in Materials Science and Engineering*, 11(2), 157.

Janssens, K.G.F., Raabe, D., Kozeschnik, E., Miodownik, M.A., Nestler, B. (2010). *Computational Materials Engineering: An Introduction to Microstructure Evolution.* Academic Press, London.

Jonas, J. (1994). Dynamic recrystallization – Scientific curiosity or industrial tool? *Materials Science and Engineering: A*, 184(2), 155–165.

Kazeminezhad, M., Taheri, A.K., Tieu, A.K. (2007). Utilization of the finite element and Monte Carlo model for simulating the recrystallization of inhomogeneous deformation of copper. *Computational Materials Science*, 38(4), 765–773.

Kim, D.-K., Jung, K., Lee, H., Im, Y.-T. (2012). Cellular automata simulation of textural evolution during primary static recrystallization of IF steel. In *Materials Science Forum*, 702. Trans Tech Publications, Wollerau.

Kim, D.-K., Woo, W., Park, W.-W., Im, Y.-T., Rollett, A. (2017). Mesoscopic coupled modeling of texture formation during recrystallization considering stored energy decomposition. *Computational Materials Science*, 129, 55–65.

Kolnoochenko, A. and Menshutina, N. (2015). CUDA-optimized cellular automata for diffusion limited processes. In *Computer Aided Chemical Engineering*, 37, 551–556.

Kowalczyk-Gajewska, K., Sztwiertnia, K., Kawałko, J., Wierzbanowski, K., Wronski, M., Frydrych, K., Stupkiewicz, S., Petryk, H. (2015). Texture evolution in titanium on complex deformation paths: Experiment and modelling. *Mater. Sci. Eng. A*, 637, 251–263.

Kruzel, F., Madej, L., Perzynski, K., Banas, K. (2014). Development of three-dimensional adaptive mesh generation for multiscale applications. *International Journal for Multiscale Computational Engineering*, 12(3), 257–269.

Kugler, G. and Turk, R. (2006). Study of the influence of initial microstructure topology on the kinetics of static recrystallization using a cellular automata model. *Computational Materials Science*, 37(3), 284–291.

Kühbach, M., Gottstein, G., Barrales-Mora, L. (2016). A statistical ensemble cellular automaton microstructure model for primary recrystallization. *Acta Materialia*, 107, 366–376.

Łach, Ł., Nowak, J., Svyetlichnyy, D. (2018). The evolution of the microstructure in AISI 304L stainless steel during the flat rolling – Modeling by frontal cellular automata and verification. *Journal of Materials Processing Technology*, 255, 488–499.

Lan, Y., Li, D., Li, Y. (2006). A mesoscale cellular automaton model for curvature-driven grain growth. *Metallurgical and Materials Transactions B*, 37, 119–129.

Li, J., Xu, H., Mattila, T.T., Kivilahti, J., Laurila, T., Paulasto-Kröckel, M. (2010). Simulation of dynamic recrystallization in solder interconnections during thermal cycling. *Computational Materials Science*, 50(2), 690–697.

Li, Z., Wang, J., Huang, H. (2019). Grain boundary curvature based 2D cellular automata simulation of grain coarsening. *Journal of Alloys and Compounds*, 791, 411–422.

Lin, Y., Liu, Y.-X., Chen, M.-S., Huang, M.-H., Ma, X., Long, Z.-L. (2016). Study of static recrystallization behavior in hot deformed Ni-based superalloy using cellular automaton model. *Materials & Design*, 99, 107–114.

Liu, J., Dai, Q., Chen, J., Chen, S., Ji, H., Dua, W., Deng, X., Wang, Z., Guo, G., Luo, H. (2017a). The two dimensional microstructure characterization of cemented carbides with an automatic image analysis process. *Ceramics International*, 43(17), 14865–14872.

Liu, Y.-X., Lin, Y., Zhou, Y. (2017b). 2D cellular automaton simulation of hot deformation behavior in a Ni-based superalloy under varying thermal-mechanical conditions. *Materials Science and Engineering: A*, 691, 88–99.

Luan, Q., Lee, J., Zheng, J.-H., Hopper, C., Jiang, J. (2020). Combining microstructural characterization with crystal plasticity and phase-field modelling for the study of static recrystallization in pure aluminium. *Computational Materials Science*, 173, 109419.

Maazi, N. and Lezzar, B. (2020). An efficient Monte Carlo Potts method for the grain growth simulation of single-phase systems. *Computer Methods in Material Science*, 20(3), 85–94.

Madej, L., Kružel, P., Cybułka, P., Perzyński, K., Banaś, B. (2012). Generation of dedicated finite element meshes for multiscale applications with Delaunay triangulation and adaptive finite element-cellular automata algorithms. *Computer Methods in Materials Science*, 12(2), 85–96.

Madej, L., Sieradzki, L., Sitko, M., Perzynski, K., Radwanski, K., Kuziak, R. (2013). Multi scale cellular automata and finite element based model for cold deformation and annealing of a ferritic – Pearlitic microstructure. *Computational Materials Science*, 77, 172–181.

Madej, L., Kwiaton, N., Kuziak, R., Sitko, M., Pietrzyk, M. (2016a). Physical and numerical simulation of cold rolling and heating during continuous annealing of DP steel strips. In *Materials Science Forum*, 854. Trans Tech Publications, Wollerau.

Madej, L., Sitko, M., Radwanski, K., Kuziak, R. (2016b). Validation and predictions of coupled finite element and cellular automata model: Influence of the degree of deformation on static recrystallization kinetics case study. *Materials Chemistry and Physics*, 179, 282–294.

Madej, L., Sitko, M., Legwand, A., Perzynski, K., Michalik, K. (2018). Development and evaluation of data transfer protocols in the fully coupled random cellular automata finite element model of dynamic recrystallization. *Journal of Computational Science*, 26, 66–77.

Madej, L., Chang, Y., Szeliga, D., Bleck, W., Pietrzyk, M. (2021). Criterion for microcrack resistance of multi-phase steels based on property gradient maps. *CIRP Annals*, 70(1), 243–246.

Mandal, S., Lao, J., Donegan, S., Rollett, A.D. (2018). Generation of statistically representative synthetic three-dimensional microstructures. *Scripta Materialia*, 146, 128–132.

Mason, J., Lind, J., Li, S., Reed, B., Kumar, M. (2015). Kinetics and anisotropy of the Monte Carlo model of grain growth. *Acta Materialia*, 82, 155–166.

Mellbin, Y., Hallberg, H., Ristinmaa, M. (2017). An extended vertex and crystal plasticity framework for efficient multiscale modeling of polycrystalline materials. *International Journal of Solids and Structures*, 125, 150–160.

von Neumann, J. (1966). *Theory of Self-Reproducing Automata*. University of Illinois Press, Urbana/London.

Niewczas, S., Sitko, M., Madej, Ł. (2022). Influence of the grain boundary curvature model on cellular automata static recrystallization simulations. *Key Engineering Materials*, 926, 1977–1985.

Ollat, M., Militzer, M., Massardier, V., Fabregue, D., Buscarlet, E., Keovilay, F., Perez, M. (2018). Mixed-mode model for ferrite-to-austenite phase transformation in dual-phase steel. *Computational Materials Science*, 149, 282–290.

Osher, S. and Fedkiw, R.P. (2001). Level set methods: An overview and some recent results. *Journal of Computational Physics*, 169(2), 463–502.

Patterson, B., Henderson, K., Gibbs, P., Imhoff, S., Clarke, A. (2014). Laboratory micro- and nanoscale x-ray tomographic investigation of Al–7 at.% Cu solidification structures. *Materials Characterization*, 95, 18–26.

Pietrzyk, M., Madej, L., Rauch, L., Szeliga, D. (2015). *Computational Materials Engineering: Achieving High Accuracy and Efficiency in Metals Processing Simulations*. Butterworth-Heinemann, Oxford and Elsevier, Waltham.

Pokharel, R., Lind, J., Li, S.F., Kenesei, P., Lebensohn, R.A., Suter, R.M., Rollett, A.D. (2015). In-situ observation of bulk 3D grain evolution during plastic deformation in polycrystalline Cu. *International Journal of Plasticity*, 67, 217–234.

Raabe, D. (1999). Introduction of a scalable three-dimensional cellular automaton with a probabilistic switching rule for the discrete mesoscale simulation of recrystallization phenomena. *Philosophical Magazine A*, 79(10), 2339–2358.

Raabe, D. and Hantcherli, L. (2005). 2D cellular automaton simulation of the recrystallization texture of an IF sheet steel under consideration of Zener pinning. *Computational Materials Science*, 34(4), 299–313.

Rauch, L. and Bachniak, D. (2017). Dynamic load balancing for cafe multiscale modelling methods for heterogeneous hardware infrastructure. *Procedia Computer Science*, 108, 1813–1822.

Rauch, L., Chmura, A., Gronostajski, Z., Polak, S., Pietrzyk, M. (2016). Cellular automata model for prediction of crack initiation and propagation in hot forging tools. *Archives of Civil and Mechanical Engineering*, 16, 437–447.

Reuther, K. and Rettenmayr, M. (2014). Perspectives for cellular automata for the simulation of dendritic solidification – A review. *Computational Materials Science*, 95, 213–220.

Rollett, A., Rohrer, G.S., Humphreys, J. (2017). *Recrystallization and Related Annealing Phenomena*, 3rd edition. Elsevier, Oxford.

Sakai, T., Ohashi, M., Chiba, K., Jonas, J. (1988). Recovery and recrystallization of polycrystalline nickel after hot working. *Acta Metallurgica*, 36(7), 1781–1790.

Salehi, M.S. and Serajzadeh, S. (2012). Simulation of static recrystallization in non-isothermal annealing using a coupled cellular automata and finite element model. *Computational Materials Science*, 53(1), 145–152.

Salvo, L., Suéry, M., Marmottant, A., Limodin, N., Bernard, D. (2010). 3D imaging in material science: Application of x-ray tomography. *Comptes Rendus Physique*, 11(9–10), 641–649.

Schoof, E., Schneider, D., Streichhan, N., Mittnacht, T., Selzer, M., Nestler, B. (2018). Multiphase-field modeling of martensitic phase transformation in a dual-phase microstructure. *Int. J. Solids Struct.*, 134, 181–194.

Sellars, C. and Zhu, Q. (2000). Microstructural modelling of aluminium alloys during thermomechanical processing. *Materials Science and Engineering: A*, 280(1), 1–7.

Shelton, R. and Dunand, D. (1996). Computer modeling of particle pushing and clustering during matrix crystallization. *Acta Materialia*, 44(11), 4571–4585.

Shen, G., Hu, B., Zheng, C., Gu, J., Li, D. (2018). Coupled simulation of ferrite recrystallization in a dual-phase steel considering deformation heterogeneity at mesoscale. *Computational Materials Science*, 149, 191–201.

Sieradzki, L. and Madej, L. (2012). Modeling of the static recrystallization with cellular automata method and digital material representation approach. *Steel Research International*, 1139–1142.

Sieradzki, L. and Madej, L. (2013). A perceptive comparison of the cellular automata and Monte Carlo techniques in application to static recrystallization modeling in polycrystalline materials. *Computational Materials Science*, 67, 156–173.

Singh, D. and Parashar, A. (2018). Effect of symmetric and asymmetric tilt grain boundaries on the tensile behaviour of BCC-niobium. *Computational Materials Science*, 143, 126–132.

Sitko, M., Chao, Q., Wang, J., Perzynski, K., Muszka, K., Madej, L. (2020a). A parallel version of the cellular automata static recrystallization model dedicated for high performance computing platforms – Development and verification. *Computational Materials Science*, 172, 109283.

Sitko, M., Mojzeszko, M., Rychlowski, L., Cios, G., Bala, P., Muszka, K., Madej, L. (2020b). Numerical procedure of three-dimensional reconstruction of ferrite-pearlite microstructure data from SEM/EBSD serial sectioning. *Procedia Manufacturing*, 47, 1217–1222.

Sitko, M., Banaś, K., Madej, L. (2021). Scaling scientific cellular automata microstructure evolution model of static recrystallization toward practical industrial calculations. *Materials*, 14(15), 4082.

Song, K., Li, Z., Fang, M., Xiao, Z., Zhu, Y., Lei, Q. (2022). Recrystallization behavior and phase transformation in a hot-rolled pure cobalt during annealing at the elevated temperature. *Materials Science and Engineering: A*, 845, 143178.

Svyetlichnyy, D.S. (2010). Modelling of the microstructure: From classical cellular automata approach to the frontal one. *Computational Materials Science*, 50(1), 92–97.

Svyetlichnyy, D.S. (2012). Reorganization of cellular space during the modeling of the microstructure evolution by frontal cellular automata. *Computational Materials Science*, 60, 153–162 [Online]. Available at: https://www.sciencedirect.com/science/article/pii/S0927025612001760.

Svyetlichnyy, D.S., Muszka, K., Majta, J. (2015). Three-dimensional frontal cellular automata modeling of the grain refinement during severe plastic deformation of microalloyed steel. *Computational Materials Science*, 102, 159–166 [Online]. Available at: https://www.sciencedirect.com/szcience/article/pii/S0927025615001354.

Takaki, T., Yamanaka, A., Higa, Y., Tomita, Y. (2007). Phase-field model during static recrystallization based on crystal-plasticity theory. *Journal of Computer-Aided Materials Design*, 14, 75–84.

Teferra, K. and Rowenhorst, D.J. (2021). Optimizing the cellular automata finite element model for additive manufacturing to simulate large microstructures. *Acta Materialia*, 213, 116930.

Wong, S., Hodgson, P., Chong, C.-J., Thomson, P. (1996). Physical modelling with application to metal working, especially to hot rolling. *Journal of Materials Processing Technology*, 62(1–3), 260–274.

Xiao, N., Hodgson, P., Rolfe, B., Li, D. (2018). Modelling discontinuous dynamic recrystallization using a quantitative multi-order-parameter phase-field method. *Computational Materials Science*, 155, 298–311.

Xu, H., Shu, X., Zuo, J., Yin, A., Wang, Y. (2022). Kinetic Monte Carlo simulation of abnormal grain growth in textured systems with anisotropic grain boundary energy and mobility. *Materials Today Communications*, 30, 103133.

Zheng, C., Xiao, N., Li, D., Li, Y. (2009). Mesoscopic modeling of austenite static recrystallization in a low carbon steel using a coupled simulation method. *Computational Materials Science*, 45(2), 568–575.

Zhu, G.-H., Mao, W.-M., Yu, Y.-N. (1999). Calculation of misorientation distribution between recrystallized grains and deformed matrix. *Scripta Materialia*, 42(1), 37–41.

List of Authors

Stéphane BERBENNI
LEM3
CNRS
Arts et Métiers ParisTech
Université de Lorraine
Metz
France

Marc BERNACKI
CNRS UMR 7635
Centre for Material Forming (CEMEF)
MINES Paris
PSL Research University
Sophia Antipolis
France

Samuel FOREST
CNRS UMR 7633
Centre des Matériaux
MINES Paris
PSL Research University
Evry
France

Dominique JEULIN
Centre de Morphologie Mathématique
MINES Paris
PSL Research University
France

Lukasz MADEJ
AGH University of Krakow
Poland

Sylvain QUEYREAU
CNRS UPR 3407
Laboratoire des Sciences des Procédés et des Matériaux
Université Sorbonne Paris Nord
Villetaneuse
France

Oleg SHCHYGLO
Interdisciplinary Centre for Advanced Materials Simulation (ICAMS)
Ruhr University Bochum
Germany

Mateusz SITKO
AGH University of Krakow
Poland

Ingo STEINBACH
Interdisciplinary Centre for Advanced
Materials Simulation (ICAMS)
Ruhr University Bochum
Germany

Index

A

abnormal grain growth, 207
anisotropic elasticity, 2, 10
anisotropy, 10
  grain boundary energy, 164, 218
apparent
  moduli, 68
  properties, 58
Arrhenius law, 5, 213
atomistic simulation, 2
austenite, 236

B

boundary conditions, 12
Burgers
  law, 4
  vector, 4, 8, 10, 11, 15, 18, 33, 39
Burke and Turnbull law, 157

C

CAFE, 264
Cahn-Hilliard, 189
cells, 262
cellular automaton, 156
Clausius–Duhem relation, 183
computational homogenization, 56
continuous
  dynamic recrystallization (CDRX), 203, 225
  nucleation, 276
crystal plasticity, 166, 273
  finite element method (CPFEM), 206

D

Darcy, 57
Debye frequency, 7
diffuse interface, 193
diffusive solid-state phase transformation (DSSPT), 204, 235
DIGIMU, 228
discontinuous dynamic recrystallization (DDRX), 203
discrete dislocation dynamics (DDD), 2
disorientation angle, 211
distance function, 215

E, F

eikonal equation, 215
elasticity, 191
fast marching method, 216
FE-LS method, 205
ferrite, 236, 269
front CA, 267

front-
capturing (FC), 155, 202
tracking (FT), 155, 202
full field, 205, 262

G

geometrically dislocation density (GND), 34, 206
Gibbs–Thomson equation, 183, 206
Ginzburg–Landau
functional, 179
theory, 177
Gourdet–Montheillet model, 225
grain growth, 166, 268, 281
graph coloration technique, 211
Green's function, 8, 10, 118, 119
grooving, 207

H, I

Hall–Petch equation, 32
Hamilton–Jacobi formulation, 216
heterogeneous media, 56
high-angle grain boundary (HAGB), 223, 227, 228
Hill's constraint tensor, 122, 124
homogenization, 56
hyperbolic term, 182
inclination, 164
interface, 183

J, K

junctions, 187
kinematic incompatibilities, 215
Kocks–Mecking law, 206
Korteveg–De Vries equation, 177

L

Laguerre–Voronoï tessellation, 208
Landau functional, 179
level-set (LS), 173, 262
Lippmann–Schwinger equation, 63
logical funtion, 263
low-angle grain boundary (LAGB), 36, 225, 275

M

Mackenzie-like DDF, 211
MacPherson Srolovitz law, 156, 170
martensitic transformation, 194
mean field, 262
metadynamic recrystallization, 203
misorientation, 167, 211, 277
mobility, 5, 211, 277
Model-
I, 160
II, 160
Monte Carlo (MC), 156, 166, 262
multi-phase fields, 173, 184
multipole expansion, 13

N

Navier, 57
nickel base superalloys, 229
nucleation, 167, 268
numerical homogenization, 56

O

OpenPhase, 185
Orowan
equations, 31
strengthening, 29
stress, 29
ortho-equilibrium, 207
Ostwald ripening (OR), 203, 242

P

para-equilibrium, 207
parabolic equation, 182
Peach and Koehler force, 8
pearlite, 276
phase field (PF), 177, 204, 262
plastic stored energy, 156, 271, 274
plasticity, 2, 194
polycrystal materials, 208

post-dynamic recrystallization, 166
potential function, 180

R

random media, 56
Read–Shockley equation, 220
recovery, 206, 274
recrystallization, 166, 196
regularization, 9
reinitialization, 216
representative volume element (RVE), 56, 69, 209
Rice's criterion, 100

S

Schmid
  factor, 18
  law, 17, 130
sigmoidal model, 220, 277
sintering, 207
site saturated nucleation, 276
Smith–Zener pinning, 164, 207, 228, 278
solidification, 198
soliton solution, 177, 179, 184
solute drag, 243
stacking fault energy, 225
static recrystallization (SRX), 203, 268
steel, 194, 218
stress singularity, 9

T

Taylor factor, 206
ThermoCalc, 239
torque terms, 164
transition rules, 263

V

Vertex, 156, 262
virtual vertices, 162
viscous drag, 169
von Neumann Mullins law, 156
Voronoï tessellation, 63, 74, 80, 95, 100, 208, 269

Y, Z

Yoshie–Laasraoui–Jonas law, 206
Zircaloy, 228